Texas

Algebra 2

Volume 2

TIMOTHY D. KANOLD

EDWARD B. BURGER

JULI K. DIXON

MATTHEW R. LARSON

STEVEN J. LEINWAND

Printed in the U.S.A.

ISBN 978-0-544-35394-7

12 2591 24

4500895464 B C D E F G

Authors

Timothy D. Kanold, Ph.D., is an award-winning international educator, author, and consultant. He is a former superintendent and director of mathematics and science at Adlai E. Stevenson High School District 125 in Lincolnshire, Illinois. He is a past president of the National Council of Supervisors of Mathematics (NCSM) and the Council for the Presidential Awardees of Mathematics (CPAM). He has served on several writing and leadership commissions for NCTM during the past decade. He presents motivational professional development seminars with a focus on developing professional learning communities (PLC's) to improve the teaching, assessing, and learning of students. He has recently authored nationally recognized articles, books, and textbooks for mathematics education and school leadership, including *What Every Principal Needs to Know about the Teaching and Learning of Mathematics*.

Edward B. Burger, Ph.D., is the President of Southwestern University, a former Francis Christopher Oakley Third Century Professor of Mathematics at Williams College, and a former vice provost at Baylor University. He has authored or coauthored more than sixty-five articles, books, and video series; delivered over five hundred addresses and workshops throughout the world; and made more than fifty radio and television appearances. He is a Fellow of the American Mathematical Society as well as having earned many national honors, including the Robert Foster Cherry Award for Great Teaching in 2010. In 2012, Microsoft Education named him a "Global Hero in Education."

Juli K. Dixon, Ph.D., is a Professor of Mathematics Education at the University of Central Florida. She has taught mathematics in urban schools at the elementary, middle, secondary, and post-secondary levels. She is an active researcher and speaker with numerous publications and conference presentations. Key areas of focus are deepening teachers' content knowledge and communicating and justifying mathematical ideas. She is a past chair of the NCTM Student Explorations in Mathematics Editorial Panel and member of the Board of Directors for the Association of Mathematics Teacher Educators.

Matthew R. Larson, Ph.D., is the K-12 mathematics curriculum specialist for the Lincoln Public Schools and served on the Board of Directors for the National Council of Teachers of Mathematics from 2010 to 2013. He is a past chair of NCTM's Research Committee and was a member of NCTM's Task Force on Linking Research and Practice. He is the author of several books on implementing the Common Core Standards for Mathematics. He has taught mathematics at the secondary and college levels and held an appointment as an honorary visiting associate professor at Teachers College, Columbia University.

Steven J. Leinwand is a Principal Research Analyst at the American Institutes for Research (AIR) in Washington, D.C., and has over 30 years in leadership positions in mathematics education. He is past president of the National Council of Supervisors of Mathematics and served on the NCTM Board of Directors. He is the author of numerous articles, books, and textbooks and has made countless presentations with topics including student achievement, reasoning, effective assessment, and successful implementation of standards.

Performance Task Consultant

Robert Kaplinsky
Teacher Specialist, Mathematics
Downey Unified School District
Downey, California

STEM Consultants
Science, Technology, Engineering, and Mathematics

Michael A. DiSpezio
Global Educator
North Falmouth, Massachusetts

Michael R. Heithaus
Executive Director, School of Environment, Arts, and Society
Professor, Department of Biological Sciences
Florida International University
North Miami, Florida

Texas Consulting Reviewers

Anne Papakonstantinou, Ed.D.
Director - Rice University School Mathematics Project
Rice University
Houston, Texas

Richard Parr
Executive Director - Rice University School Mathematics Project
Rice University
Houston, Texas

Susan Troutman
Director for Secondary Programs - Rice University School Mathematics Project
Rice University
Houston, Texas

Carolyn White
Director for Elementary and Intermediate Programs - Rice University School Mathematics Project
Rice University
Houston, Texas

Alice Fisher
Director of Technology Applications and Integration - Rice University School Mathematics Project
Rice University
Houston, Texas

Texas Reviewers

Ashley Alford
Flower Mound High School
Flower Mound, Texas

Sydney Bentz
Flower Mound High School
Flower Mound, Texas

David Surdovel
Manor ISD
Manor, Texas

Matthew J. Waldmann, M.S.,
Educator
Arlington High School
Arlington, Texas

Megan A. Uken
Flower Mound High School
Flower Mound, Texas

Rational Functions, Expressions, and Equations

UNIT 4

Volume 2

MODULE 9 Rational Functions

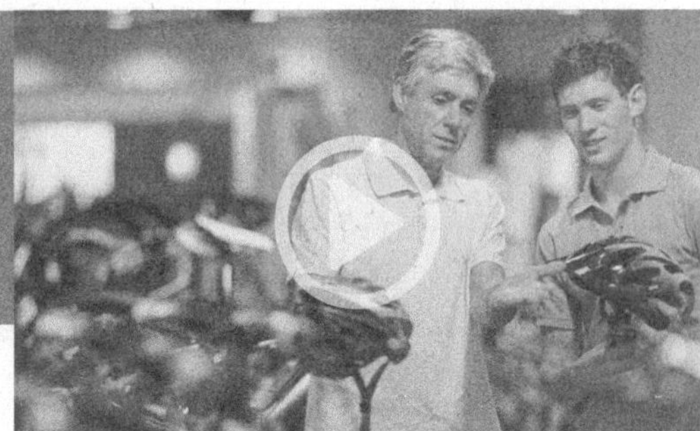

TEKS

MODULE 10 Rational Expressions and Equations

TEKS

UNIT 5

Radical Functions, Expressions, and Equations

Volume 2

MODULE 11 Radical Functions

TEKS

MODULE 12 Radical Expressions and Equations

TEKS

Exponential and Logarithmic Functions and Equations

MODULE 13 Exponential Functions

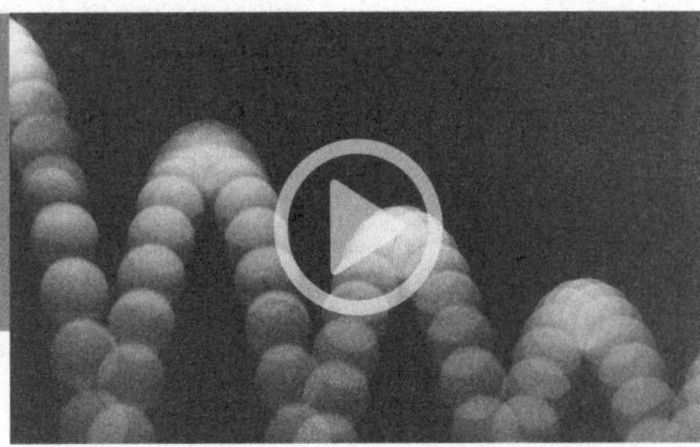

TEKS

MODULE 14 Modeling with Exponential and Other Functions

TEKS

MODULE 15 Logarithmic Functions

TEKS

MODULE 16 Logarithmic Properties and Exponential and Logarithmic Equations

TEKS

UNIT 4

Rational Functions, Expressions, and Equations

MODULE 9

Rational Functions

TEKS TEKS A2.2.A, A2.6.G, A2.6.L, A2.6.K, A2.7.I

MODULE 10

Rational Expressions and Equations

TEKS TEKS A2.6.I, A2.7.F

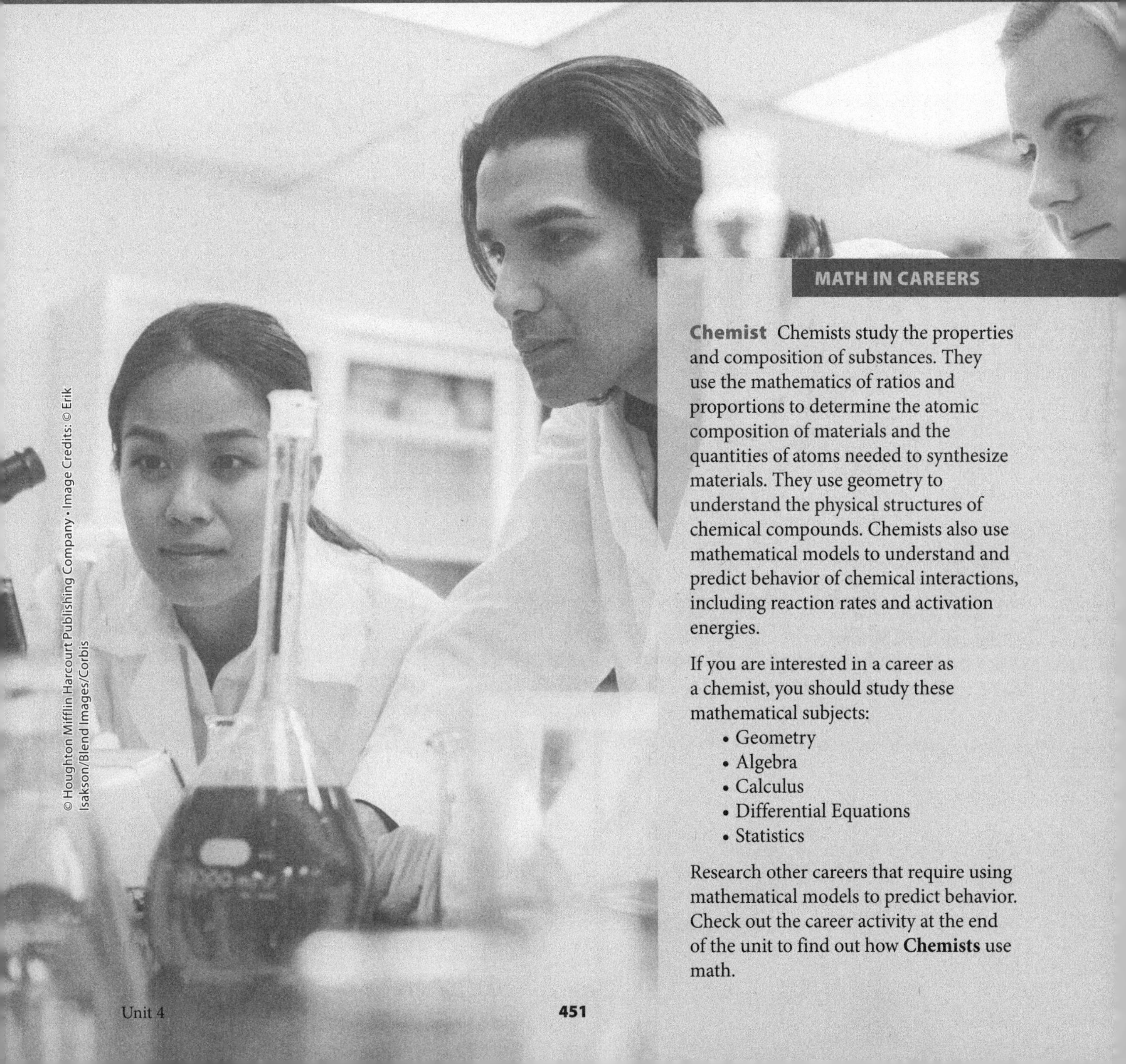

MATH IN CAREERS

Chemist Chemists study the properties and composition of substances. They use the mathematics of ratios and proportions to determine the atomic composition of materials and the quantities of atoms needed to synthesize materials. They use geometry to understand the physical structures of chemical compounds. Chemists also use mathematical models to understand and predict behavior of chemical interactions, including reaction rates and activation energies.

If you are interested in a career as a chemist, you should study these mathematical subjects:

- Geometry
- Algebra
- Calculus
- Differential Equations
- Statistics

Research other careers that require using mathematical models to predict behavior. Check out the career activity at the end of the unit to find out how **Chemists** use math.

Reading Start-Up

Vocabulary

Review Words

✔ direct variation (variación directa)

✔ domain (dominio)

✔ parent function (función madre)

✔ range (rango)

✔ rational expression (expresión racional)

✔ reciprocal (recíproco)

Preview Words

asymptote (asíntota)

closure (cerradura)

inverse variation (variación inversa)

rational function (función racional)

Visualize Vocabulary

Use the ✓ words to complete the chart.

	the set of output values of a function or relation
	a linear relationship between two variables, x and y, that can be written in the form $y = kx$, where k is a nonzero constant
	the multiplicative inverse of a number; the product of a number and its reciprocal is 1
	the simplest function with the defining characteristics of the function family
	the set of all possible input values of a relation or function
	an algebraic expression whose numerator and denominator are polynomials and whose denominator has a degree ≥ 1.

Understand Vocabulary

To become familiar with some of the vocabulary terms in the module, consider the following. You may refer to the module, the glossary, or a dictionary.

1. ______________ is a relationship between two variables, x and y, that can be written in the form $y = \frac{k}{x}$, where k is a nonzero constant and $x \neq 0$.
2. A line that a graph approaches as the value of the variable becomes extremely large or small is an __________.
3. A function whose rule can be written as a rational expression is a ______________.

Active Reading

Double-Door Fold Before beginning each lesson, create a double-door fold to compare the characteristics of two expressions, functions, or variations. This can help you identify the similarities and differences between the topics.

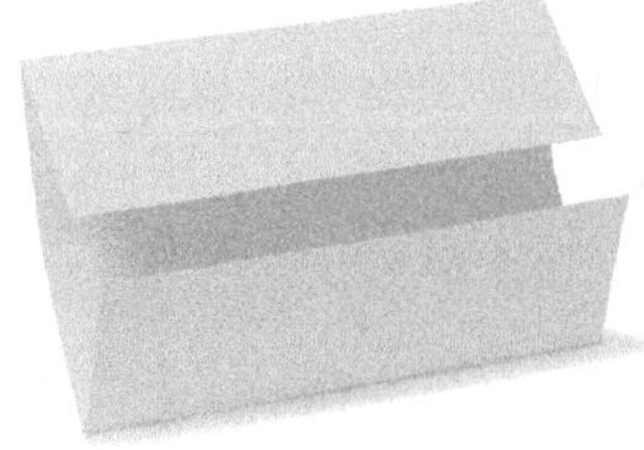

Rational Functions

MODULE 9

Essential Question: How can you use rational functions to solve real-world problems?

TEKS

REAL WORLD VIDEO
As a consumer, you may shop around for the best deal on a new bike helmet. Check out the video to see how sporting goods manufacturers can use rational functions to help set pricing and sales goals.

MODULE PERFORMANCE TASK PREVIEW

What Is the Profit?

Like any business, a manufacturer of bike helmets must pay attention to ways to minimize costs and maximize profit. Businesses use mathematical functions to calculate and predict various quantities, including profits, costs, and revenue. What are some of the ways a business can use a profit function? Let's find out!

Are YOU Ready?

Complete these exercises to review skills you will need for this module.

Personal Math Trainer
- Online Homework
- Hints and Help
- Extra Practice

Graphing Linear Nonproportional Relationships

Example 1 Graph $y = -\frac{1}{2}x - 3$

Plot the y-intercept $(0, -3)$

The slope is $-\frac{1}{2}$, so from $(0, -3)$, plot the next point up 1 and left 2.

Draw a line through the two points.

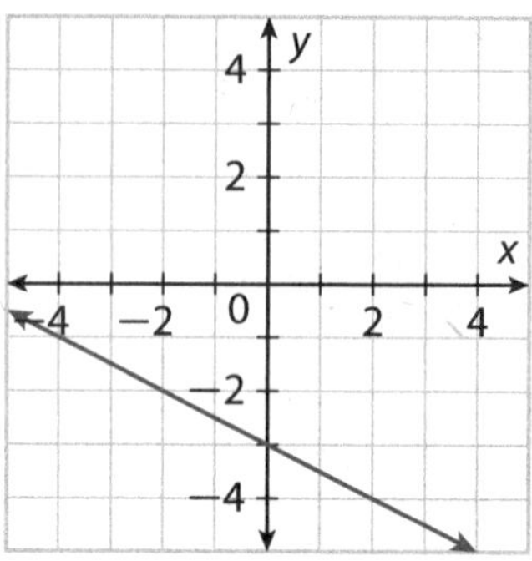

Graph each relationship.

1. $y = 3x - 4$

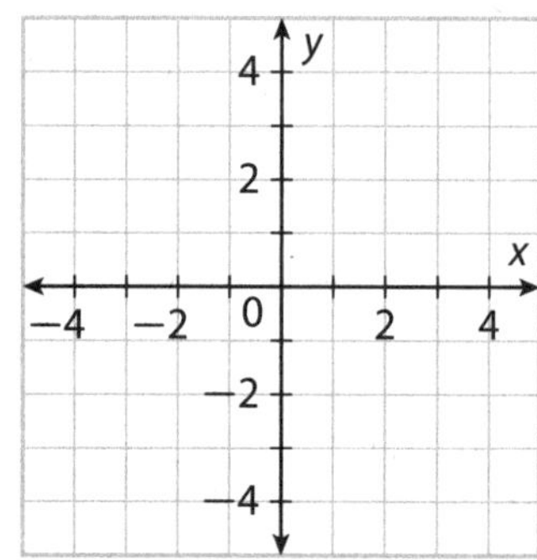

2. $y = -\frac{3}{4}x + 1$

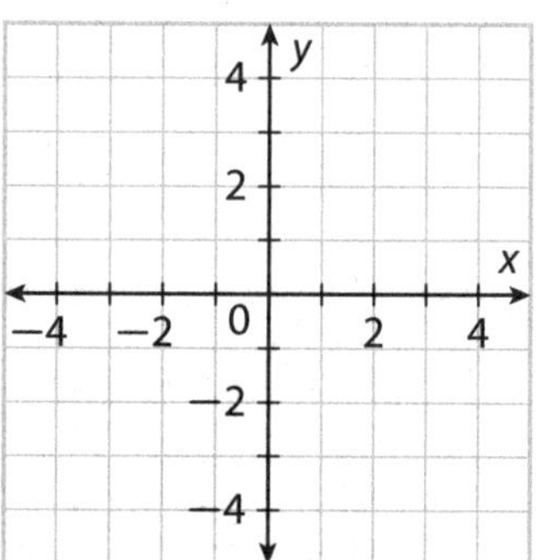

Direct and Inverse Variation

Example 2 The graph of a direct variation function passes through $(1, 10)$. Write the equation for the function.

$k = \frac{y}{x} = \frac{10}{1} = 10$, so the equation is $y = 10x$.

Example 3 The graph of an inverse variation function passes through $(4, 3)$. Write the equation for the function.

$k = xy = 4 \cdot 3 = 12$, so the equation is $y = \frac{12}{x}$.

Write a direct variation equation for the graph that passes through the point.

3. $(2, 5)$ ________ **4.** $(3, 6)$ ________ **5.** $(4, 1)$ ________

Write an inverse variation equation for the graph that passes through the point.

6. $(8, 1)$ ________ **7.** $(5, 10)$ ________ **8.** $(2, 6)$ ________

Name________________________ Class______________ Date__________

9.1 Inverse Variation

Resource Locker

Essential Question: What does it mean for one variable to vary inversely as another variable?

A2.6.L Formulate and solve equations involving inverse variation.

Explore Investigating Inverse Variation

You know that the area A of a rectangle with length ℓ and width w is given by the formula $A = \ell w$. If you hold the width constant and allow the length to vary, the area will vary also. For instance, if $w = 2$, then the equation relating A and ℓ is $A = 2\ell$, A is said to *vary directly* as ℓ and 2 is called the *constant of variation*. In general, any equation of the form $y = ax$, where a is a nonzero constant, is a *direct variation*, and its graph is a line that passes through the origin. For instance, the graph of $A = 2\ell$ is shown. The graph is a ray rather than a line because length and area are nonnegative quantities.

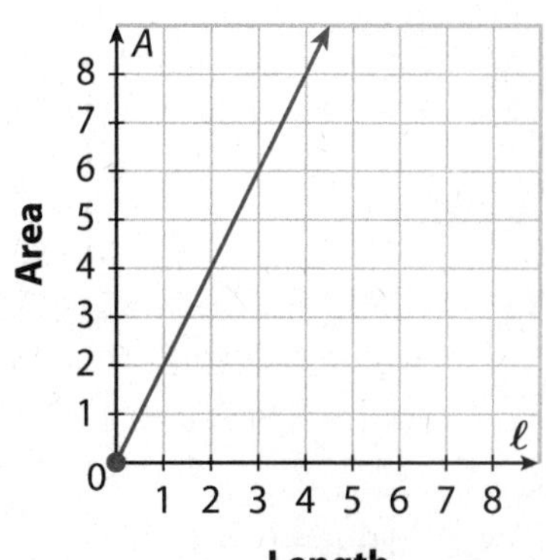

In this Explore, you will consider what happens to the length and width of a rectangle if you hold the area constant.

(A) Let the area of a rectangle be 12 square units. Complete the table, which lists possible positive-integer lengths and widths for the rectangle. Note that it doesn't matter whether the width is less than, equal to, or greater than the length.

Length	Width
1	12
2	
3	
4	
6	
12	

Ⓑ On the coordinate plane shown, draw each of the rectangles having the lengths and widths in the table from Step A. Each rectangle should have its lower-left corner at the origin. The first rectangle, having a length of 1 and a width of 12, is already drawn for you.

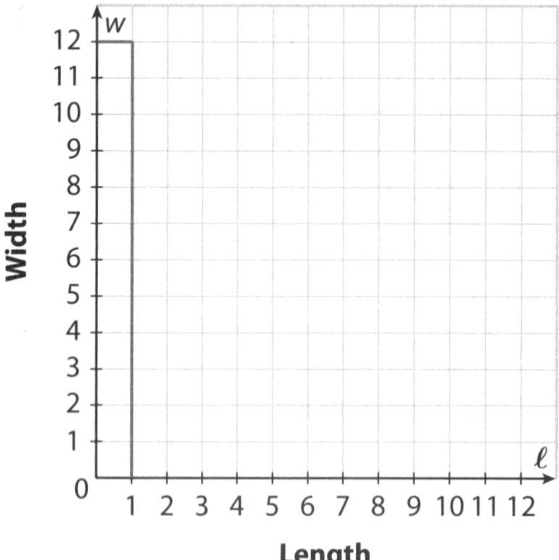

Ⓒ Using the coordinate plane in Step B, draw a smooth curve through the upper-right corners of the rectangles.

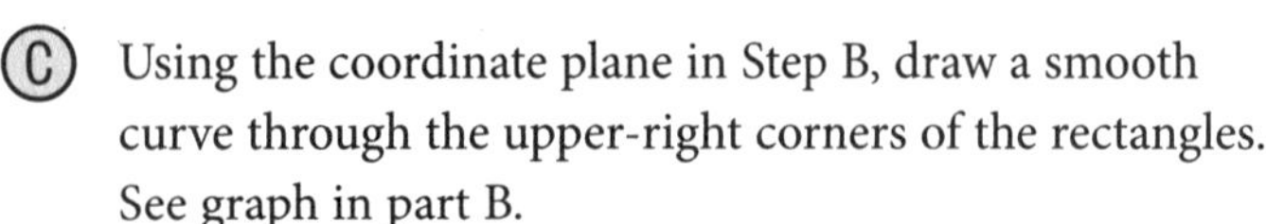

See graph in part B.

Ⓓ Write an equation that gives w in terms of ℓ. ______

Reflect

1. In the direct variation equation $A = 2\ell$, the value of A increases as the value of ℓ increases. For the equation that you wrote in Step D, what can you say about the value of w as the value of ℓ increases?

______

Explain 1 Formulating and Solving Inverse Variation Equations

An **inverse variation** is a relationship between two variables x and y that can be written in the form $y = \frac{a}{x}$ where $a \neq 0$. In this relationship, y is said to vary inversely as x, and a is called the **constant of variation**.

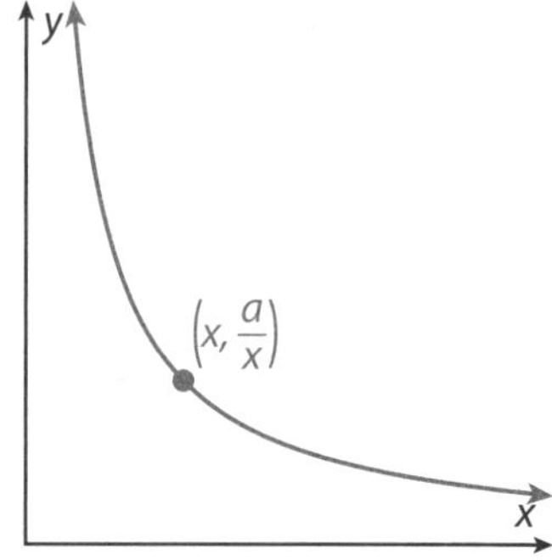

If x and y represent positive real-world quantities, then the graph of $y = \frac{a}{x}$ is a Quadrant I curve that passes through points of the form $\left(x, \frac{a}{x}\right)$ and has the following end behavior:

- 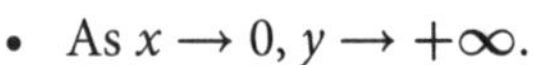As $x \rightarrow 0$, $y \rightarrow +\infty$.
- As $x \rightarrow +\infty$, $y \rightarrow 0$.

To use the equation $y = \frac{a}{x}$ to solve problems, you may find it helpful to rewrite it as $xy = a$.

Example 1 **Write an equation relating the variables and use it to answer the question.**

(A) The time t (in days) that it takes a theater crew to set up a stage for a musical varies inversely as the number of workers w. If 30 workers can set up the stage in 4 days, how many days would it take if only 24 workers are available?

Write an equation relating w and t by finding the constant of variation, a.

Write the general equation. $t = \frac{a}{w}$ Substitute known values. $4 = \frac{a}{30}$

Solve for a. $120 = a$ So, an equation is $t = \frac{120}{w}$.

Find the amount of time needed to set up the stage with 24 workers.

Using the equation $t = \frac{120}{w}$, substitute 24 for w to get $t = \frac{120}{24} = 5$.

So, 24 workers can set up the stage in 5 days.

(B) The time t (in hours) that it takes a group of volunteers to clean up a city park varies inversely as the number of volunteers v. If 10 volunteers can clean up the park in 6 hours, how many volunteers would be needed to clean up the park in 4 hours?

Write an equation relating v and t by finding the constant of variation, a.

Write the general equation. $t = \frac{a}{v}$

Substitute known values. $10 = \frac{a}{\square}$

Solve for a. $\square = a$ So, an equation is $t = \frac{\square}{v}$.

Find the number of volunteers needed to clean up the park in 4 hours.

Using the equivalent equation $vt = \square$, substitute 4 for t and solve for v to get $v = \square$.

So, ______ volunteers are needed to clean up the park in 4 hours.

Reflect

2. **Discussion** Using the equivalent equation $xy = a$ for an inverse variation, explain why the point (q, p) must be on the graph of the equation if the point (p, q) is.

______

______

______

3. **Discussion** What does the fact that for every point (p, q) on the graph of $y = \frac{a}{x}$ the point (q, p) is also on the graph tell you about the symmetry of the graph?

______

Your Turn

Write an equation relating the variables and use it to answer the question.

4. Kim lives in a suburb and drives to work in a city. The time t (in hours) it takes her to get to work varies inversely with her average driving speed s (in miles per hour). When she averages 20 miles per hour in heavy traffic, it takes her 1.5 hours to get to work. How long would the trip take if her average driving speed is 50 miles per hour in light traffic?

5. Boyle's law says that the volume V of a gas held in a container at a constant temperature varies inversely with the pressure P on the gas. The volume of a particular gas is 8 liters at a pressure of 3 atmospheres. What is the pressure when the volume is 6 liters?

Explain 2 Distinguishing between Inverse Variation and Direct Variation

As you know, you can rewrite the inverse variation equation $y = \frac{a}{x}$ as $xy = a$. The alternative form of the equation gives you a way to check for inverse variation in a table of data: If the products of the paired values of the variables are constant (or nearly constant), then inverse variation exists.

A direct variation equation has the form $y = ax$, which you can rewrite as $\frac{y}{x} = a$. So, to check for direct variation in a table of data, see if the ratios of the paired values of the variables are constant.

Example 2 **Determine whether the two variables vary inversely or directly. Then write an equation and use it to answer the question.**

(A) The table gives the total cost (in dollars) of various numbers of tickets to a school play. What is the cost of 12 tickets?

Tickets, t	3	7	9
Total cost, C	12	28	36

Check both the products tC and the ratios $\frac{C}{t}$ to see which are constant.

t	C	tC	$\frac{C}{t}$
3	12	36	4
7	28	196	4
9	36	324	4

Because the ratios $\frac{C}{t}$ are constant, C varies directly as t.

Write the equation. $C = 4t$

Find the cost for 12 tickets. $C = 4(12) = 48$

So, 12 tickets cost $48.

(B) The table gives the time (in hours) needed to drive to a destination at various average speeds (in miles per hour). What is the time needed to get to the destination when driving at an average speed of 48 miles per hour?

Average Speed, s	12	15	20
Time, t	50	40	30

Check both the products st and the ratios $\frac{t}{s}$ to see which are constant.

s	t	st	$\frac{t}{s}$
12	50		$4.1\overline{6}$
15	40	600	
20	30		

Because the __________ are constant, s varies __________ as t.

Write the equation.

Find the time needed to get to the destination when driving at an average speed of 48 miles per hour.

$t = \frac{\square}{48} = \square$

So, it takes ______ hours to get to the destination when driving at an average speed of 48 miles per hour.

Reflect

6. State the real-world significance of the constant of variation in Part A and Part B.

Your Turn

Determine whether the two variables vary inversely or directly. Then write an equation and use it to answer the question.

7. The table gives the number of pages read from a book for various amounts of time (in hours) spent reading. How many pages can be read in 6 hours?

Time, t	2	3	5
Pages Read, p	40	60	100

8. The table gives the amount (in dollars) that each person owes when a group of people buys a birthday present for a friend. How much does each person owe when 8 people chip in for the present?

Number of People, p	2	4	5
Amount each Owes, A	40	20	16

Elaborate

9. How do direct variation and inverse variation differ?

10. What is another way to write the inverse variation equation $y = \frac{a}{x}$? How is this equation helpful?

11. **Essential Question Check-In** If x and y represent positive real-world quantities that vary inversely, what happens to y as x approaches 0? What happens to y as x increases without bound?

✪ Evaluate: Homework and Practice

- Online Homework
- Hints and Help
- Extra Practice

1. Let the area of a rectangle be 16 square units. On the coordinate plane shown, draw all rectangles having positive-integer lengths and widths. (Note that it doesn't matter whether the width is less than, equal to, or greater than the length.) Each rectangle should have its lower-left corner at the origin. The first rectangle, having a length of 1 and a width of 16, is already drawn for you. After drawing the rectangles, draw a smooth curve through their upper-right corners.

2. Write an equation of the curve that you drew in Evaluate 1. The equation should give the width w of a rectangle in terms of the length ℓ.

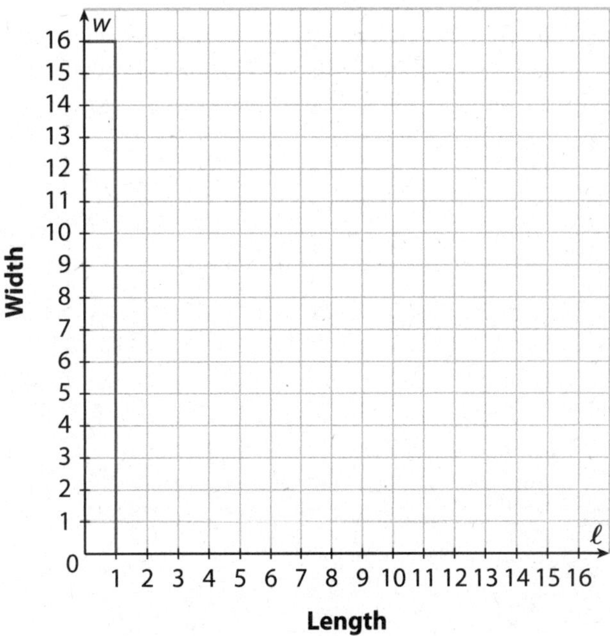

Write an equation relating the variables and use it to answer the question.

3. Given that y varies inversely as x, and $y = 8$ when $x = 3$, what is the value of y when $x = 12$?

4. Given that y varies inversely as x, and $y = 10$ when $x = 4$, what is the value of x when $y = 8$?

5. The time t (in hours) that it takes a pump to empty a tank of water varies inversely with the pumping rate r (in gallons per hour). If it takes 3 hours to empty a tank of water when the pumping rate is 80 gallons per hour, how long does it take to empty the tank when the pumping rate is 60 gallons per hour?

6. The number of flowers f that a gardener can plant along a border of a garden varies inversely with the distance d (in inches) between the flowers. If the gardener can fill the border with 30 flowers planted 12 inches apart, how far apart should the gardener plant 36 flowers?

7. The number of presents p that Tim can afford to buy varies inversely with their average cost C (in dollars). If Tim can afford 5 presents when their average cost is $12, what average cost would 3 presents have?

8. A club rents a bus for a trip. The cost C (in dollars) that each person pays to cover the cost of the bus varies inversely with the number of people p who go on the trip. It will cost $30 per person if 50 people go on the trip. How much will it cost per person if 40 people go on the trip?

9. For a fundraiser, members of the booster club wash cars by hand. The time t (in minutes) it takes to wash a car varies inversely as the number of people p who are washing the car. If 2 people can wash a car in 20 minutes, how many people would be needed to wash a car in 8 minutes?

10. A gear with 32 teeth meshes with a gear with 40 teeth so that when one gear revolves, the other one does as well. The number of revolutions r that each gear makes varies inversely with the gear's number of teeth t. When the gear with 32 teeth makes 5 revolutions, how many revolutions does the gear with 40 teeth make?

11. **Music** The frequency f (in hertz) of a vibrating guitar string varies inversely as its length ℓ (in centimeters). If a guitar string 65 centimeters long vibrates with a frequency of 110 hertz, at what frequency would the guitar string vibrate when the guitarist reduces the string's length to 22 centimeters?

12. **Physics** When a lever is placed on a fulcrum and a force is applied at each end, the lever will be in balance as long as the magnitude m (in pounds) of each force and the distance d (in feet) of each force from the fulcrum satisfy an inverse variation relationship. If a force of 50 pounds is applied to one end of a lever at a distance of 2 feet from the fulcrum, what force must be applied to other end, which is 5 feet from the fulcrum, to bring the lever into balance?

Determine whether the two variables vary inversely or directly. Then write an equation and use it to answer the question.

13. Given the table of data, what is y when $x = 15$?

x	1	3	4
y	30	10	7.5

14. Given the table of data, what is y when $x = 10$?

x	2	5	6
y	30	75	90

15. The table gives the cost (in dollars) per person when friends share in renting a mountain cabin for a weekend. What is the cost per person when 6 friends rent the cabin?

Number of People, p	2	3	5
Cost per Person, C	90	60	36

16. The table gives the amount of gas (in gallons) used when driving a car various distances (in miles) on highways. What amount of gas is used when driving 336 miles on highways?

Distance Driven, d	112	140	224
Amount of Gas, g	4	5	8

17. The table gives the cost (in dollars) of renting a rowboat at a lake for various amounts of time (in hours). What is the cost of renting a rowboat for 3.5 hours?

Time, t	2	2.5	3
Rental Cost, C	28	35	42

18. The table gives the total working time (in hours) that it takes a crew of painters to paint a house. How much time does it take 6 painters to paint the house?

Number of Painters, p	2	3	5
Total Working Time, t	48	32	19.2

19. The table gives the speed of a bicycle (in miles per hour) when a cyclist pedals at various rates (in revolutions per minute of the pedals) with the bicycle in a particular gear. What is the bicycle's speed when the cyclist pedals at a rate of 72 revolutions per minute?

Pedaling Rate, r	30	60	90
Bicycle's Speed, s	5	10	15

20. The table gives the number of laptops sold in a month when a store sells a particular model of laptop at various prices (in dollars). How many laptops would be sold in a month when the store sells the laptop for $500?

Price of Laptop, p	600	720	800
Number of Laptops Sold, ℓ	60	50	45

21. The table gives the number of small figurines that can be placed on a display shelf for various distances (in centimeters) between them. How many figurines can be placed on the shelf when they are 24 centimeters apart?

Distance between Figurines, s	10	15	20
Number of Figurines, f	12	8	6

22. The table gives the amount of water (in gallons) coming out of a garden hose for various amounts of time (in minutes). How much water comes out of the garden hose in 20 minutes?

Time, t	4	9	16
Amount of Water, w	72	162	288

23. Determine whether each of the following situations represents direct variation or inverse variation. Select the correct answer for each lettered part.

a. When traveling a fixed distance, the travel time is a function of average speed. ○ Direct variation ○ Inverse variation

b. For items that are priced the same, the total cost is a function of the number of items purchased. ○ Direct variation ○ Inverse variation

c. When traveling at a fixed speed, the distance traveled is a function of the travel time. ○ Direct variation ○ Inverse variation

d. For a fixed amount of money, the number of identical items that can be purchased is a function of the cost per item. ○ Direct variation ○ Inverse variation

H.O.T. Focus on Higher Order Thinking

24. Justify Reasoning If y varies directly as x, what happens to the value of y when the value of x is doubled? If y varies inversely as x, what happens to the value of y when the value of x is doubled? Use the general equations for direct and inverse variation to justify your reasoning.

25. Explain the Error Boyle's law says that the volume V of a gas held in a container at a constant temperature varies inversely with the pressure P on the gas. The volume of a particular gas is 6 liters at a pressure of 2 atmospheres. Jerry says that if the pressure is changed to 3 atmospheres, the volume will be 9 liters. Explain and correct his error.

26. Represent Real-World Situations The volume V of a gas varies inversely as the pressure P and directly as the temperature T. A particular gas has a volume of 10 liters, a temperature of 300 kelvins, and a pressure of 1.5 atmospheres. If the gas is compressed to a volume of 7.5 liters and is heated to a temperature of 360 kelvins, what will the pressure be? Write an equation and use it to answer the question.

Lesson Performance Task

You have a collection of CDs whose songs you want to transfer to your new MP3 player. The MP3 player has 32,000 megabytes (MB) of storage. An average song lasts 4 minutes and requires 40 MB of storage on a CD.

a. Write a function that gives the number S of songs that your MP3 player can store if the average file size of a song is s MB.

b. If you were to transfer the songs from your CDs in their current size, how many songs could you expect to store on your MP3 player?

c. The MP3 file format compresses songs without much loss in the quality of the sound. Typically, the MP3 format compresses a CD file to a tenth of its size. How many songs from your CDs can you expect to store as MP3 files on your MP3 player?

d. In general, if you could use a file format that compresses the songs on your CDs by a factor of $\frac{1}{f}$ where $f > 1$, how many songs in that format can you expect to store on your MP3 player?

Name________________________ Class____________ Date________

9.2 Graphing Simple Rational Functions

Resource Locker

Essential Question: How are the graphs of $f(x) = a\left(\frac{1}{x-h}\right) + k$ and $f(x) = \frac{1}{\frac{1}{b}(x-h)} + k$ related to the graph of $f(x) = \frac{1}{x}$?

TEKS **A2.6.G** Analyze the effect on the graphs of $f(x) = \frac{1}{x}$ when $f(x)$ is replaced by $af(x)$, $f(bx)$, $f(x - c)$, and $f(x) + d$ for specific positive and negative real values of a, b, c, and d. Also A2.2.A, A2.6.K, A2.7.I

Explore 1 Graphing and Analyzing $f(x) = \frac{1}{x}$

A **rational function** is a function of the form $f(x) = \frac{p(x)}{q(x)}$ where $p(x)$ and $q(x)$ are polynomials. The most basic rational function with a variable expression in the denominator is $f(x) = \frac{1}{x}$.

(A) State the domain of $f(x) = \frac{1}{x}$.

The function accepts all real numbers except ___, because division by ___ is undefined. So, the function's domain is as follows:

- As an inequality: $x <$ ☐ or $x >$ ☐
- In set notation: $\{x \mid x \neq$ ☐ $\}$
- In interval notation (where the symbol ∪ means *union*):

$(-\infty,$ ☐ $) \cup ($ ☐ $, +\infty)$

(B) Determine the end behavior of $f(x) = \frac{1}{x}$.

First, complete the tables.

x Increases without Bound	
x	$f(x) = \frac{1}{x}$
100	
1000	
10,000	

x Decreases without Bound	
x	$f(x) = \frac{1}{x}$
−100	
−1000	
−10,000	

Next, summarize the results.

- As $x \to +\infty$, $f(x) \to$ ☐.
- As $x \to -\infty$, $f(x) \to$ ☐.

Ⓒ Be more precise about the end behavior of $f(x) = \frac{1}{x}$, and determine what this means for the graph of the function.

You can be more precise about the end behavior by using the notation $f(x) \to 0^+$, which means that the value of $f(x)$ approaches 0 from the positive direction (that is, the value of $f(x)$ is positive as it approaches 0), and the notation $f(x) \to 0^-$, which means that the value of $f(x)$ approaches 0 from the negative direction. So, the end behavior of the function is more precisely summarized as follows:

- As $x \to +\infty$, $f(x) \to$ ☐.
- As $x \to -\infty$, $f(x) \to$ ☐.

The end behavior indicates that the graph of $f(x)$ approaches, but does not cross, the [x-axis/y-axis], so that axis is an asymptote for the graph.

Ⓓ Examine the behavior of $f(x) = \frac{1}{x}$ near $x = 0$, and determine what this means for the graph of the function.

First, complete the tables.

x Approaches 0 from the Positive Direction	
x	$f(x) = \frac{1}{x}$
0.01	
0.001	
0.0001	

x Approaches 0 from the Negative Direction	
x	$f(x) = \frac{1}{x}$
−0.01	
−0.001	
−0.0001	

Next, summarize the results.

- As $x \to 0^+$, $f(x) \to$ ☐.
- As $x \to 0^-$, $f(x) \to$ ☐.

The behavior of $f(x) = \frac{1}{x}$ near $x = 0$ indicates that the graph of $f(x)$ approaches, but does not cross, the [x-axis/y-axis], so that axis is also an asymptote for the graph.

Ⓔ Graph $f(x) = \frac{1}{x}$.

First, determine the sign of $f(x)$ on the two parts of its domain.

- When x is a negative number, $f(x)$ is a [positive/negative] number.
- When x is a positive number, $f(x)$ is a [positive/negative] number.

Next, complete the tables.

Negative Values of *x*	
x	$f(x) = \frac{1}{x}$
−2	
−1	
−0.5	

Positive Values of *x*	
x	$f(x) = \frac{1}{x}$
0.5	
1	
2	

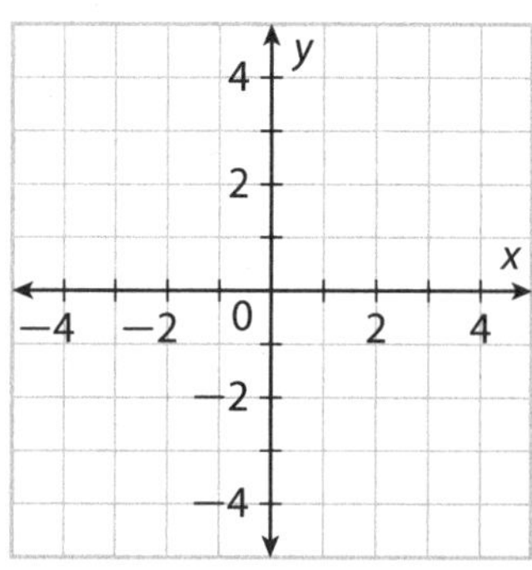

Finally, use the information from this step and all previous steps to draw the graph. Draw asymptotes as dashed lines.

(F) State the range of $f(x) = \frac{1}{x}$.

The function takes on all real numbers except ________, so the function's range is as follows:

- As an inequality: $y <$ ☐ or $y >$ ☐
- In set notation: $\{y \mid y \neq \square\}$
- In interval notation (where the symbol ∪ means *union*): $(-\infty, \square) \cup (\square, +\infty)$

(G) Identify the intervals where the function is increasing and where it is decreasing.

____

____

(H) Determine whether $f(x) = \frac{1}{x}$ is an even function, an odd function, or neither.

____

Reflect

1. How does the graph of $f(x) = \frac{1}{x}$ show that the function has no zeros?

____

2. **Discussion** A graph is said to be *symmetric about the origin* (and the origin is called the graph's *point of symmetry*) if for every point (x, y) on the graph, the point $(-x, -y)$ is also on the graph. Is the graph of $f(x) = \frac{1}{x}$ symmetric about the origin? Explain.

3. Give any line(s) of symmetry for the graph of $f(x) = \frac{1}{x}$.

____

Explore 2 Predicting Transformations of the Graph of $f(x) = \frac{1}{x}$

In an earlier lesson, you learned how to transform the graph of a function using reflections across the x- and y-axes, vertical and horizontal stretches and compressions, and translations. Here, you will apply those transformations to the graph of the function $f(x) = \frac{1}{x}$.

(A) Predict the effect of the parameter h on the graph of $g(x) = \frac{1}{x-h}$ for each function.

a. The graph of $g(x) = \frac{1}{x-3}$ is a ________ of the graph of $f(x)$ [right/left/up/down] 3 units.

b. The graph of $g(x) = \frac{1}{x+3}$ is a ________ of the graph of $f(x)$ [right/left/up/down] 3 units.

Check your predictions using a graphing calculator.

Ⓑ Predict the effect of the parameter k on the graph of $g(x) = \frac{1}{x} + k$ for each function.

a. The graph of $g(x) = \frac{1}{x} + 3$ is a ______________ of the graph of $f(x)$ [right/left/up/down] 3 units.

b. The graph of $g(x) = \frac{1}{x} - 3$ is a ______________ of the graph of $f(x)$ [right/left/up/down] 3 units.

Check your predictions using a graphing calculator.

Ⓒ Predict the effect of the parameter a on the graph of $g(x) = a\left(\frac{1}{x}\right)$ for each function.

a. The graph of $g(x) = 2\left(\frac{1}{x}\right)$ is a ______________ of the graph of $f(x)$ by a factor of _____.

b. The graph of $g(x) = 0.5\left(\frac{1}{x}\right)$ is a ______________ of the graph of $f(x)$ by a factor of _____.

c. The graph of $g(x) = -0.5\left(\frac{1}{x}\right)$ is a ______________ of the graph of $f(x)$ by a factor of _____ as well as a ______________ across the __________.

d. The graph of $g(x) = -2\left(\frac{1}{x}\right)$ is a ______________ of the graph of $f(x)$ by a factor of _____ as well as a ______________ across the __________.

Check your predictions using a graphing calculator.

Ⓓ Predict the effect of the parameter b on the graph of $g(x) = \frac{1}{\frac{1}{b}x}$ for each function.

a. The graph of $g(x) = \frac{1}{0.5x}$ is a ______________ of the graph of $f(x)$ by a factor of _____.

b. The graph of $g(x) = \frac{1}{2x}$ is a ______________ of the graph of $f(x)$ by a factor of _____.

c. The graph of $g(x) = \frac{1}{-0.5x}$ is a ______________ of the graph of $f(x)$ by a factor of _____ as well as a ______________ across the __________.

d. The graph of $g(x) = \frac{1}{-2x}$ is a ______________ of the graph of $f(x)$ by a factor of _____ as well as a ______________ across the __________.

Reflect

4. For each function from Step D, there is a function from Step C that has an identical graph even though the graphs of the two functions involve different transformations of the graph of $f(x)$. Using the letters and numbers in the lists below, match each function from Step D with a function from Step C and show why the functions are equivalent.

Functions from Step D	Functions from Step C
1. $g(x) = \frac{1}{0.5x}$	a. $g(x) = 2\left(\frac{1}{x}\right)$
2. $g(x) = \frac{1}{2x}$	b. $g(x) = 0.5\left(\frac{1}{x}\right)$
3. $g(x) = \frac{1}{-0.5x}$	c. $g(x) = -0.5\left(\frac{1}{x}\right)$
4. $g(x) = \frac{1}{-2x}$	d. $g(x) = -2\left(\frac{1}{x}\right)$

Explain 1 Graphing Simple Rational Functions

When graphing transformations of $f(x) = \frac{1}{x}$, it helps to consider the effect of the transformations on the following features of the graph of $f(x)$: the vertical asymptote, $x = 0$; the horizontal asymptote, $y = 0$; and two reference points, $(-1, -1)$ and $(1, 1)$. The table lists these features of the graph of $f(x)$ and the corresponding features of the graph of $g(x) = a\left(\frac{1}{\frac{1}{b}(x-h)}\right) + k$. Note that the asymptotes are affected only by the parameters h and k, while the reference points are affected by all four parameters.

Feature	$f(x) = \frac{1}{x}$	$g(x) = a\left(\frac{1}{\frac{1}{b}(x-h)}\right) + k$
Vertical asymptote	$x = 0$	$x = h$
Horizontal asymptote	$y = 0$	$y = k$
Reference point	$(-1, -1)$	$(-b + h, -a + k)$
Reference point	$(1, 1)$	$(b + h, a + k)$

Example 1 **Identify the transformations of the graph of $f(x) = \frac{1}{x}$ that produce the graph of the given function $g(x)$. Then graph $g(x)$ on the same coordinate plane as the graph of $f(x)$ by applying the transformations to the asymptotes $x = 0$ and $y = 0$ to the reference points $(-1, -1)$ and $(1, 1)$. Also state the domain and range of $g(x)$ using inequalities, set notation, and interval notation.**

Ⓐ $g(x) = 3\left(\frac{1}{x - 1}\right) + 2$

The transformations of the graph of $f(x)$ that produce the graph of $g(x)$ are:

- a vertical stretch by a factor of 3
- a translation of 1 unit to the right and 2 units up

Note that the translation of 1 unit to the right affects only the x-coordinates, while the vertical stretch by a factor of 3 and the translation of 2 units up affect only the y-coordinates.

Feature	$f(x) = \frac{1}{x}$	$g(x) = 3\left(\frac{1}{x - 1}\right) + 2$
Vertical asymptote	$x = 0$	$x = 1$
Horizontal asymptote	$y = 0$	$y = 2$
Reference point	$(-1, -1)$	$(-1 + 1, 3(-1) + 2) = (0, -1)$
Reference point	$(1, 1)$	$(1 + 1, 3(1) + 2) = (2, 5)$

Domain of $g(x)$:

Inequality: $x < 1$ or $x > 1$

Set notation: $\{x \mid x \neq 1\}$

Interval notation: $(-\infty, 1) \cup (1, +\infty)$

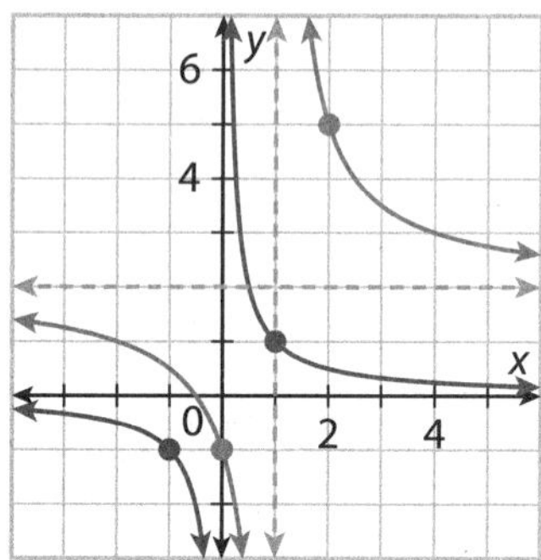

Range of $g(x)$:

Inequality: $y < 2$ or $y > 2$

Set notation: $\{y \mid y \neq 2\}$

Interval notation: $(-\infty, 2) \cup (2, +\infty)$

Ⓑ $g(x) = \frac{1}{2(x+3)} - 1$

The transformations of the graph of $f(x)$ that produce the graph of $g(x)$ are:

- a horizontal compression by a factor of $\frac{1}{2}$
- a translation of 3 units to the left and 1 unit down

Note that the horizontal compression by a factor of $\frac{1}{2}$ and the translation of 3 units to the left affect only the x-coordinates of points on the graph of $f(x)$, while the translation of 1 unit down affects only the y-coordinates.

Feature	$f(x) = \frac{1}{x}$	$g(x) = \frac{1}{2(x+3)} - 1$
Vertical asymptote	$x = 0$	$x = \square$
Horizontal asymptote	$y = 0$	$y = \square$
Reference point	$(-1, -1)$	$\left(\frac{1}{2}(\square) - 3, \square - 1\right) = (\square, \square)$
Reference point	$(1, 1)$	$\left(\frac{1}{2}(\square) - 3, \square - 1\right) = (\square, \square)$

Domain of $g(x)$:

Inequality: $x < \square$ or $x > \square$

Set notation: $\{x \mid x \neq \square\}$

Interval notation: $(-\infty, \square) \cup (\square, +\infty)$

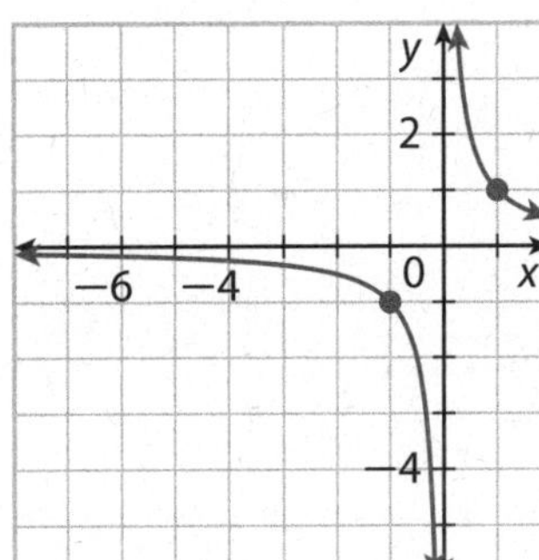

Range of $g(x)$:

Inequality: $y < \square$ or $y > \square$

Set notation: $\{y \mid y \neq \square\}$

Interval notation: $(-\infty, \square) \cup (\square, +\infty)$

Your Turn

Identify the transformations of the graph of $f(x) = \frac{1}{x}$ that produce the graph of the given function $g(x)$. Then graph $g(x)$ on the same coordinate plane as the graph of $f(x)$ by applying the transformations to the asymptotes $x = 0$ and $y = 0$ to the reference points $(-1, -1)$ and $(1, 1)$. Also state the domain and range of $g(x)$ using inequalities, set notation, and interval notation.

5. $g(x) = -0.5\left(\frac{1}{x+1}\right) - 3$

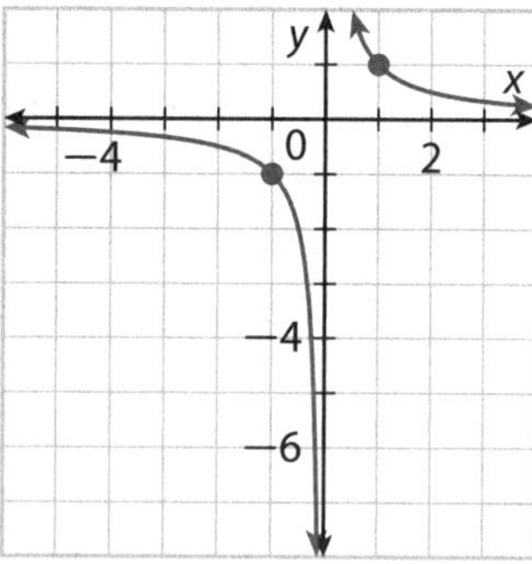

6. $g(x) = \frac{1}{-0.5(x-2)} + 1$

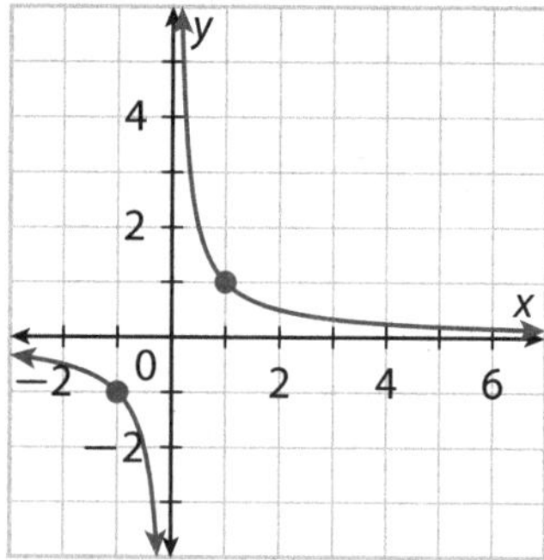

Explain 2 Rewriting Simple Rational Functions in Order to Graph Them

When given a rational function of the form $g(x) = \frac{mx + n}{px + q}$, where $m \neq 0$ and $p \neq 0$, you can carry out the division of the numerator by the denominator to write the function in the form $g(x) = a\left(\frac{1}{x - h}\right) + k$ or $g(x) = \frac{1}{\frac{1}{b}(x - h)} + k$ in order to graph it.

Example 2 **Rewrite the function in the form $g(x) = a\left(\frac{1}{x - h}\right) + k$ or $g(x) = \frac{1}{\frac{1}{b}(x - h)} + k$ and graph it. Also state the domain and range using inequalities, set notation, and interval notation.**

(A) $g(x) = \frac{3x - 4}{x - 1}$

Use long division.

$$x - 1 \overline{\smash{\big)}\, 3x - 4} \quad \text{quotient } 3$$
$$3x - 3$$
$$-1$$

So, the quotient is 3, and the remainder is -1. Using the fact that $dividend = quotient + \frac{remainder}{divisor}$, you have $g(x) = 3 + \frac{-1}{x - 1}$, or $g(x) = -\frac{1}{x - 1} + 3$.

The graph of $g(x)$ has vertical asymptote $x = 1$, horizontal asymptote $y = 3$, and reference points $(-1 + 1, -(-1) + 3) = (0, 4)$ and $(1 + 1, -(1) + 3) = (2, 2)$.

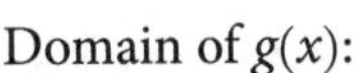

Domain of $g(x)$:

Inequality: $x < 1$ or $x > 1$

Set notation: $\{x \mid x \neq 1\}$

Interval notation: $(-\infty, 1) \cup (1, +\infty)$

Domain of $g(x)$:

Inequality: $y < 3$ or $y > 3$

Set notation: $\{y \mid y \neq 3\}$

Interval notation: $(-\infty, 3) \cup (3, +\infty)$

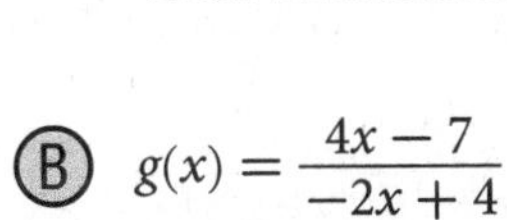

(B) $g(x) = \frac{4x - 7}{-2x + 4}$

Use long division.

$$-2x + 4 \overline{\smash{\big)}\, 4x - 7} \quad \text{quotient } -2$$
$$4x - 8$$
$$\square$$

So, the quotient is -2, and the remainder is ______. Using the fact that $dividend = quotient + \frac{remainder}{divisor}$, you have

$g(x) = -2 + \frac{\square}{-2x + 4}$, or $g(x) = \frac{\square}{-2(x - \square)} - 2$.

The graph of $g(x)$ has vertical asymptote $x = \square$, horizontal asymptote $y = -2$, and reference points

$\left(-\frac{1}{2}(-1) + \square, -1 - 2\right) = \left(\square, -3\right)$ and $\left(-\frac{1}{2}(1) + \square, 1 - 2\right) = \left(\square, -1\right)$.

Domain of $g(x)$:

Inequality: $x < \square$ or $x > \square$

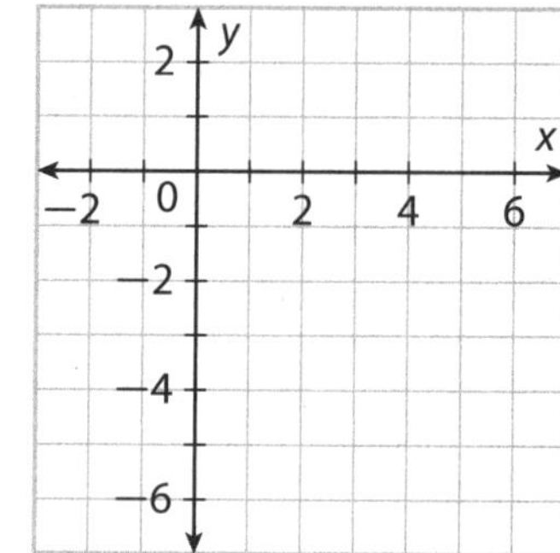

Set notation: $\{x | x \neq \square\}$

Interval notation: $(-\infty, \square) \cup (\square, +\infty)$

Range of $g(x)$:

Inequality: $y < \square$ or $y > \square$

Set notation: $\{y | y \neq \square\}$

Interval notation: $(-\infty, \square) \cup (\square, +\infty)$

Reflect

7. In Part A, the graph of $g(x)$ is the result of what transformations of the graph of $f(x) = \frac{1}{x}$?

8. In Part B, the graph of $g(x)$ is the result of what transformations of the graph of $f(x) = \frac{1}{x}$?

Your Turn

Rewrite the function in the form $g(x) = a\left(\frac{1}{x-h}\right) + k$ or $g(x) = \frac{1}{\frac{1}{b}(x-h)} + k$ and graph it. Also state the domain and range using inequalities, set notation, and interval notation.

9. $g(x) = \dfrac{3x+8}{x+2}$

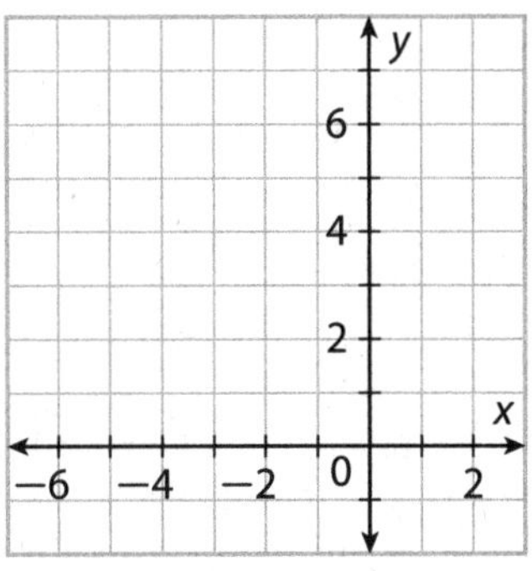

 Explain 3 Writing Simple Rational Functions

When given the graph of a simple rational function, you can write its equation using one of the general forms $g(x) = a\left(\frac{1}{x-h}\right) + k$ and $g(x) = \frac{1}{\frac{1}{b}(x-h)} + k$ after identifying the values of the parameters using information obtained from the graph.

Example 3

Ⓐ **Write the equation of the function whose graph is shown. Use the form $g(x) = a\left(\frac{1}{x-h}\right) + k$.**

Since the graph's vertical asymptote is $x = 3$, the value of the parameter h is 3. Since the graph's horizontal asymptote is $y = 4$, the value of the parameter k is 4.

Substitute these values into the general form of the function.

$$g(x) = a\left(\frac{1}{x-3}\right) + 4$$

Now use one of the points, such as $(4, 6)$, to find the value of the parameter a.

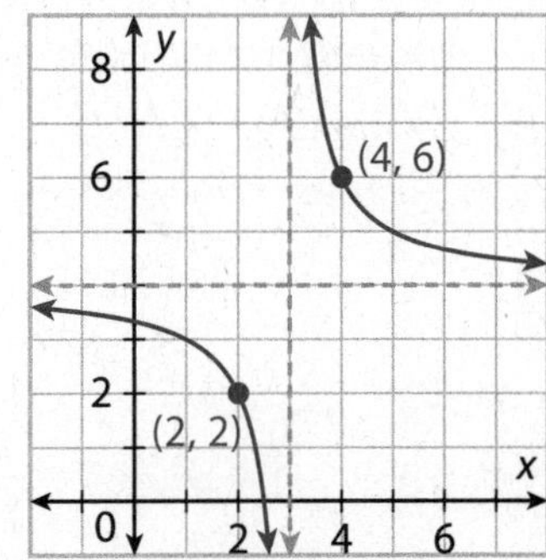

$$g(x) = a\left(\frac{1}{x-3}\right) + 4$$

$$6 = a\left(\frac{1}{4-3}\right) + 4$$

$$6 = a + 4$$

$$2 = a$$

So, $g(x) = 2\left(\frac{1}{x-3}\right) + 4$.

(B) **Write the equation of the function whose graph is shown. Use the form $g(x) = \frac{1}{\frac{1}{b}(x-h)} + k$.**

Since the graph's vertical asymptote is $x = -3$, the value of the parameter h is -3. Since the graph's horizontal asymptote is $y = \square$, the value of the parameter k is ____.

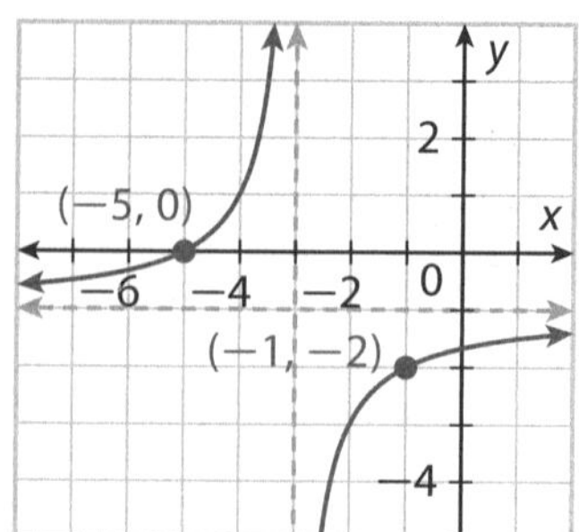

Substitute these values into the general form of the function.

$$g(x) = \frac{1}{\frac{1}{b}(x+3)} + \square$$

Now use one of the points, such as $(-5, 0)$, to find the value of the parameter a.

$$g(x) = \frac{1}{\frac{1}{b}(x+3)} + \square$$

$$0 = \frac{1}{\frac{1}{b}(-5+3)} + \square$$

$$\square = \frac{1}{\frac{1}{b}(-2)}$$

$$\frac{1}{b}(-2) \cdot \square = 1$$

$$\frac{1}{b} = \square$$

$$b = \square$$

So, $g(x) = \frac{1}{\frac{1}{\square}(x+3)} + \square$, or $g(x) = \frac{1}{-0.5(x+3)} - 1$.

Reflect

10. Discussion In Parts A and B, the coordinates of a second point on the graph of $g(x)$ are given. In what way can those coordinates be useful?

Your Turn

11. Write the function whose graph is shown.
Use the form $g(x) = a\left(\frac{1}{x-h}\right) + k$.

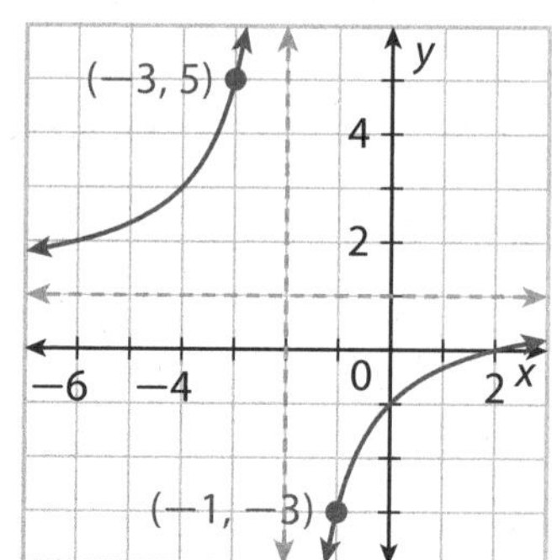

12. Write the equation of the function whose graph is shown. Use the form $g(x) = \frac{1}{\frac{1}{b}(x - h)} + k$.

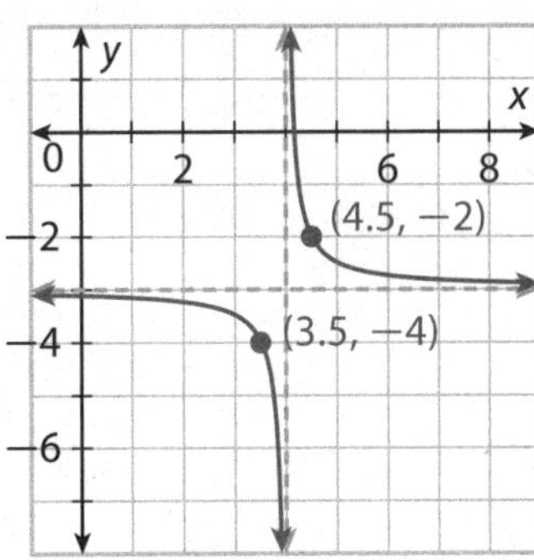

Explain 4 Modeling with Simple Rational Functions

In a real-world situation where there is a shared cost and a per-person or per-item cost, you can model the situation using a rational function that has the general form $f(x) = \frac{a}{x - h} + k$ where $f(x)$ is the total cost for each person or item.

Example 4

(A) **Mary and some of her friends are thinking about renting a car while staying at a beach resort for a vacation. The cost per person for staying at the beach resort is \$300, and the cost of the car rental is \$220. If the friends agree to share the cost of the car rental, what is the minimum number of people who must go on the trip so that the total cost for each person is no more than \$350?**

Analyze Information

Identify the important information.

- The cost per person for the resort is ______.
- The cost of the car rental is ______.
- The most that each person will spend is ______

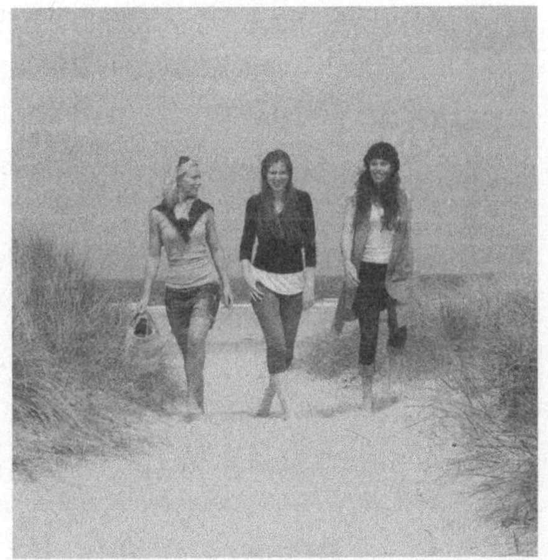

Formulate a Plan

Create a rational function that gives the total cost for each person. Graph the function, and use the graph to answer the question.

Solve

Let p be the number of people who agree to go on the trip. Let $C(p)$ be the cost (in dollars) for each person.

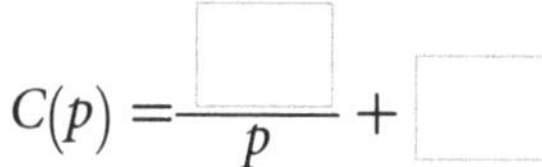

$C(p) = \frac{\square}{p} + \square$

Graph the function, recognizing that the graph involves two transformations of the graph of the parent rational function:

- a vertical stretch by a factor of ______
- a vertical translation of ______ units up

Also draw the line $C(p) = 350$.

The graphs intersect between $p = \square$ and $p = \square$, so the minimum number of people who must go on the trip in order for the total cost for each person to be no more than \$350 is ____.

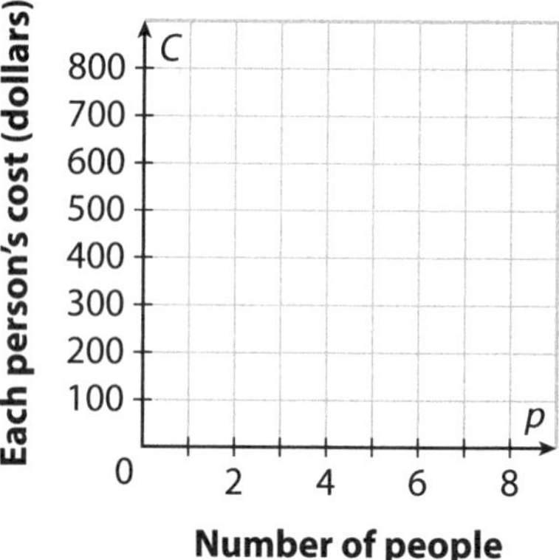

Justify and Evaluate

Check the solution by evaluating the function $C(p)$. Since $C(4) = \square > 350$ and $C(5) = \square < 350$, the minimum number of people who must go on the trip is ______.

Your Turn

13. Justin has purchased a basic silk screening kit for applying designs to fabric. The kit costs \$200. He plans to buy T-shirts for \$10 each, apply a design that he creates to them, and then sell them. Model this situation with a rational function that gives the average cost of a silk-screened T-shirt when the cost of the kit is included in the calculation. Use the graph of the function to determine the minimum number of T-shirts that brings the average cost below \$17.50.

C

40
35
30
25
20
15
10
5
0

10 20 30 40

s

Average cost (dollars)

Number of T-shirts

Elaborate

14. Compare and contrast the attributes of $f(x) = \frac{1}{x}$ and the attributes of $g(x) = -\frac{1}{x}$.

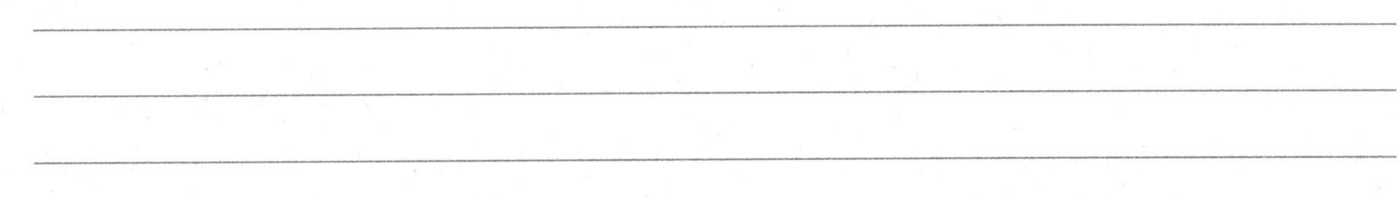

15. State the domain and range of $f(x) = a\left(\frac{1}{x - h}\right) + k$ using inequalities, set notation, and interval notation.

16. Given that the model $C(p) = \frac{100}{p} + 50$ represents the total cost C (in dollars) for each person in a group of p people when there is a shared expense and an individual expense, describe what the expressions $\frac{100}{p}$ and 50 represent.

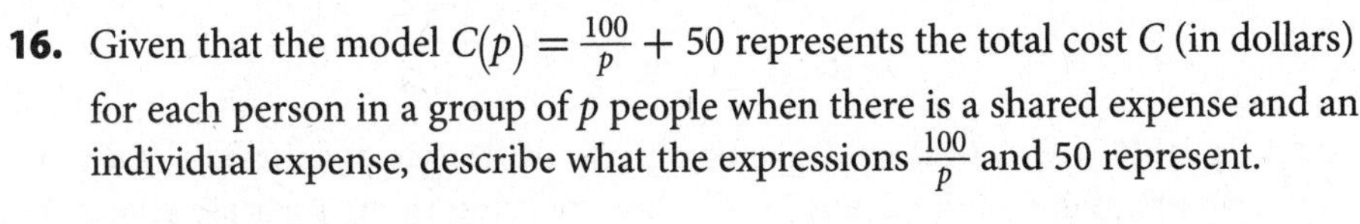

17. Essential Question Check-In Describe the transformations you must perform on the graph of $f(x) = \frac{1}{x}$ to obtain the graph of $f(x) = a\left(\frac{1}{x - h}\right) + k$.

Evaluate: Homework and Practice

- Online Homework
- Hints and Help
- Extra Practice

Describe how the graph of $g(x)$ is related to the graph of $f(x) = \frac{1}{x}$.

1. $g(x) = \frac{1}{x} + 4$

2. $g(x) = 5\left(\frac{1}{x}\right)$

3. $g(x) = \frac{1}{x + 3}$

4. $g(x) = \frac{1}{0.1x}$

5. $g(x) = \frac{1}{x} - 7$

6. $g(x) = \frac{1}{x - 8}$

7. $g(x) = -0.1\left(\frac{1}{x}\right)$

8. $g(x) = \frac{1}{-3x}$

Identify the transformations of the graph of $f(x)$ that produce the graph of the given function $g(x)$. Then graph $g(x)$ on the same coordinate plane as the graph of $f(x)$ by applying the transformations to the asymptotes $x = 0$ and $y = 0$ and to the reference points $(-1, -1)$ and $(1, 1)$. Also state the domain and range of $g(x)$ using inequalities, set notation, and interval notation.

9. $g(x) = 3\left(\frac{1}{x + 1}\right) - 2$

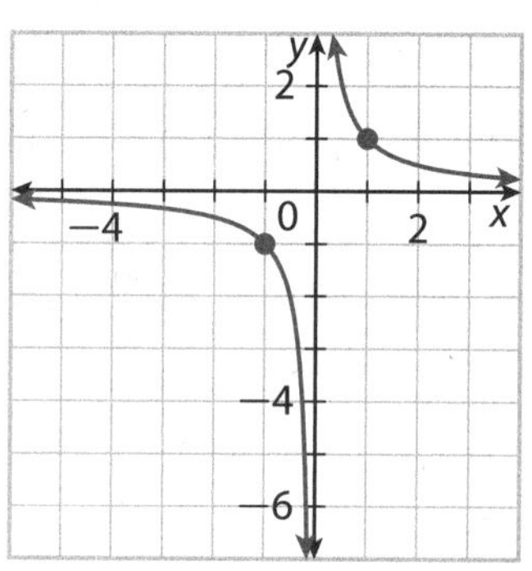

10. $g(x) = \frac{1}{-0.5(x - 3)} + 1$

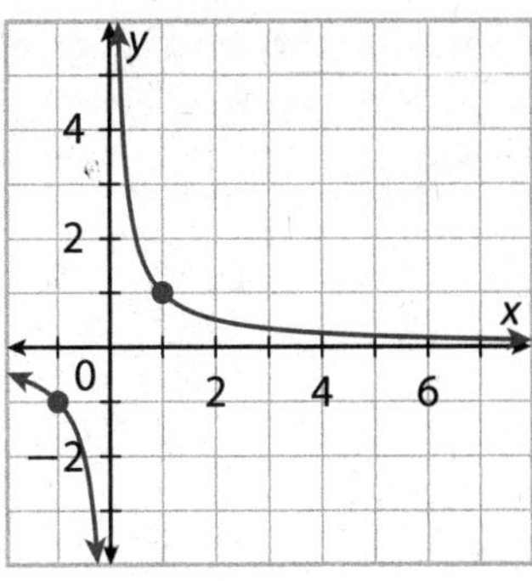

11. $g(x) = -0.5\left(\frac{1}{x - 1}\right) - 2$

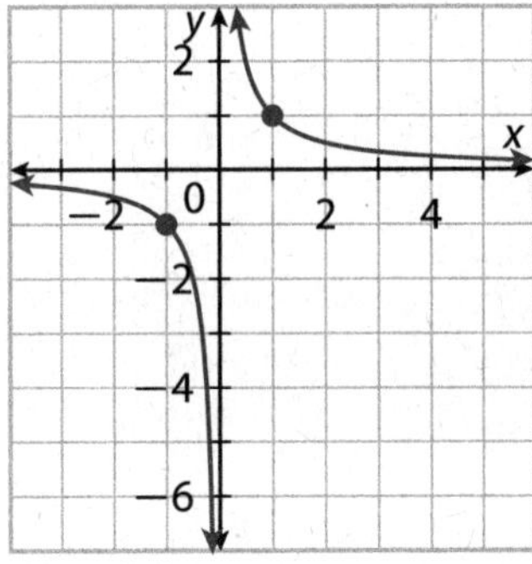

12. $g(x) = \frac{1}{2(x + 2)} + 3$

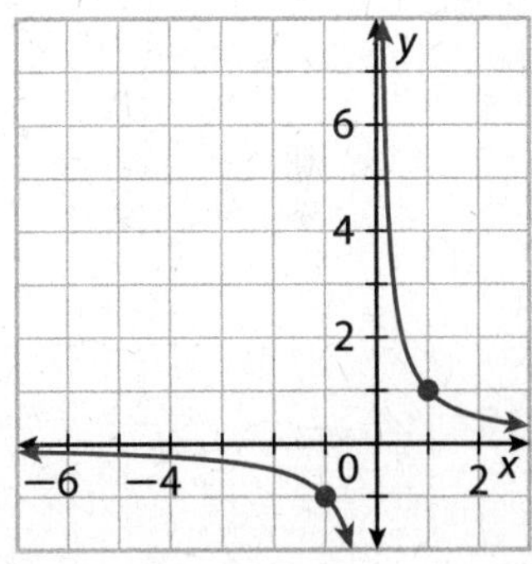

Rewrite the function in the form $g(x) = a\frac{1}{(x-h)} + k$ or $g(x) = \frac{1}{\frac{1}{b}(x-h)} + k$ and graph it. Also state the domain and range using inequalities, set notation, and interval notation.

13. $g(x) = \dfrac{3x - 5}{x - 1}$

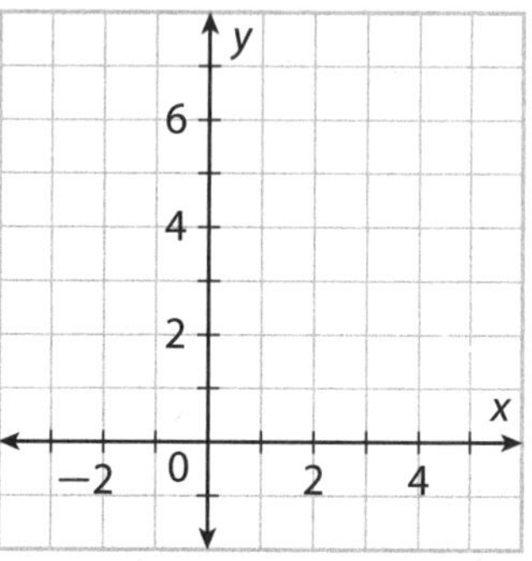

14. $g(x) = \dfrac{x + 5}{0.5x + 2}$

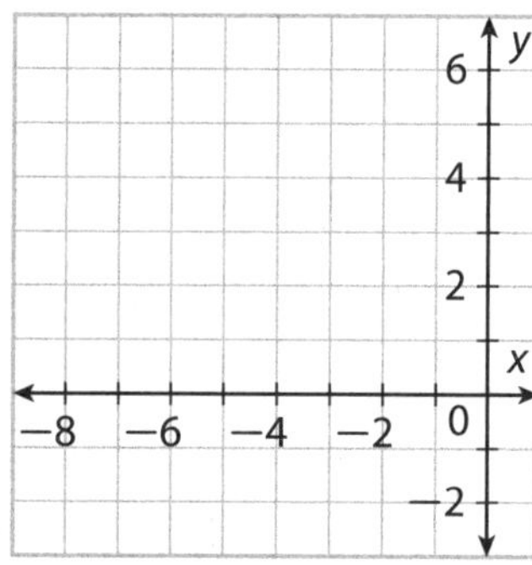

15. $g(x) = \dfrac{-4x + 11}{x - 2}$

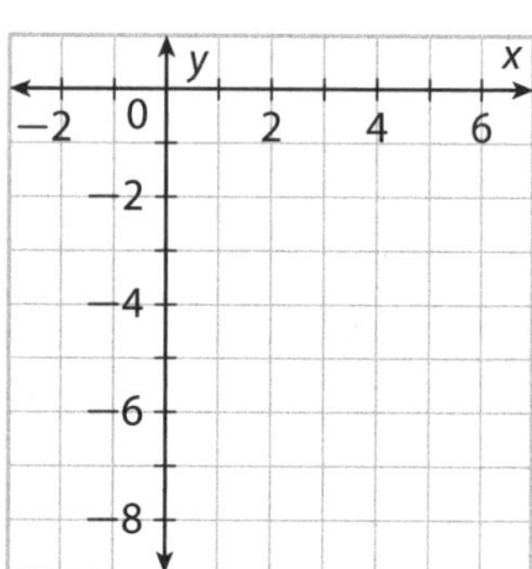

16. $g(x) = \dfrac{4x + 13}{-2x - 6}$

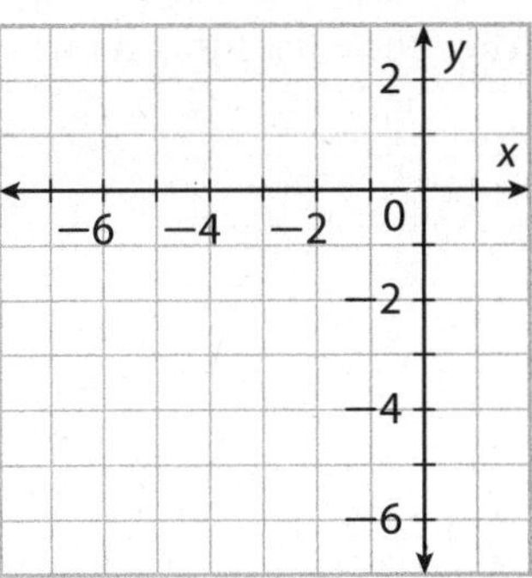

17. Write the function whose graph is shown. Use the form $g(x) = a\left(\frac{1}{x - h}\right) + k$.

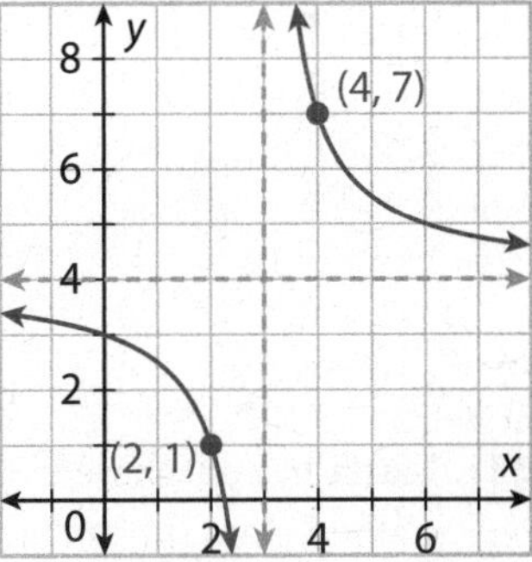

18. Write the function whose graph is shown. Use the form $g(x) = \dfrac{1}{\frac{1}{b}(x - h)} + k$

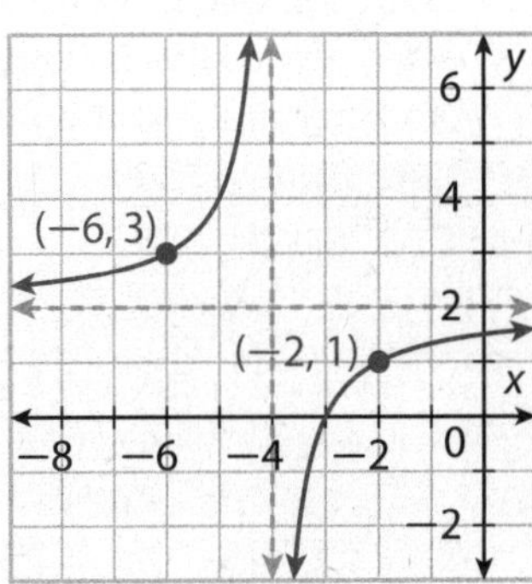

19. Write the function whose graph is shown. Use the form $g(x) = \frac{1}{\frac{1}{b}(x - h)} + k$.

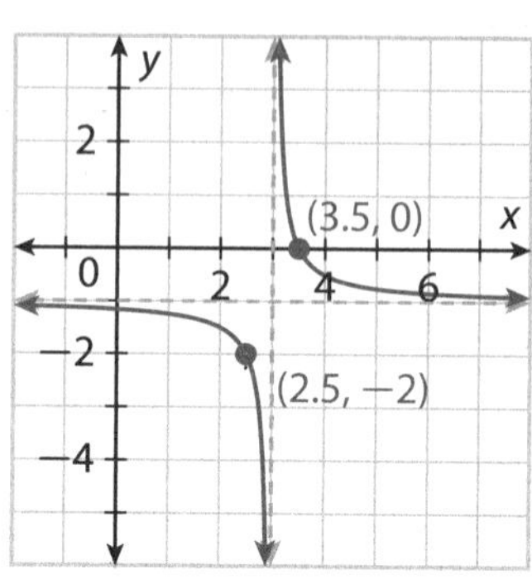

20. Write the function whose graph is shown. Use the form $g(x) = a\left(\frac{1}{x - h}\right) + k$

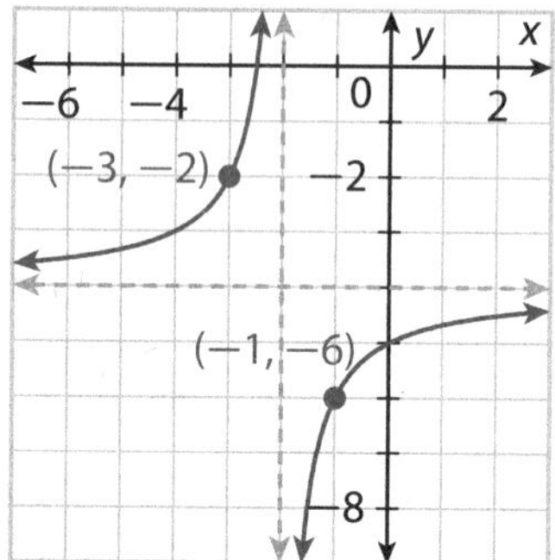

21. Maria has purchased a basic stained glass kit for \$100. She plans to make stained glass suncatchers and sell them. She estimates that the materials for making each suncatcher will cost \$15. Model this situation with a rational function that gives the average cost of a stained glass suncatcher when the cost of the kit is included in the calculation. Use the graph of the function to determine the minimum number of suncatchers that brings the average cost below \$22.50.

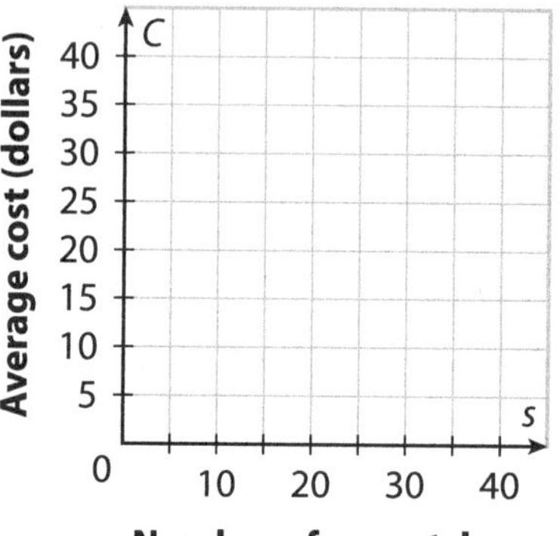

22. Amy has purchased a basic letterpress kit for \$140. She plans to make wedding invitations. She estimates that the cost of the paper and envelope for each invitation is \$2. Model this situation with a rational function that gives the average cost of a wedding invitation when the cost of the kit is included in the calculation. Use the graph of the function to determine the minimum number of invitations that brings the average cost below \$5.

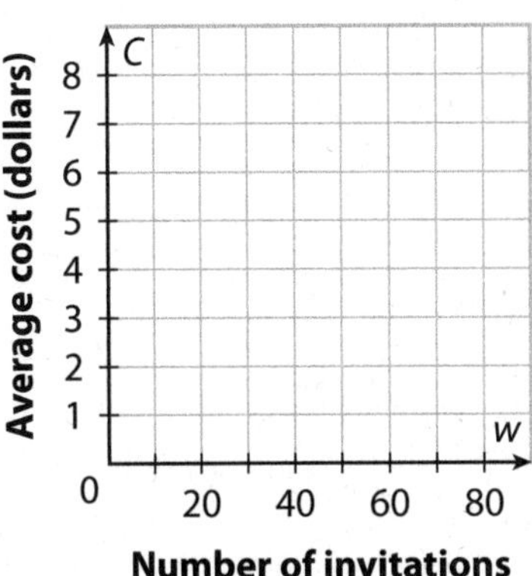

23. Multiple Response Select the transformations of the graph of the parent rational function that result in the graph of $g(x) = \frac{1}{2(x-3)} + 1$

A. Horizontal stretch by a factor of 2

B. Horizontal compression by a factor of $\frac{1}{2}$

C. Vertical stretch by a factor of 2

D. Vertical compression by a factor of $\frac{1}{2}$

E. Translation 1 unit up

F. Translation 1 unit down

G. Translation 3 units right

H. Translation 3 units left

H.O.T. Focus on Higher Order Thinking

24. Justify Reasoning Explain why, for positive numbers a and b, a vertical stretch or compression of the graph of $f(x) = \frac{1}{x}$ by a factor of a and, separately, a horizontal stretch or compression of the graph of $f(x)$ by a factor of b result in the same graph when a and b are equal.

25. Communicate Mathematical Ideas Determine the domain and range of the rational function $g(x) = \frac{mx+n}{px+q}$ where $p \neq 0$. Give your answer in set notation, and explain your reasoning.

Lesson Performance Task

Graham wants to take snowboarding lessons at a nearby ski resort that charges $40 per week for a class that meets once during the week for 1 hour and gives him a lift ticket valid for 4 hours. The resort also charges a one-time equipment-rental fee of $99 for uninterrupted enrollment in classes. The resort requires that learners pay for three weeks of classes at a time.

a. Write a model that gives Graham's average weekly enrollment cost (in dollars) as a function of the time (in weeks) that Graham takes classes.

b. How much would Graham's average weekly enrollment cost be if he took classes only for the minimum of three weeks?

c. For how many weeks would Graham need to take classes for his average weekly enrollment cost to be at most $60? Describe how you can use a graphing calculator to graph the function from part a in order to answer this question, and then state the answer.

Name_______________ Class_______________ Date_______________

9.3 Graphing More Complicated Rational Functions

Resource Locker

Essential Question: **What features of the graph of a rational function should you identify in order to sketch the graph? How do you identify those features?**

A2.6.K Determine the asymptotic restrictions on the domain of a rational function and represent domain and range using interval notation, inequalities, and set notation. Also A2.7.I

Explore 1 Investigating Domains and Vertical Asympotes of More Complicated Rational Functions

You know that the rational function $f(x) = \frac{1}{x-2} + 3$ has the domain $\{x | x \neq 2\}$ because the function is undefined at $x = 2$. Its graph has the vertical asymptote $x = 2$ because as $x \to 2^+$ (x approaches 2 from the right), $f(x) \to +\infty$, and as $x \to 2^-$ (x approaches 2 from the left), $f(x) \to -\infty$. In this Explore, you will investigate the domains and vertical asymptotes of other rational functions.

(A) Complete the table by identifying each function's domain based on the x-values for which the function is undefined. Write the domain using an inequality, set notation, and interval notation. Then state the equations of what you think the vertical asymptotes of the function's graph are.

Function	Domain	Possible Vertical Asymptotes
$f(x) = \frac{x+3}{x-1}$		
$f(x) = \frac{(x+5)(x-1)}{x+1}$		
$f(x) = \frac{x-4}{(x+1)(x-1)}$		
$f(x) = \frac{2x^2-3x+9}{x^2-x-6}$		

(B) Using a graphing calculator, graph each of the functions from Step A, and check to see if vertical asymptotes occur where you expect. Are there any unexpected results?

(C) Examine the behavior of $f(x) = \dfrac{x+3}{x-1}$ near $x = 1$.

First, complete the tables.

x approaches 1 from the right	
x	$f(x) = \dfrac{x+3}{x-1}$
1.1	
1.01	
1.001	

x approaches 1 from the left	
x	$f(x) = \dfrac{x+3}{x-1}$
0.9	
0.99	
0.999	

Next, summarize the results.

- As $x \to 1^+$, $f(x) \to$ ______.
- As $x \to 1^-$, $f(x) \to$ ______.

The behaviour of $f(x) = \dfrac{x+3}{x-1}$ near $x = 1$ shows that the graph of $f(x)$ [has/does not] have a vertical asymptote at $x = 1$.

(D) Examine the behavior of $f(x) = \dfrac{(x+5)(x-1)}{(x+1)}$ near $x = -1$.

First, complete the tables.

x approaches -1 from the right	
x	$f(x) = \dfrac{(x+5)(x-1)}{x+1}$
-0.9	
-0.99	
-0.999	

x approaches -1 from the left	
x	$f(x) = \dfrac{(x+5)(x-1)}{x+1}$
-1.1	
-1.01	
-1.001	

Next, summarize the results.

- As $x \to -1^+$, $f(x) \to$ ______.
- As $x \to -1^-$, $f(x) \to$ ______.

The behavior of $f(x) = \dfrac{(x+5)(x-1)}{(x+1)}$ near $x = -1$ shows that the graph of $f(x)$ [has/does not have] a vertical asymptote at $x = -1$.

 Examine the behavior of $f(x) = \dfrac{x - 4}{(x + 1)(x - 1)}$ near $x = -1$ and $x = 1$.

First, complete the tables. Round results to the nearest tenth.

x approaches −1 from the right	
x	$f(x) = \dfrac{x - 4}{(x + 1)(x - 1)}$
−0.9	
−0.99	
−0.999	

x approaches −1 from the left	
x	$f(x) = \dfrac{x - 4}{(x + 1)(x - 1)}$
−1.1	
−1.01	
−1.001	

x approaches 1 from the right	
x	$f(x) = \dfrac{x - 4}{(x + 1)(x - 1)}$
1.1	
1.01	
1.001	

x approaches 1 from the left	
x	$f(x) = \dfrac{x - 4}{(x + 1)(x - 1)}$
0.9	
0.99	
0.999	

Next, summarize the results.

- As $x \to -1^+$, $f(x) \to$ ☐.
- As $x \to -1^-$, $f(x) \to$ ☐.
- As $x \to 1^+$, $f(x) \to$ ☐.
- As $x \to 1^-$, $f(x) \to$ ☐.

The behavior of $f(x) = \frac{x - 4}{(x + 1)(x - 1)}$ near $x = -1$ shows that the graph of $f(x)$ has/does not have a vertical asymptote at $x = -1$. The behavior of $f(x) = \frac{x - 4}{(x + 1)(x - 1)}$ near $x = 1$ shows that the graph of $f(x)$ has/does not have a vertical asymptote at $x = 1$.

 Examine the behavior of $f(x) = \dfrac{2x^2 - 3x - 9}{x^2 - x - 6}$ near $x = -2$ and $x = 3$.

First, complete the tables. Round results to the nearest ten thousandth if necessary.

x approaches −2 from the right	
x	$f(x) = \dfrac{2x^2 - 3x - 9}{x^2 - x - 6}$
−1.9	
−1.99	
−1.999	

x approaches −2 from the left	
x	$f(x) = \dfrac{2x^2 - 3x - 9}{x^2 - x - 6}$
−2.1	
−2.01	
−2.001	

x approaches 3 from the right	
x	$f(x) = \dfrac{2x^2 - 3x - 9}{x^2 - x - 6}$
3.1	
3.01	
3.001	

x approaches 3 from the left	
x	$f(x) = \dfrac{2x^2 - 3x - 9}{x^2 - x - 6}$
2.9	
2.99	
2.999	

Next, summarize the results.

- As $x \to -2^+$, $f(x) \to$ ______.
- As $x \to -2^-$, $f(x) \to$ ______.
- As $x \to 3^+$, $f(x) \to$ ______.
- As $x \to 3^-$, $f(x) \to$ ______.

The behavior of $f(x) = \frac{2x^2 - 3x - 9}{x^2 - x - 6}$ near $x = -2$ shows that the graph of $f(x)$ has/does not have a vertical asymptote at $x = -2$. The behavior of $f(x) = \frac{2x^2 - 3x - 9}{x^2 - x - 6}$ near $x = 3$ shows that the graph of $f(x)$ has/does not have a vertical asymptote at $x = 3$.

Reflect

1. Rewrite $f(x) = \frac{2x^2 - 3x - 9}{x^2 - x - 6}$ so that its numerator and denominator are factored. How does this form of the function explain the behavior of the function near $x = 3$?

2. **Discussion** When you graph $f(x) = \dfrac{2x^2 - 3x - 9}{x^2 - x - 6}$ on a graphing calculator, you can't tell that the function is undefined for $x = 3$. How does using the calculator's table feature help? What do you think the graph should look like to make it clear that the function is undefined at $x = 3$?

Explore 2 Investigating End Behaviors and Ranges of More Complicated Rational Functions

You can determine the end behavior of a rational function by using polynomial division. When you divide a rational function's numerator (dividend) by its denominator (divisor), you obtain a quotient and a remainder. You can then rewrite the function in the form $f(x) = \text{quotient} + \frac{\text{remainder}}{\text{divisor}}$. Because the degree of the remainder must be less than the degree of the divisor, the expression $\frac{\text{remainder}}{\text{divisor}}$ approaches 0 as x increases or decreases without bound. This means that the function has the same end behavior as the quotient.

(A) The table lists the same rational functions as in the previous Explore. Complete the table. When writing a function in the form $f(x) = \text{quotient} + \frac{\text{remainder}}{\text{divisor}}$, you will need to write numerator and denominator as polynomials in standard form before carrying out the division.

Function	Function in the form $f(x) = \text{quotient} + \frac{\text{remainder}}{\text{divisor}}$	End behavior
$f(x) = \frac{x+3}{x-1}$		
$f(x) = \frac{(x+5)(x-1)}{x+1}$		
$f(x) = \frac{x-4}{(x+1)(x-1)}$		
$f(x) = \frac{2x^2-3x-9}{x^2-x-6}$		

(B) Identify the functions in Step A whose graphs have horizontal asymptotes.

Ⓒ Using a graphing calculator, graph each of the functions in Step A to check its end behavior. Then examine the graph to determine the function's range, and record the range in the table using an inequality, set notation, and interval notation. (For the third function, you should examine a table of values using an x-value increment of 0.1. Once you have a sense of where the function's minimum and maximum values occur, select **3:minimum** from the **CALC** menu to determine the minimum value and **4:maximum** from the **CALC** menu to determine the maximum value. Round these values to the nearest hundredth, recognizing that the range you give is an approximation of the actual range.)

Function	Range
$f(x) = \frac{x+3}{x-1}$	
$f(x) = \frac{(x+5)(x-1)}{x+1}$	
$f(x) = \frac{x-4}{(x+1)(x-1)}$	
$f(x) = \frac{2x^2-3x-9}{x^2-x-6}$	

Reflect

3. What line does the graph of $f(x) = \frac{(x+5)(x-1)}{x+1}$ approach as x increases or decreases without bound? Explain.

4. If you rewrite the function $f(x) = \frac{x-4}{(x+1)(x-1)}$ as $f(x) = \frac{x-4}{x^2-1}$ and then multiply the numerator and denominator by $\frac{1}{x^2}$, you get $f(x) = \frac{\frac{1}{x} - \frac{4}{x^2}}{1 - \frac{1}{x^2}}$. Using this form of the function, describe what happens to each term in the numerator and denominator as x increases without bound and as x decreases without bound. Then explain how these results give you the function's end behavior.

5. Discussion Which of the four functions in this Explore has a graph that crosses its horizontal asymptote? How do you reconcile the fact that a graph is supposed to approach a line but not cross it in order for the line to be an asymptote with the fact that the graph of the function you identified crosses its horizontal asymptote?

Explain 1 Sketching the Graphs of More Complicated Rational Functions

As you have seen, there can be breaks in the graph of a rational function. These breaks are called *discontinuities*, and there are two kinds:

1. When a rational function has a factor in the denominator that is not also in the numerator, an *infinite discontinuity* occurs at the value of x for which the factor equals 0. On the graph of the function, an infinite discontinuity appears as a vertical asymptote.

2. When a rational function has a factor in the denominator that is also in the numerator, a *point discontinuity* occurs at the value of x for which the factor equals 0. On the graph of the function, a point discontinuity appears as a "hole."

The graph of a rational function can also have a horizontal asymptote. It is determined by the degrees and leading coefficients of the function's numerator and denominator. If the numerator is a polynomial $p(x)$ in standard form with leading coefficient a and the denominator is a polynomial $q(x)$ in standard form with leading coefficient b, then an examination of the function's end behavior gives the following results.

Relationship between Degree of $p(x)$ and Degree of $q(x)$	Equation of Horizontal Asymptote (if one exists)
Degree of $p(x) <$ degree of $q(x)$	$y = 0$
Degree of $p(x) =$ degree of $q(x)$	$y = \frac{a}{b}$
Degree of $p(x) >$ degree of $q(x)$	There is no horizontal asymptote. The function instead increases or decreases without bound as x increases or decreases without bound. In particular, when the degree of the numerator is 1 more than the degree of the denominator, the function's graph approaches a slanted line, called a *slant asymptote*, as x increases or decreases without bound.

You can sketch the graph of a rational function by identifying where vertical asymptotes, "holes," and horizontal asymptotes occur. Using the factors of the numerator and denominator, you can also establish intervals on the x-axis where either an x-intercept or a discontinuity occurs and then check the signs of the factors on those intervals to determine whether the graph lies above or below the x-axis.

Example 1 **Sketch the graph of the given rational function. (If the degree of the numerator is 1 more than the degree of the denominator, find the graph's slant asymptote by dividing the numerator by the denominator.) Also state the function's domain and range using inequalities, set notation, and interval notation. (If your sketch indicates that the function has maximum or minimum values, use a graphing calculator to find those values to the nearest hundredth when determining the range.)**

Ⓐ $f(x) = \frac{x+1}{x-2}$

Identify vertical asymptotes and "holes."

The function is undefined when $x - 2 = 0$, or $x = 2$. Since $x - 2$ is not a factor of the numerator, there is a vertical asymptote rather than a "hole" at $x = 2$.

Identify horizontal asymptotes and slant asymptotes.

The numerator and denominator have the same degree and the leading coefficient of each is 1, so there is a horizontal asymptote at $y = \frac{1}{1} = 1$.

Identify x-intercepts.

An x-intercept occurs when $x + 1 = 0$, or $x = -1$.

Check the sign of the function on the intervals $x < -1$, $-1 < x < 2$, and $x > 2$.

Interval	Sign of $x+1$	Sign of $x-2$	Sign of $f(x) = \frac{x+1}{x-2}$
$x < -1$	$-$	$-$	$+$
$-1 < x < 2$	$+$	$-$	$-$
$x > 2$	$+$	$+$	$+$

Sketch the graph using all this information. Then state the domain and range.

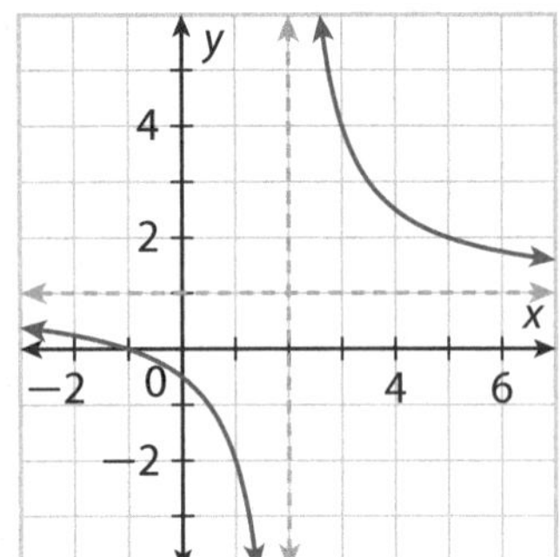

Domain:

Inequality: $x < 2$ or $x > 2$

Set notation: $\{x \mid x \neq 2\}$

Interval notation: $(-\infty, 2) \cup (2, +\infty)$

Range:

Inequality: $y < 1$ or $y > 1$

Set notation: $\{y \mid y \neq 1\}$

Interval notation: $(-\infty, 1) \cup (1, +\infty)$

 $f(x) = \dfrac{x^2 + x - 2}{x + 3}$

Factor the function's numerator.

$$f(x) = \frac{x^2 + x - 2}{x + 3} = \frac{(x - 1)(x + 2)}{x + 3}$$

Identify vertical asymptotes and "holes."

The function is undefined when $x + 3 = 0$, or $x = \square$. Since $x + 3$ is not a factor of the numerator, there is a vertical asymptote rather than a "hole" at $x = \square$.

Identify horizontal asymptotes and slant asymptotes.

Because the degree of the numerator is 1 more than the degree of the denominator, there is no horizontal asymptote, but there is a slant asymptote. Divide the numerator by the denominator to identify the slant asymptote.

$$\begin{array}{r} x - \square \\ x + 3 \overline{)\, x^2 + x - 2} \\ \underline{x^2 + 3x} \\ -2x - 2 \\ \underline{-2x - 6} \\ 4 \end{array}$$

So, the line $y = x - \square$ is the slant asymptote.

Identify x-intercepts.

There are two x-intercepts: when $x - 1 = 0$, or $x = \square$, and when $x + 2 = 0$, or $x = \square$.

Check the sign of the function on the intervals $x < -3$, $-3 < x < -2$, $-2 < x < 1$, and $x > 1$.

Interval	Sign of $x + 3$	Sign of $x + 2$	Sign of $x - 1$	Sign of $f(x) = \frac{(x-1)(x+2)}{x+3}$
$x < -3$	−	−	−	−
$-3 < x < -2$				
$-2 < x < 1$				
$x > 1$				

Sketch the graph using all this information. Then state the domain and range.

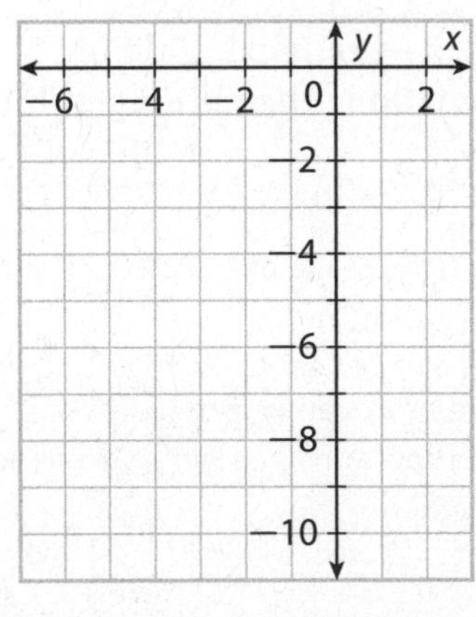

Domain:

Inequality: $x < \square$ or $x > \square$

Set notation: $\{x \mid x \neq \square\}$

Interval notation: $(-\infty, \square) \cup (\square, +\infty)$

The sketch indicates that the function has a maximum value and a minimum value. Using **3:minimum** from the **CALC** menu on a graphing calculator gives −1 as the minimum value. Using **4:maximum** from the **CALC** menu on a graphing calculator gives −5 as the maximum value.

Range: Inequality: $y <$ ☐ or $y >$ ☐

Set notation: $\{y \mid y <$ ☐ or $y >$ ☐$\}$ Interval notation: $(-\infty,$ ☐$) \cup ($☐$, +\infty)$

Your Turn

Sketch the graph of the given rational function. (If the degree of the numerator is 1 more than the degree of the denominator, find the graph's slant asymptote by dividing the numerator by the denominator.) Also state the function's domain and range using inequalities, set notation, and interval notation. (If your sketch indicates that the function has maximum or minimum values, use a graphing calculator to find those values to the nearest hundredth when determining the range.)

6. $f(x) = \dfrac{x + 1}{x^2 + 3x - 4}$

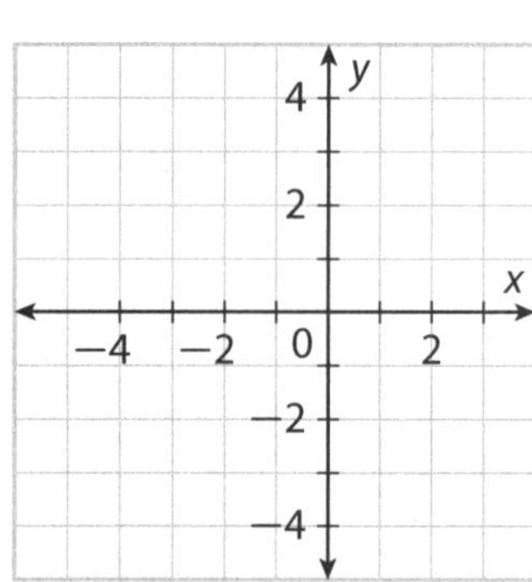

Explain 2 Modeling with More Complicated Rational Functions

When two real-world variable quantities are compared using a ratio or rate, the comparison is a rational function. You can solve problems about the ratio or rate by graphing the rational function.

Example 2 **Write a rational function to model the situation, or use the given rational function. State a reasonable domain and range for the function using set notation. Then use a graphing calculator to graph the function and answer the question.**

(A) A baseball team has won 32 games out of 56 games played, for a winning percentage of $\frac{32}{56} \approx 0.571$. How many consecutive games must the team win to raise its winning percentage to 0.600?

Let w be the number of consecutive games to be won. Then the total number of games won is the function $T_{\text{won}}(w) = 32 + w$, and the total number of games played is the function $T_{\text{played}}(w) = 56 + w$.

The rational function that gives the team's winning percentage p (as a decimal) is
$p(w) = \frac{T_{\text{won}}(w)}{T_{\text{played}}(w)} = \frac{32 + w}{56 + w}$.

The domain of the rational function is $\left\{w \middle| w \geq 0 \text{ and } w \text{ is a whole number}\right\}$. Note that you do not need to exclude -56 from the domain, because only nonnegative whole-number values of w make sense in this situation.

Since the function models what happens to the team's winning percentage from consecutive wins (no losses), the values of $p(w)$ start at 0.571 and approach 1 as w increases without bound. So, the range is $\left\{p \middle| 0.571 \leq p < 1\right\}$.

Graph $y = \frac{32 + x}{56 + x}$ on a graphing calculator using a viewing window that shows 0 to 10 on the x-axis and 0.5 to 0.7 on the y-axis. Also graph the line $y = 0.6$. To find where the graphs intersect, select **5: intersect** from the **CALC** menu.

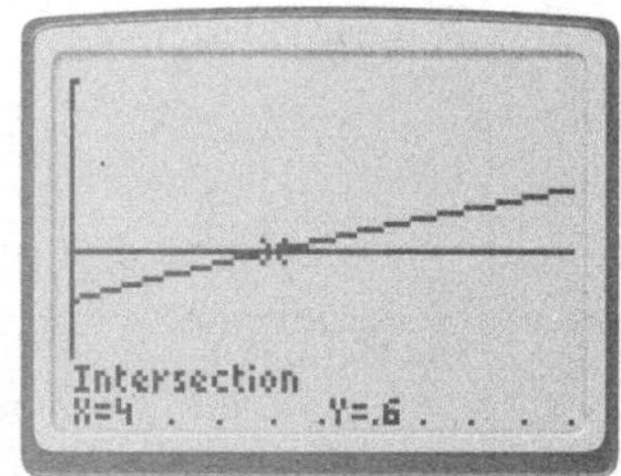

So, the team's winning percentage (as a decimal) will be 0.600 if the team wins 4 consecutive games.

(B) Two friends decide spend an afternoon canoeing on a river. They travel 4 miles upstream and 6 miles downstream. In still water, they know that their average paddling speed is 5 miles per hour. If their canoe trip takes 4 hours, what is the average speed of the river's current? To answer the question, use the rational function $t(c) = \frac{4}{5 - c} + \frac{6}{5 + c} = \frac{50 - 2c}{(5 - c)(5 + c)}$ where c is the average speed of the current (in miles per hour) and t is the time (in hours) spent canoeing 4 miles against the current at a rate of $5 - c$ miles per hour and 6 miles with the current at a rate of $5 + c$ miles per hour.

In order for the friends to travel upstream, the speed of the current must be less than their average paddling speed, so the domain of the function is $\left\{c \middle| 0 \leq c < \square\right\}$. If the friends canoed in still water ($c = 0$), the trip would take a total of $\frac{4}{5} + \frac{6}{5} = \square$ hours.

As c approaches 5 from the left, the value of $\frac{6}{5 + c}$ approaches $\frac{6}{10} = 0.6$ hour, but the value of $\frac{4}{5 - c}$ ______________________. So, the range of the function is $\left\{t \middle| t \geq \square\right\}$.

Graph $y = \frac{50 - 2x}{(5 - x)(5 + x)}$ on a graphing calculator using a viewing window that shows 0 to 5 on the x-axis and 2 to 5 on the y-axis. Also graph the line $y = \square$. To find where the graphs intersect, select **5:intersect** from the **CALC** menu. The calculator shows that the average speed of the current is about ______ miles per hour.

Your Turn

Write a rational function to model the situation, or use the given rational function. State a reasonable domain and range for the function using set notation. Then use a graphing calculator to graph the function and answer the question.

7. A saline solution is a mixture of salt and water. A $p\%$ saline solution contains $p\%$ salt and $(100 - p)\%$ water by mass. A chemist has 300 grams of a 4% saline solution that needs to be strengthened to a 6% solution by adding salt. How much salt should the chemist add?

Elaborate

8. How can you show that the vertical line $x = c$, where c is a constant, is an asymptote for the graph of a rational function?

9. How can you determine the end behavior of a rational function?

10. **Essential Question Check-In** How do you identify any vertical asymptotes and "holes" that the graph of a rational function has?

Evaluate: Homework and Practice

- Online Homework
- Hints and Help
- Extra Practice

State the domain using an inequality, set notation, and interval notation. For any x-value excluded from the domain, state whether the graph has a vertical asymptote or a "hole" at that x-value. Use a graphing calculator to check your answer.

1. $f(x) = \dfrac{x+5}{x+1}$

2. $f(x) = \dfrac{x^2+2x-3}{x^2-4x+3}$

Divide the numerator by the denominator to write the function in the form $f(x) = \text{quotient} + \dfrac{\text{remainder}}{\text{divisor}}$ and determine the function's end behavior. Then, using a graphing calculator to examine the function's graph, state the range using an inequality, set notation, and interval notation.

3. $f(x) = \dfrac{3x+1}{x-2}$

4. $f(x) = \dfrac{x}{(x-2)(x+3)}$

5. $f(x) = \dfrac{x^2-5x+6}{x-1}$

6. $f(x) = \dfrac{4x^2-1}{x^2+x-2}$

Sketch the graph of the given rational function. (If the degree of the numerator is 1 more than the degree of the denominator, find the graph's slant asymptote by dividing the numerator by the denominator.) Also state the function's domain and range using inequalities, set notation, and interval notation. (If your sketch indicates that the function has maximum or minimum values, use a graphing calculator to find those values to the nearest hundredth when determining the range.)

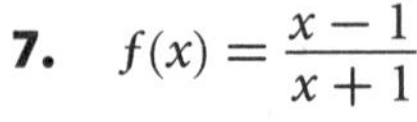

7. $f(x) = \dfrac{x-1}{x+1}$

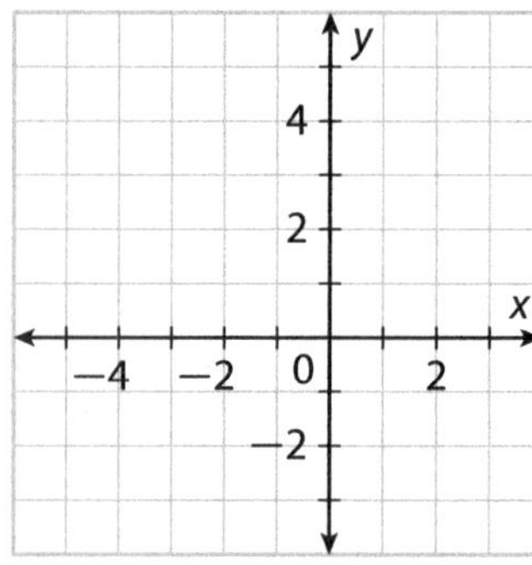

8. $f(x) = \dfrac{x-1}{(x-2)(x+3)}$

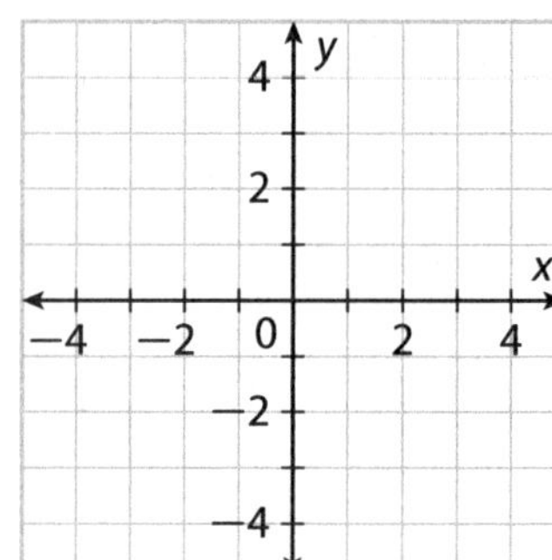

9. $f(x) = \dfrac{(x+1)(x-1)}{x+2}$

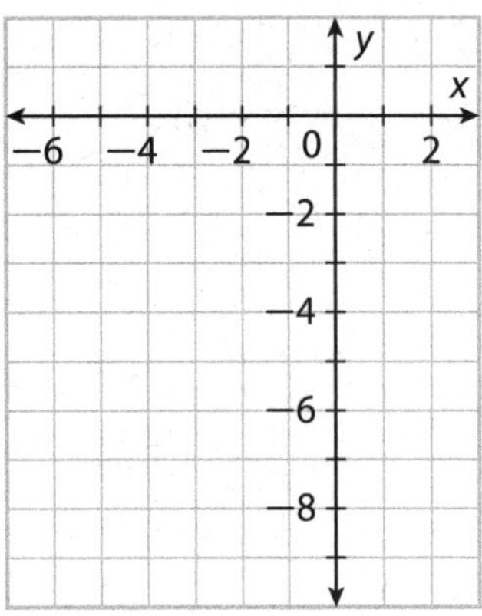

10. $f(x) = \dfrac{-3x(x-2)}{(x-2)(x+2)}$

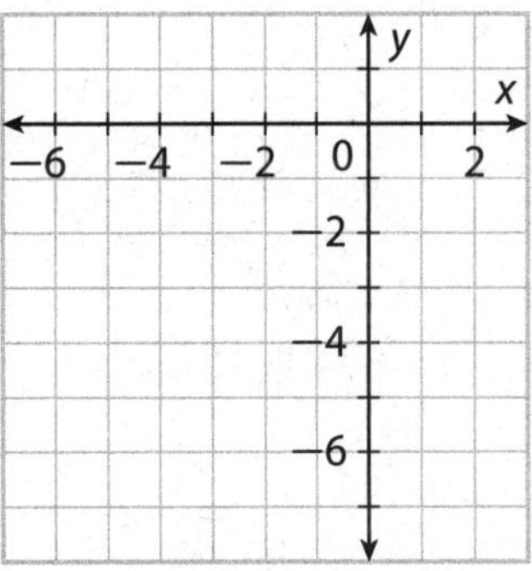

11. $f(x) = \dfrac{x^2 + 2x - 4}{x - 1}$

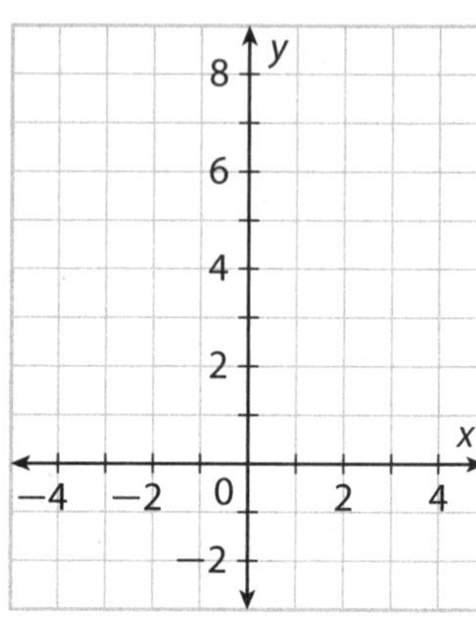

12. $f(x) = \dfrac{2x^2 - 4x}{x^2 + 4x + 4}$

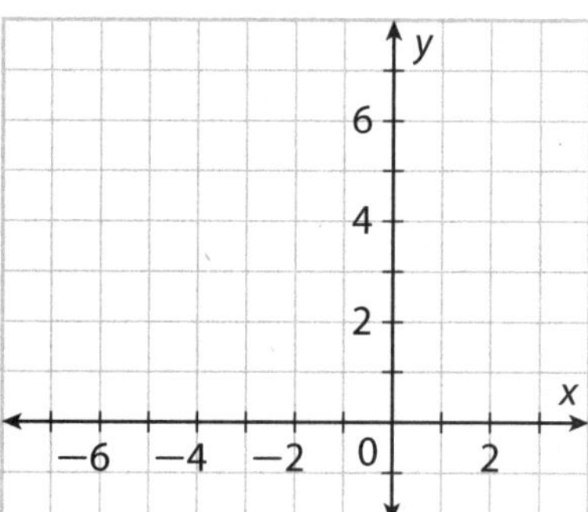

Write a rational function to model the situation, or use the given rational function. State a reasonable domain and range for the function using set notation. Then use a graphing calculator to graph the function and answer the question.

13. A basketball team has won 16 games out of 23 games played, for a winning percentage (expressed as a decimal) of $\frac{16}{23} \approx 0.696$. How many consecutive games must the team win to raise its winning percentage to 0.750?

14. So far this season, a baseball player has had 84 hits in 294 times at bat, for a batting average of $\frac{84}{294} \approx 0.286$. How many consecutive hits must the player get to raise his batting average to 0.300?

15. A kayaker traveled 5 miles upstream and then 8 miles downstream on a river. The average speed of the current was 3 miles per hour. If the kayaker was paddling for 5 hours, what was the kayaker's average paddling speed? To answer the question, use the rational function $t(s) = \frac{5}{s-3} + \frac{8}{s+3} = \frac{13s-9}{(s-3)(s+3)}$ where s is the kayaker's average paddling speed (in miles per hour) and t is the time (in hours) spent kayaking 5 miles against the current at a rate of $s - 3$ miles per hour and 8 miles with the current at a rate of $s + 3$ miles per hour.

16. In aviation, *air speed* refers to a plane's speed in still air. A small plane maintains a certain air speed when flying to and from a city that is 400 miles away. On the trip out, the plane flies against a wind, which has an average speed of 40 miles per hour. On the return trip, the plane flies with the wind. If the total flight time for the round trip is 3.5 hours, what is the plane's average air speed? To answer this question, use the rational function $t(s) = \frac{400}{s-40} + \frac{400}{s+40} = \frac{800}{(s-40)(s+40)}$ air speed (in miles per hour) and t is the total flight time (in hours) for the round trip.

17. Multiple Response Select the statements that apply to the rational function $f(x) = \dfrac{x-2}{x^2 - x - 6}$.

A. The function's domain is $\left\{x \middle| x \neq -2 \text{ and } x \neq 3\right\}$.

B. The function's domain is $\left\{x \middle| x \neq -2 \text{ and } x \neq -3\right\}$.

C. The function's range is $\left\{y \middle| y \neq 0\right\}$.

D. The function's range is $\left\{y \middle| -\infty \; y < +\infty\right\}$.

E. The function's graph has vertical asymptotes at $x = -2$ and $x = 3$.

F. The function's graph has a vertical asymptote at $x = -3$ and a "hole" at $x = 2$.

G. The function's graph has a horizontal asymptote at $y = 0$.

H. The function's graph has a horizontal asymptote at $y = 1$.

H.O.T. Focus on Higher Order Thinking

18. Draw Conclusions For what value(s) of a does the graph of $f(x) = \dfrac{x + a}{x^2 + 4x + 3}$ have a "hole"? Explain. Then, for each value of a, state the domain and the range of $f(x)$ using interval notation.

19. Critique Reasoning A student claims that the functions $f(x) = \dfrac{4x^2 - 1}{4x + 2}$ and $g(x) = \dfrac{4x + 2}{4x^2 - 1}$ have different domains but identical ranges. Which part of the student's claim is correct, and which is false? Explain.

Lesson Performance Task

In professional baseball, the smallest allowable volume of a baseball is 92.06% of the largest allowable volume, and the range of allowable radii is 0.04 inch.

a. Let r be the largest allowable radius (in inches) of a baseball. Write expressions, both in terms of r, for the largest allowable volume of the baseball and the smallest allowable volume of the baseball. (Use the formula for the volume of a sphere, $V = \frac{4}{3}\pi r^3$.)

b. Write and simplify a function that gives the ratio R of the smallest allowable volume of a baseball to the largest allowable volume.

c. Use a graphing calculator to graph the function from part b, and use the graph to find the smallest allowable radius and the largest allowable radius of a baseball. Round your answers to the nearest hundredth.

STUDY GUIDE REVIEW

Rational Functions

MODULE 9

Essential Question: How can you use rational functions to solve real-world problems?

Key Vocabulary
asymptote *(asíntota)*
constant of variation *(constante de variación)*
inverse variation *(variación inversa)*
parent function *(función madre)*
rational function *(función racional)*

KEY EXAMPLE *(Lesson 9.1)*

The variables x and y vary inversely, and $y = 12$ when $x = 3$. What is the value of y when $x = -4$?

$y = \frac{k}{x}$ Find the value of the constant of variation.

$12 = \frac{k}{3}$

$36 = k$

The inverse variation equation is $y = \frac{36}{x}$. When $x = -4$, $y = \frac{36}{-4}$, $= -9$.

KEY EXAMPLE *(Lesson 9.3)*

Graph $y = \frac{2x^2}{x+2}$. State the domain, range, x-intercept, and identify any asymptotes.

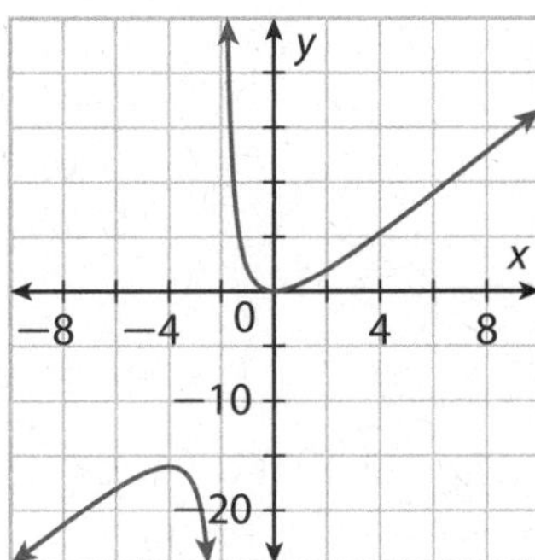

Domain: $x < -2$ or $x > -2$

Range: $y \leq -16$ or $y \geq 0$

The numerator has 0 as its only zero, so the x-intercept is $(0, 0)$.

The denominator has -2 as its only zero, so the graph has a vertical asymptote at $x = -2$.

EXERCISES

Write an equation relating the variables and use it to answer the question. *(Lesson 9.1)*

1. The variables x and y vary inversely. When $x = 6$, $y = 2$. What is y when $x = 15$?

2. For a fixed voltage, the current I flowing in a wire varies inversely as the resistance R of the wire. If the current is 8 amperes when the resistance is 15 ohms, what will the resistance be when the current is 5 amperes?

Describe how the graph of $g(x)$ is related to the graph of $f(x) = \frac{1}{x}$. *(Lesson 9.2)*

3. $g(x) = \frac{1}{x + 4}$

4. $g(x) = \frac{1}{x - 2} + 3$

Graph the function using a graphing calculator. State the domain and, x-intercept(s), and identify asymptotes. *(Lesson 9.3)*

5. $f(x) = \frac{x^2 - 3x}{x + 4}$

6. $f(x) = \frac{x - 3}{x^2 + 6x + 5}$

MODULE PERFORMANCE TASK

What Is the Profit?

A sporting goods store sells two styles of bike helmets: Style A for $30 each and Style B for $40 each. The store is trying to calculate its average profit on each style of helmet, using the rational function $A(x) = \frac{P(x)}{x}$, where $P(x)$ is the profit on the sale of x helmets. The helmet supplier offers a of volume discount for orders up to 500 helmets, using the cost formulas shown in the table. For each style, how does per-helmet profit change as the number of helmets increases? What is the maximum per-helmet profit?

Number of Helmets (x)	100	200	300	400	500
Style A					
Revenue					
Cost: $100 + (20 - 0.01x)x$					
Profit					
Style B					
Revenue					
Cost: $250 + (30 - 0.03x)x$					
Profit					

Start by organizing your data in the table. Then use your own paper to complete the task. Use graphs, numbers, words, or algebra to explain how you reached your conclusion.

Ready to Go On?

9.1–9.3 Rational Expressions and Equations

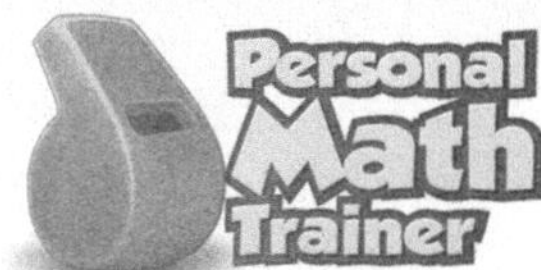

- Online Homework
- Hints and Help
- Extra Practice

Write an equation relating the variables and use it to answer the question. *(Lesson 9.1)*

1. The variables x and y vary inversely. When $x = -12$, $y = \frac{2}{3}$. What is y when $x = -3$?

2. The number of hours h needed to pick a field of berries varies inversely as the number of people working p, and $h = 20$ hours when $p = 10$ people. Find h when $p = 25$ people.

Describe how the graph of $g(x)$ is related to the graph of $f(x) = \frac{1}{x}$. *(Lesson 9.2)*

3. $g(x) = \frac{5}{x} - 3$

4. $g(x) = \frac{1}{-0.5(x-2)} + 4$

Graph the function using a graphing calculator. State the domain, x-intercept(s), and identify asymptotes. *(Lessons 9.2, 9.3)*

5. $f(x) = \frac{2x-4}{x+3}$

6. $f(x) = \frac{x^2-9}{x-2}$

ESSENTIAL QUESTION

7. How do you identify the asymptotes of a rational function?

Assessment Readiness

1. Which statement about the graph of $y = \dfrac{2x + 5}{x - 1}$ is false?
   A. The line $x = 1$ is an asymptote.
   B. The line $y = 2$ is an asymptote.
   C. The function is undefined for $x = -1$.
   D. The point $(0, -5)$ lies on the graph.

2. What is the solution set of $x^2 - 64 < 0$?
   A. $x < -8$ or $x > -8$
   B. $-8 < x < 8$
   C. $-64 < x < 64$
   D. $x < -64$ or $x > -64$

3. Which is a factor of $x^3 + 2x^2 - 5x - 6$?
   A. $(x + 2)$
   B. $(x - 2)$
   C. $(x - 5)$
   D. $(x + 6)$

4. The time t in days that it takes a movie crew to build a set varies inversely as the number of workers w. If 25 workers can build the set in 3 days, how long would it take 15 workers to build the set?
   A. 5 days
   B. 15 days
   C. 2 days
   D. 25 days

5. You have subscribed to a cable television service. The cable company charges a one-time installation fee of \$30 and a monthly fee of \$50. Write a model that gives the average cost per month as a function of months subscribed to the service. After how many months will the average cost be \$56? Explain your thinking and identify any asymptotes on the graph of the function.

MODULE 10

Rational Expressions and Equations

TEKS

Essential Question: How can you use rational expressions and equations to solve real-world problems?

REAL WORLD VIDEO
Robotic arms and other prosthetic devices are among the wonders of robotics. Check out some of the other cutting-edge applications of modern robotics.

MODULE PERFORMANCE TASK PREVIEW

Robots and Resistors

Robotics engineers design robots and develop applications for them, such as executing high-precision tasks in factories, cleaning toxic waste, and locating and defusing bombs. People who work in robotics are skilled in areas such as electronics and computer programming. How can a rational expression be used to help design the circuitry for a robot? Let's find out!

Are YOU Ready?

Complete these exercises to review skills you will need for this module.

Personal Math Trainer
- Online Homework
- Hints and Help
- Extra Practice

Graphing Linear Proportional Relationships

Example 1

Graph $y = \frac{1}{2}x$.

When $x = 0$, $y = 0$, so plot $(0, 0)$ on the graph.

The slope is $\frac{1}{2}$, so from $(0, 0)$, plot the next point up 1 and over 2.

Draw a line through the two points.

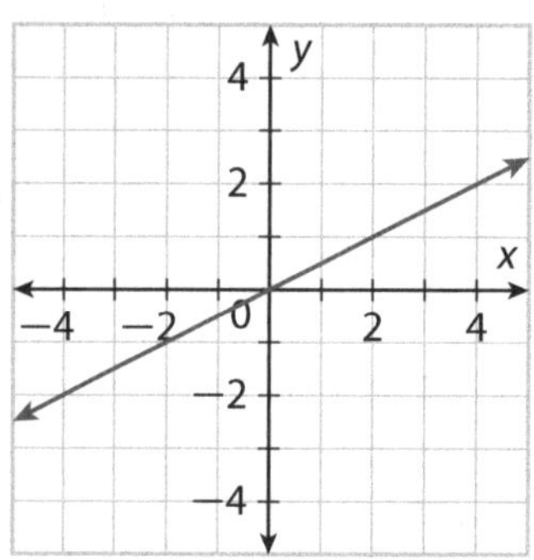

Graph each proportional relationship.

1. $y = 2x$

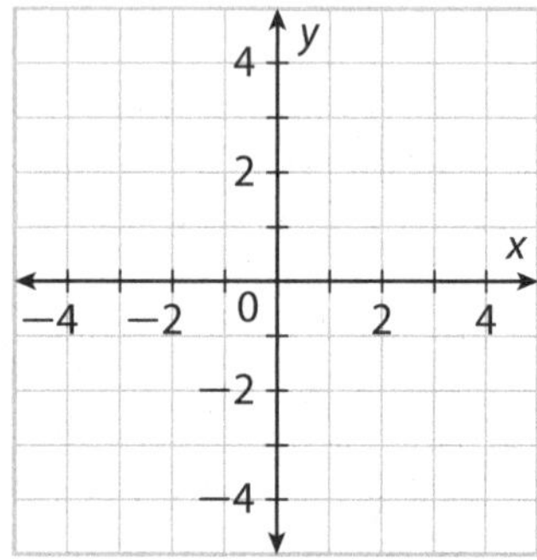

2. $y = \frac{2}{3}x$

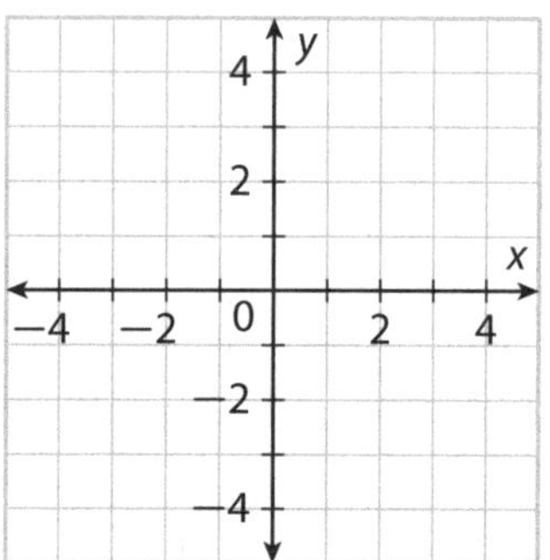

Direct and Inverse Variation

Example 2 In a direct variation the constant is 4 and its graph passes through $(x, 10)$. Find x.

$y = kx \rightarrow 10 = 4x \rightarrow x = 2.5$

Example 3 In an inverse variation the constant is 2.4 and its graph passes through $(6, y)$. Find y.

$k = xy \rightarrow 2.4 = 6y \rightarrow y = 0.4$

Find the missing variable for each direct variation.

3. $k = -1$; $(x, -5)$ ______

4. $k = 3$; $(9, y)$ ______

5. $k = \frac{1}{3}$; $(x, -2)$ ______

Find the missing variable for each inverse variation.

6. $k = 8$; $(x, -10)$ ______

7. $k = -2$; $(5, y)$ ______

8. $k = 6$; $(x, 1.5)$ ______

Name________________________ Class____________ Date________

10.1 Adding and Subtracting Rational Expressions

Resource Locker

Essential Question: How can you add and subtract rational expressions?

A2.7.F Determine the sum, difference…of rational expressions with integral exponents of degree one and of degree two.

Explore Identifying Excluded Values

Given a rational expression, identify the excluded values by finding the zeroes of the denominator. If possible, simplify the expression.

(A) $\frac{(1 - x^2)}{x - 1}$

The denominator of the expression is ______.

(B) Since division by 0 is not defined, the excluded values for this expression are all the values that would make the denominator equal to 0.

$x - 1 = 0$

$x = \square$

(C) Begin simplifying the expression by factoring the numerator.

$$\frac{(1 - x^2)}{x - 1} = \frac{(\square)(\square)}{x - 1}$$

(D) Divide out terms common to both the numerator and the denominator.

$$\frac{(1 - x^2)}{x - 1} = \frac{(\square)(\square)}{-(1 - x)} = \square = \square$$

(E) The simplified expression is

$\frac{(1 - x^2)}{x - 1} = \square$, whenever $x \neq \square$

(F) What is the domain for this function? What is its range?

____

Reflect

1. What factors can be divided out of the numerator and denominator?

____

Explain 1 Writing Equivalent Rational Expressions

Given a rational expression, there are different ways to write an equivalent rational expression. When common terms are divided out, the result is an equivalent but simplified expression.

Example 1 **Simplify the expressions.**

(A) Write $\frac{3x}{(x+3)}$ as an equivalent rational expression that has a denominator of $(x+3)(x+5)$.

The expression $\frac{3x}{(x+3)}$ has a denominator of $(x+3)$.

The factor missing from the denominator is $(x+5)$.

Introduce a common factor, $(x+5)$.

$$\frac{3x}{(x+3)} = \frac{3x(x+5)}{(x+3)(x+5)}$$

$\frac{3x}{(x+3)}$ is equivalent to $\frac{3x(x+5)}{(x+3)(x+5)}$.

(B) Simplify the expression $\frac{(x^2+5x+6)}{(x^2+3x+2)(x+3)}$.

Write the expression. $\frac{(x^2+5x+6)}{(x^2+3x+2)(x+3)}$

Factor the numerator and denominator. ________________

Divide out like terms. ______

Your Turn

2. Write $\frac{5}{5x-25}$ as an equivalent expression with a denominator of $(x-5)(x+1)$.

3. Simplify the expression $\frac{(x+x^3)(1-x^2)}{(x^2-x^6)}$.

Explain 2 Adding and Subtracting Rational Expressions

Adding and subtracting rational expressions is similar to adding and subtracting fractions.

Example 2 **Add or subtract. Identify any excluded values and simplify your answer.**

 $\frac{x^2 + 4x + 2}{x^2} + \frac{x^2}{x^2 + x}$

Factor the denominators. $\frac{x^2 + 4x + 2}{x^2} + \frac{x^2}{x(x + 1)}$

Identify where the expression is not defined. The first expression is undefined when $x = 0$. The second expression is undefined when $x = 0$ and when $x = -1$.

Find a common denominator. The LCM for x^2 and $x(x + 1)$ is $x^2(x + 1)$.

Write the expressions with a common denominator by multiplying both by the appropriate form of 1. $\frac{(x + 1)}{(x + 1)} \cdot \frac{x^2 + 4x + 2}{x^2} + \frac{x^2}{x(x + 1)} \cdot \frac{x}{x}$

Simplify each numerator. $= \frac{x^3 + 5x^2 + 6x + 2}{x^2(x + 1)} + \frac{x^3}{x^2(x + 1)}$

Add. $= \frac{2x^3 + 5x^2 + 6x + 2}{x^2(x + 1)}$

It is clear that x and x^2 are not factors of the numerator, and you can show by synthetic division that $x + 1$ is not a factor of the numerator. Since none of the factors of the denominator are factors of the numerator, the expression cannot be further simplified.

 $\frac{2x^2}{x^2 - 5x} - \frac{x^2 + 3x - 4}{x^2}$

Factor the denominators. $\frac{2x^2}{\boxed{}} - \frac{x^2 + 3x - 4}{x^2}$

Identify where the expression is not defined. The first expression is undefined when $x = 0$ and when $x = 5$. The second expression is undefined when $x = 0$.

Find a common denominator. The LCM for $x(x - 5)$ and x^2 is ______.

Write the expressions with a common denominator by multiplying both by the appropriate form of 1. $\boxed{} \cdot \frac{2x^2}{x(x - 5)} - \frac{x^2 + 3x - 4}{x^2} \cdot \frac{x - 5}{x - 5}$

Simplify each numerator. $= \frac{2x^3}{x^2(x - 5)} - \frac{x^3 - 2x^2 - 19x + 20}{x^2(x - 5)}$

Subtract. $= \frac{\boxed{} + 2x^2 + 19x - 20}{x^2(x - 5)}$

It is clear that x and x^2 are not factors of the numerator, and you can show by synthetic division that $x - 5$ is not a factor of the numerator. Since none of the factors of the denominator are factors of the numerator, the expression cannot be further simplified.

Your Turn

Add each pair of expressions, simplifying the result and noting the combined excluded values. Then subtract the second expression from the first, again simplifying the result and noting the combined excluded values.

4. $-x^2$ and $\frac{1}{(1 - x^2)}$

5. $\frac{x^2}{(4 - x^2)}$ and $\frac{1}{(2 - x)}$

Explain 3 Adding and Subtracting with Rational Models

Rational expressions can model real-world phenomena, and can be used to calculate measurements of those phenomena.

Example 3 **Find the sum or difference of the models to solve the problem.**

(A) Two groups have agreed that each will contribute \$2000 for an upcoming trip. Group A has 6 more people than group B. Let x represent the number of people in group A. Write and simplify an expression in terms of x that represents the difference between the number of dollars each person in group A must contribute and the number each person in group B must contribute.

$$\frac{2000}{x} - \frac{2000}{x-6} = \frac{2000(x-6)}{x(x-6)} - \frac{2000x}{(x-6)x}$$

$$= \frac{2000x - 12{,}000 - 2000x}{x(x-6)}$$

$$= -\frac{12{,}000}{x(x-6)}$$

(B) A freight train averages 30 miles per hour traveling to its destination with full cars and 40 miles per hour on the return trip with empty cars. Find the total time in terms of d. Use the formula $t = \frac{d}{r}$.

Let d represent the one-way distance.

Total time: $\frac{d}{30} + \frac{d}{40} = \frac{d \cdot \square}{30 \cdot \square} + \frac{d \cdot \square}{40 \cdot \square}$

$$= \frac{d \cdot \square + d \cdot \square}{\square}$$

$$= \frac{\square}{\square}d$$

Your Turn

6. A hiker averages 1.4 miles per hour when walking downhill on a mountain trail and 0.8 miles per hour on the return trip when walking uphill. Find the total time in terms of d. Use the formula $t = \frac{d}{r}$.

7. Yvette ran at an average speed of 6.20 feet per second during the first two laps of a race and an average speed of 7.75 feet per second during the second two laps of a race. Find her total time in terms of d, the distance around the racecourse.

Elaborate

8. Why do rational expressions have excluded values?

9. How can you tell if your answer is written in simplest form?

10. **Essential Question Check-In** Why must the excluded values of each expression in a sum or difference of rational expressions also be excluded values for the simplified expression?

Evaluate: Homework and Practice

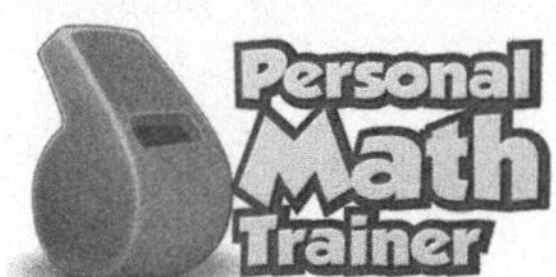

- Online Homework
- Hints and Help
- Extra Practice

Given a rational expression, identify the excluded values by finding the zeroes of the denominator.

1. $\dfrac{x-1}{x^2+3x-4}$

2. $\dfrac{4}{x(x+17)}$

Write the given expression as an equivalent rational expression that has the given denominator.

3. Expression: $\dfrac{x-7}{x+8}$

   Denominator: x^3+8x^2

4. Expression: $\dfrac{3x^3}{3x-6}$

   Denominator: $(2-x)(x^2+9)$

Simplify the given expression.

5. $\dfrac{(-4-4x)}{(x^2-x-2)}$

6. $\dfrac{-x-8}{x^2+9x+8}$

7. $\dfrac{6x^2+5x+1}{3x^2+4x+1}$

8. $\dfrac{x^4-1}{x^2+1}$

Add or subtract the given expressions, simplifying each result and noting the combined excluded values.

9. $\dfrac{1}{1+x}+\dfrac{1-x}{x}$

10. $\dfrac{x+4}{x^2-4}+\dfrac{-2x-2}{x^2-4}$

11. $\frac{1}{2+x} - \frac{2-x}{x}$

12. $\frac{4x^4+4}{x^2+1} - \frac{8}{x^2+1}$

13. $\frac{x^4-2}{x^2-2} + \frac{2}{-x^2+2}$

14. $\frac{1}{x^2+3x-4} - \frac{1}{x^2-3x+2}$

15. $\frac{3}{x^2-4} - \frac{x+5}{x+2}$

16. $\frac{-3}{9x^2-4} + \frac{1}{3x^2+2x}$

17. $\dfrac{x-2}{x+2} + \dfrac{1}{x^2-4} - \dfrac{x+2}{2-x}$

18. $\dfrac{x-3}{x+3} - \dfrac{1}{x-3} + \dfrac{x+2}{3-x}$

19. A company has two factories, factory A and factory B. The cost per item to produce q items in factory A is $\dfrac{200+13q}{q}$. The cost per item to produce q items in factory B is $\dfrac{300+25q}{2q}$. Find an expression for the sum of these costs per item. Then divide this expression by 2 to find an expression for the average cost per item to produce q items in each factory.

20. An auto race consists of 8 laps. A driver completes the first 3 laps at an average speed of 185 miles per hour and the remaining laps at an average speed of 200 miles per hour. Let d represent the length of one lap. Find the time in terms of d that it takes the driver to complete the race.

21. The junior and senior classes of a high school are cleaning up a beach. Each class has pledged to clean 1600 meters of shoreline. The junior class has 12 more students than the senior class. Let s represent the number of students in the senior class. Write and simplify an expression in terms of s that represents the difference between the number of meters of shoreline each senior must clean and the number of meters each junior must clean.

22. Architecture The Renaissance architect Andrea Palladio believed that the height of a room with vaulted ceilings should be the harmonic mean of the length and width. The harmonic mean of two positive numbers a and b is equal to $\frac{2}{\frac{1}{a}+\frac{1}{b}}$. Simplify this expression. What are the excluded values? What do they mean in this problem?

23. Match each expression with the correct excluded value(s).

a. $\dfrac{3x+5}{x+2}$ ______ no excluded values

b. $\dfrac{1+x}{x^2-1}$ ______ $x \neq 0, -2$

c. $\dfrac{3x^4-12}{x^2+4}$ ______ $x \neq 1, -1$

d. $\dfrac{3x+6}{x^2(x+2)}$ ______ $x \neq -2$

H.O.T. Focus on Higher Order Thinking

24. **Explain the Error** George was asked to write three different expressions equivalent to $2x - 3$, but with successive excluded values of $x = 1$, $x = 2$, and $x = -3$. He wrote the following expressions:

a. $\frac{2x - 3}{x - 1}$

b. $\frac{2x - 3}{x - 2}$

c. $\frac{2x - 3}{x + 3}$

What error did George make? Write the correct expressions, then write an expression that has all three excluded values.

25. **Communicate Mathematical Ideas** Write a rational expression with excluded values at $x = 0$ and $x = 17$.

26. **Critical Thinking** Sketch the graph of the rational equation $y = \frac{x^2 + 3x + 2}{x + 1}$. Think about how to show graphically that a graph exists over a domain except at one point.

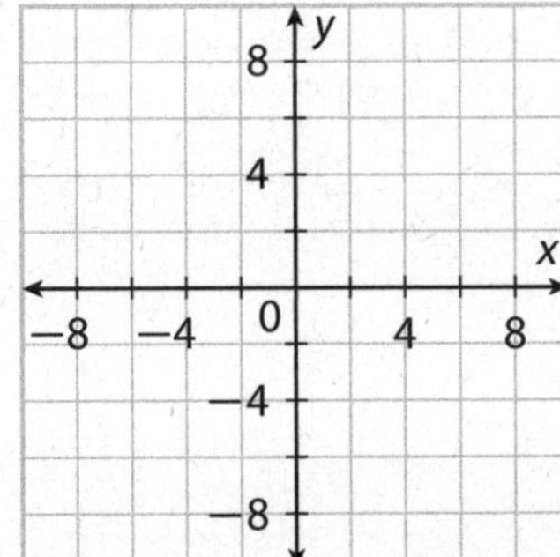

Lesson Performance Task

A kayaker spends an afternoon paddling on a river. She travels 3 miles upstream and 3 miles downstream in a total of 4 hours. In still water, the kayaker can travel at an average speed of 2 miles per hour. Based on this information, can you estimate the average speed of the river's current? Is your answer reasonable?

Next, assume the average speed of the kayaker is an unknown, k, and not necessarily 2 miles per hour. What is the range of possible average kayaker speeds under the rest of the constraints?

Name____________________ Class____________ Date________

10.2 Multiplying and Dividing Rational Expressions

Resource Locker

Essential Question: How can you multiply and divide rational expressions?

A2.7.F Determine the ... product, and quotient of rational expressions with integral exponents of degree one and of degree two.

Explore Relating Multiplication Concepts

Use the facts you know about multiplying rational numbers to determine how to multiply rational expressions.

(A) How do you multiply $\frac{4}{5} \cdot \frac{5}{6}$?

Multiply 4 by ___ to find the __________ of the product, and multiply 5 by ___ to find the __________.

(B) $\frac{4}{5} \cdot \frac{5}{6} = \frac{\square}{\square}$

(C) To simplify, factor the numerator and denominator.

$20 = \square$

$30 = \square$

(D) Cancel common factors in the numerator and denominator to simplify the product.

$\frac{4}{5} \cdot \frac{5}{6} = \frac{20}{30} = \frac{2 \cdot 2 \cdot 5}{2 \cdot 3 \cdot 5} = \frac{\square}{\square}$

(E) Based on the steps used for multiplying rational numbers, how can you multiply the rational expression $\frac{x+1}{x-1} \cdot \frac{3}{2(x+1)}$?

____

____

Reflect

1. **Discussion** Multiplying rational expressions is similar to multiplying rational numbers. Likewise, dividing rational expressions is similar to dividing rational numbers. How could you use the steps for dividing rational numbers to divide rational expressions?

____

____

____

Explain 1 Multiplying Rational Expressions

To multiply rational expressions, multiply the numerators to find the numerator of the product, and multiply the denominators to find the denominator. Then, simplify the product by cancelling common factors.

Note the excluded values of the product, which are any values of the variable for which the expression is undefined.

Example 1 **Find the products and any excluded values.**

(A) $\frac{3x^2}{x^2-2x-8} \cdot \frac{2x^2-6x-20}{x^2-3x-10}$

$$\frac{3x^2}{x^2-2x-8} \cdot \frac{2x^2-6x-20}{x^2-3x-10} = \frac{3x^2}{(x+2)(x-4)} \cdot \frac{2(x+2)(x-5)}{(x+2)(x-5)}$$

Factor the numerators and denominators.

$$= \frac{6x^2(x+2)(x-5)}{(x+2)(x-4)(x+2)(x-5)}$$

Multiply the numerators and multiply the denominators.

$$= \frac{6x^2\cancel{(x+2)}\cancel{(x-5)}}{\cancel{(x+2)}(x-4)(x+2)\cancel{(x-5)}}$$

Cancel the common factors in the numerator and denominator.

$$= \frac{6x^2}{(x+2)(x-4)}$$

Determine what values of x make each expression undefined.

$\frac{3x^2}{x^2-2x-8}$: The denominator is 0 when $x = -2$ and $x = 4$.

$\frac{2x^2-6x-20}{x^2-3x-10}$: The denominator is 0 when $x = -2$ and $x = 5$.

Excluded values: $x = -2$, $x = 4$, and $x = 5$

(B) $\frac{x^2-8x}{14(x^2+8x+15)} \cdot \frac{7x+35}{x+8}$

$$\frac{x^2-8x}{14(x^2+8x+15)} \cdot \frac{7x+35}{x+8} = \frac{\boxed{}(x-8)}{14\left(\boxed{}\right)(x+5)} \cdot \frac{7\left(\boxed{}\right)}{x+8}$$

Factor the numerators and denominators.

$$= \frac{7x(x-8)\left(\boxed{}\right)}{14\left(\boxed{}\right)(x+5)(x+8)}$$

Multiply the numerators and multiply the denominators.

$$= \frac{\boxed{}}{\boxed{}}$$

Cancel the common factors in the numerator and denominator.

Determine what values of x make each expression undefined.

$\frac{x^2-8x}{14(x^2+8x+15)}$: The denominator is 0 when $\boxed{}$.

$\frac{7x+35}{x+8}$: The denominator is 0 when $\boxed{}$.

Excluded values: $\boxed{}$

Your Turn

Find the products and any excluded values.

2. $\dfrac{x^2-9}{x^2-5x-24} \cdot \dfrac{x-8}{2x^2-18x}$

3. $\dfrac{x}{x-9} \cdot \dfrac{3x-27}{x+1}$

Explain 2 Dividing Rational Expressions

To divide rational expressions, change the division problem to a multiplication problem by multiplying by the reciprocal. Then, follow the steps for multiplying rational expressions.

Example 2 **Find the quotients and any excluded values.**

(A) $\dfrac{x+2}{x-4} \div \dfrac{x}{3x-12} = \dfrac{x+2}{x-4} \cdot \dfrac{3x-12}{x}$ Multiply by the reciprocal.

$= \dfrac{3(x+2)(x-4)}{x(x-4)}$ Factor and multiply the numerators and the denominators.

$= \dfrac{3(x+2)\cancel{(x-4)}}{x\cancel{(x-4)}} = \dfrac{3(x+2)}{x}$ Divide out common factors and simplify.

Determine the values that make each expression undefined.

$\dfrac{x+2}{x-4}$ and $\dfrac{x}{3x-12}$ are undefined when $x = 4$. $\dfrac{3x-12}{x}$ is undefined when $x = 0$.

Excluded values: $x = 4$ and $x = 0$.

(B) $\dfrac{(x+7)^2}{x^2} \div \dfrac{x^2+9x+14}{x^2+x-2} = \dfrac{(x+7)^2}{x^2} \cdot \dfrac{x^2+x-2}{x^2+9x+14}$ Multiply by the reciprocal.

$= \dfrac{(x+7)(x+7)}{x^2} \cdot \dfrac{(x+2)(x-1)}{(x+7)(x+2)}$ Factor the numerators and denominators.

$= \dfrac{(x+7)(x+7)(x+2)(x-1)}{x^2(x+7)(x+2)}$ Multiply the numerators and multiply the denominators.

$= \dfrac{\cancel{(x+7)}(x+7)\cancel{(x+2)}(x-1)}{x^2\cancel{(x+7)}\cancel{(x+2)}}$ Divide out common factors in the numerator and denominator.

$= \dfrac{(x+7)(x-1)}{x^2}$

Determine what values of x make each expression undefined.

$\frac{(x+7)^2}{x}$: $x = 0$;

$\frac{x^2+9x+14}{x^2+x-2}$: $x = -2$ and $x = 1$;

$\frac{x^2+x-2}{x^2+9x+14}$: $x = -7$ and $x = -2$;

Excluded values: $x = 0$, $x = -7$, $x = 1$, and $x = -2$

C $\frac{6x}{3x-30} \div \frac{9x^2-27x-36}{x^2-10x}$

$\frac{6x}{3x-30} \div \frac{9x^2-27x-36}{x^2-10x} = \frac{6x}{3x-30} \cdot \frac{\square}{\square}$ Multiply by the reciprocal.

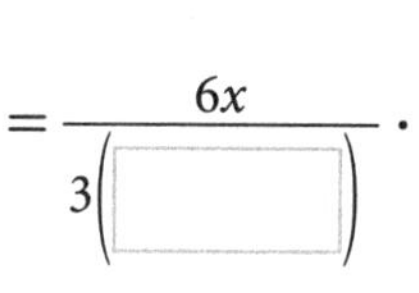

$= \frac{6x}{3(\square)} \cdot \frac{x(\square)}{9(x+1)(\square)}$ Factor the numerators and denominators.

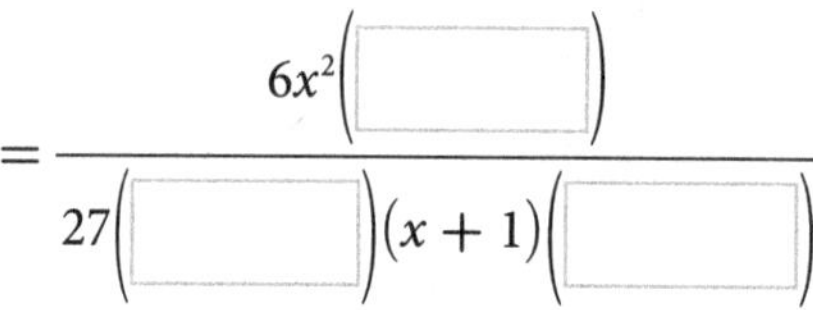

$= \frac{6x^2(\square)}{27(\square)(x+1)(\square)}$ Multiply the numerators and multiply the denominators.

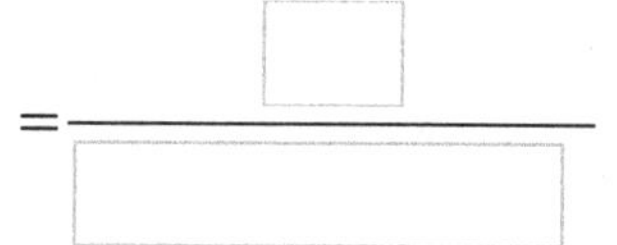

$= \frac{\square}{\square}$ Cancel the common factors in the numerator and denominator.

Determine what values of x make each expression undefined.

$\frac{6x}{3x-30}$: The denominator is 0 when $\square$.

$\frac{9x^2-27x-36}{x^2-10x}$: The denominator is 0 when $\square$.

$\frac{x^2-10x}{9x^2-27x-36}$: The denominator is 0 when $\square$.

Excluded values: $\square$

Your Turn

Find the quotients and any excluded values.

4. $\frac{x+11}{4x} \div \frac{2x+6}{x^2+2x-3}$

5. $\frac{20}{x^2-7x} \div \frac{5x^2-40x}{x^2-15x+56}$

Explain 3 Activity: Investigating Closure

A set of numbers is said to be closed, or to have **closure**, under a given operation if the result of the operation on any two numbers in the set is also in the set.

(A) Recall whether the set of whole numbers, the set of integers, and the set of rational numbers are closed under each of the four basic operations.

	Addition	Subtraction	Multiplication	Division
Whole Numbers				
Integers				
Rational Numbers				

(B) Look at the set of rational expressions. Use the rational expressions $\frac{p(x)}{q(x)}$ and $\frac{r(x)}{s(x)}$ where $p(x)$, $q(x)$, $r(x)$ and $s(x)$ are nonzero. Add the rational expressions.

$$\frac{p(x)}{q(x)} + \frac{r(x)}{s(x)} = \square$$

(C) Is the set of rational expressions closed under addition? Explain.

(D) Subtract the rational expressions.

$$\frac{p(x)}{q(x)} - \frac{r(x)}{s(x)} = \square$$

(E) Is the set of rational expressions closed under subtraction? Explain.

(F) Multiply the rational expressions.

$$\frac{p(x)}{q(x)} \cdot \frac{r(x)}{s(x)} = \square$$

(G) Is the set of rational expressions closed under multiplication? Explain.

(H) Divide the rational expressions.

$$\frac{p(x)}{q(x)} \div \frac{r(x)}{s(x)} = \square$$

(I) Is the set of rational expressions closed under division? Explain.

Reflect

6. Are rational expressions most like whole numbers, integers, or rational numbers? Explain.

Explain 4 Multiplying and Dividing with Rational Models

Models involving rational expressions can be solved using the same steps to multiply or divide rational expressions.

Example 3 **Solve the problems using rational expressions.**

(A) Leonard drives 40 miles to work every day. One-fifth of his drive is on city roads, where he averages 30 miles per hour. The other part of his drive is on a highway, where he averages 55 miles per hour. The expression $\frac{d_c r_h + d_h r_c}{r_c r_h}$ represents the total time spent driving, in hours. In the expression, d_c represents the distance traveled on city roads, d_h represents the distance traveled on the highway, r_c is the average speed on city roads, and r_h is the average speed on the highway. Use the expression to find the average speed of Leonard's drive.

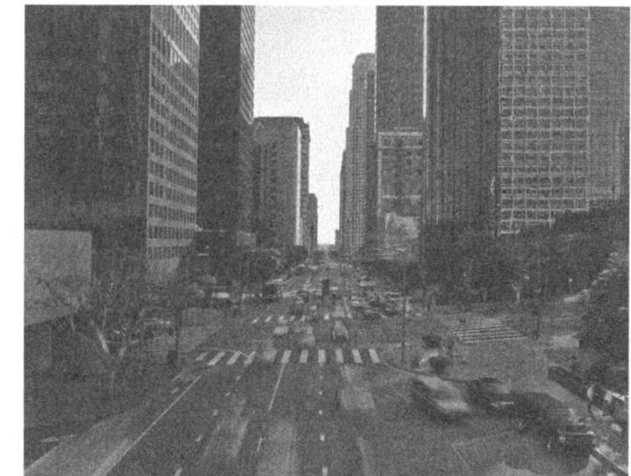

The total distance traveled is 40 miles. Find an expression for the average speed, r, of Leonard's drive.

$$r = \text{Total distance traveled} \div \text{Total time}$$
$$= 40 \div \frac{d_c r_h + d_h r_c}{r_c r_h}$$
$$= 40 \cdot \frac{r_c r_h}{d_c r_h + d_h r_c}$$
$$= \frac{40 r_c r_h}{d_c r_h + d_h r_c}$$

Find the values of d_c and d_h.

$$d_c = \frac{1}{5}(40) = 8 \text{ miles}$$
$$d_h = 40 - 8 = 32 \text{ miles}$$

Solve for r by substituting in the given values from the problem.

$$r = \frac{d r_c r_h}{d_c r_h + d_h r_c}$$
$$= \frac{40 \cdot 55 \cdot 30}{8 \cdot 55 + 32 \cdot 30}$$
$$\approx 47 \text{ miles per hour}$$

The average speed of Leonard's drive is about 47 miles per hour.

(B) The fuel efficiency of Tanika's car at highway speeds is 35 miles per gallon. The expression $\frac{48E - 216}{E(E - 6)}$ represents the total gas consumed, in gallons, when Tanika drives 36 miles on a highway and 12 miles in a town to get to her relative's house. In the expression, E represents the fuel efficiency, in miles per gallon, of Tanika's car at highway speeds. Use the expression to find the average rate of gas consumed on her trip.

The total distance traveled is $\square$ miles. Find an expression for the average rate of gas consumed, g, on Tanika's trip.

$$g = \text{Total gas consumed} \div \text{Total distance traveled}$$

$$= \frac{48E - 216}{E(E-6)} \div \square$$

$$= \frac{48E - 216}{\square\, E(E-6)}$$

The value of E is $\square$.

Solve for g by substituting in the value of E.

$$g = \frac{48\left(\square\right) - 216}{48\left(\square\right)\left(\square - 6\right)}$$

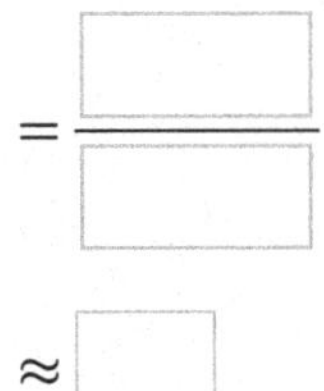

$$\approx \square$$

The average rate of gas consumed on Tanika's trip is about $\square$ gallon per mile.

Your Turn

7. The distance traveled by a car undergoing constant acceleration, a, for a time, t, is given by $d = v_0t + \frac{1}{2}at^2$ where v_0 is the initial velocity of the car. Two cars are side by side with the same initial velocity. One car accelerates and the other car does not. Write an expression for the ratio of the distance traveled by the accelerating car to the distance traveled by the nonaccelerating car as a function of time.

Elaborate

8. Explain how finding excluded values when dividing one rational expression by another is different from multiplying two rational expressions.

9. **Essential Question Check-In** How is dividing rational expressions related to multiplying rational expressions?

Evaluate: Homework and Practice

- Online Homework
- Hints and Help
- Extra Practice

1. Explain how to multiply the rational expressions.

$$\frac{x-3}{2} \cdot \frac{x^2-3x+4}{x^2-2x}$$

Find the products and any excluded values.

2. $\frac{x}{3x-6} \cdot \frac{x-2}{x+9}$

3. $\frac{5x^2+25x}{2} \cdot \frac{4x}{x+5}$

4. $\frac{x^2-2x-15}{10x+30} \cdot \frac{3}{x^2-3x-10}$

5. $\frac{x^2-1}{x^2+5x+4} \cdot \frac{x^2}{x^2-x}$

6. $\dfrac{x^2 + 14x + 33}{4x} \cdot \dfrac{x^2 - 3x}{x + 3} \cdot \dfrac{8x - 56}{x^2 + 4x - 77}$

7. $\dfrac{9x^2}{x - 6} \cdot \dfrac{x^2 - 36}{3x - 6} \cdot \dfrac{3}{4x^2 + 24x}$

Find the quotients and any excluded values.

8. $\dfrac{5x^2 + 10x}{x^2 + 2x + 1} \div \dfrac{20x + 40}{x^2 - 1}$

9. $\dfrac{x^2 - 9x + 18}{x^2 + 9x + 18} \div \dfrac{x^2 - 36}{x^2 - 9}$

10. $\dfrac{-x^2 + x + 20}{5x^2 - 25x} \div \dfrac{x + 4}{2x - 14}$

11. $\dfrac{x + 3}{x^2 + 8x + 15} \div \dfrac{x^2 - 25}{x - 5}$

12. $\dfrac{x^2 - 10x + 9}{3x} \div \dfrac{x^2 - 7x - 18}{x^2 + 2x}$

13. $\dfrac{8x + 32}{x^2 + 8x + 16} \div \dfrac{x^2 - 6x}{x^2 - 2x - 24}$

Let $p(x) = \frac{1}{x+1}$ and $q(x) = \frac{1}{x-1}$. Perform the operation, and show that it results in another rational expression.

14. $p(x) + q(x)$

15. $p(x) - q(x)$

16. $p(x) \cdot q(x)$

17. $p(x) \div q(x)$

18. The distance a race car travels is given by the equation $d = v_0t + \frac{1}{2}at^2$ where v_0 is the initial speed of the race car, a is the acceleration, and t is the time travelled. Near the beginning of a race, the driver accelerates for 9 seconds at a rate of 4 m/s². The driver's initial speed was 75 m/s. Find the driver's average speed during the acceleration.

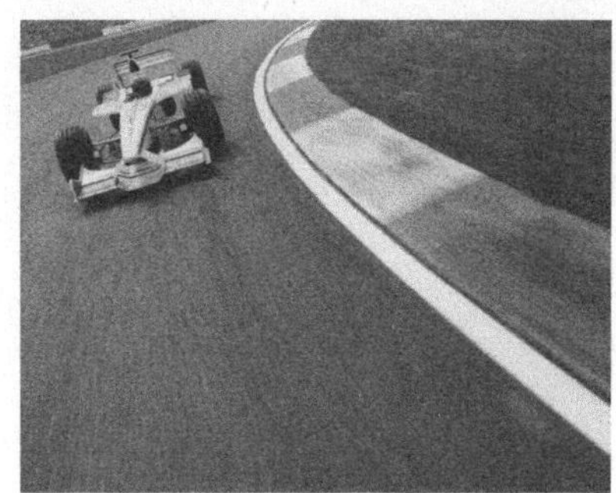

19. Julianna is designing a circular track that will consist of three concentric rings. The radius of the middle ring is 6 meters greater than that of the inner ring and 6 meters less than that of the outer ring. Find an expression for the ratio of the length of the outer ring to the length of the middle ring and another for the ratio of the length of the outer ring to length of the inner ring. If the radius of the inner ring is set at 90 meters, how many times longer is the outer ring than the middle ring and the inner ring?

20. Geometry Find a rational expression for the ratio of the surface area of a cylinder to the volume of a cylinder. Then find the ratio when the radius is 3 inches and the height is 10 inches.

H.O.T. Focus on Higher Order Thinking

21. Explain the Error Maria finds an expression equivalent to $\frac{x^2 - 4x - 45}{3x - 15} \div \frac{6x^2 - 150}{x^2 - 5x}$. Her work is shown. Find and correct Maria's mistake.

$$\frac{x^2 - 4x - 45}{3x - 15} \div \frac{6x^2 - 150}{x^2 - 5x} = \frac{(x-9)(x+5)}{3(x-5)} \div \frac{6(x+5)(x-5)}{x(x-5)}$$

$$= \frac{6(x-9)(x+5)(x+5)(x-5)}{3x(x-5)(x-5)}$$

$$= \frac{2(x-9)(x+5)^2}{x(x-5)}$$

22. Critical Thinking Multiply the expressions. What do you notice about the resulting expression?

$$\left(\frac{3}{x-4} + \frac{x^3 - 4x}{8x^2 - 32}\right)\left(\frac{3x+18}{x^2 + 2x - 24} - \frac{x}{8}\right)$$

23. **Multi-Step** Jordan is making a garden with an area of $x^2 + 13x + 30$ square feet and a length of $x + 3$ feet.

a. Find an expression for the width of Jordan's garden.

b. If Karl makes a garden with an area of $3x^2 + 48x + 180$ square feet and a length of $x + 6$, how many times larger is the width of Jon's garden than Jordan's?

c. If x is equal to 4, what are the dimensions of both Jordan's and Karl's gardens?

Lesson Performance Task

Who has the advantage, taller or shorter runners? Almost all of the energy generated by a long-distance runner is released in the form of heat. For a runner with height H and speed V, the rate h_g of heat generated and the rate h_r of heat released can be modeled by $h_g = k_1H^3V^2$ and $h_r = k_2H^2$, k_1 and k_2 being constants. So, how does a runner's height affect the amount of heat she releases as she increases her speed?

Name ______________ Class ______________ Date ______________

10.3 Solving Rational Equations

Essential Question: What methods are there for solving rational equations?

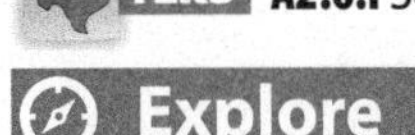

A2.6.I Solve rational equations that have real solutions. Also A2.6.H, A2.6.J

Explore Solving Rational Equations Graphically

A rational equation is an equation that contains one or more rational expressions. The time t in hours it takes to travel d miles can be determined by using the equation $t = \frac{d}{r}$, where r is the average rate of speed. This equation is an example of a rational equation. One method to solving rational equations is by graphing.

Solve the rational equation $\frac{x}{x-3} = 2$ by graphing.

(A) First, identify any excluded values. A number is an excluded value of a rational expression if substituting the number into the expression results in a division by 0, which is undefined. Solve $x - 3 = 0$ for x.

$x - 3 = 0$

$x = \square$

(B) So, 3 is an excluded value of the rational equation. Rewrite the equation with 0 on one side.

$\frac{x}{x-3} = 2$

$\square = 0$

(C) Graph the left side of the equation as a function. Substitute y for 0 and complete the table below.

x	y	(x, y)
0		
1		
2		
4		
5		
9		

(D) Use the table to graph the function.

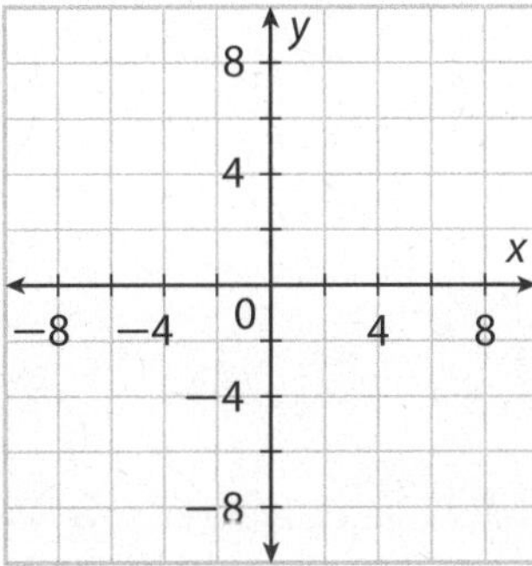

(E) Identify any x-intercepts of the graph.

There is an x-intercept at $\square$.

(F) Is the value of x an excluded value? What is the solution of $\frac{x}{x-3} = 2$?

____

____

Reflect

1. **Discussion** Why does rewriting a rational equation with 0 on one side help with solving the equation?

______

______

Explain 1 Identifying the LCD of Two Rational Expressions

Given two or more rational expressions, the least common denominator (LCD) is found by factoring each denominator and finding the least common multiple (LCM) of the factors. This technique is useful for the addition and subtraction of expressions with unlike denominators.

Least Common Denominator (LCD) of Rational Expressions

To find the LCD of rational expressions:

1. Factor each denominator completely. Write any repeated factors as powers.
2. List the different factors. If the denominators have common factors, use the highest power of each common factor.

Example 2 **Find the LCD for each set of rational expressions.**

(A) $\frac{-2}{3x - 15}$ and $\frac{6x}{4x + 28}$

Factor each denominator completely.

$3x - 15 = 3(x - 5)$

$4x + 28 = 4(x + 7)$

List the different factors.

$3, 4, x - 5, x + 7$

The LCD is $3 \cdot 4(x - 5)(x + 7)$,

or $12(x - 5)(x + 7)$.

(B) $\frac{-14}{x^2 - 11x + 14}$ and $\frac{9}{x^2 - 6x + 9}$

Factor each denominator completely.

$x^2 - 11x + 24 =$ ______

$x^2 - 6x + 9 =$ ______

List the different factors.

______ and ______

Taking the highest power of $(x - 3)$,

the LCD is ______.

Reflect

2. **Discussion** When is the LCD of two rational expressions not equal to the product of their denominators?

______

Your Turn

Find the LCD for each set of rational expressions.

3. $\dfrac{x+6}{8x-24}$ and $\dfrac{14x}{10x-30}$

4. $\dfrac{12x}{15x+60} = \dfrac{5}{x^2+9x+20}$

Explain 2 Solving Rational Equations Algebraically

Rational equations can be solved algebraically by multiplying through by the LCD and solving the resulting polynomial equation. However, this eliminates the information about the excluded values of the original equation. Sometimes an excluded value of the original equation is a solution of the polynomial equation, and in this case the excluded value will be an **extraneous solution** of the polynomial equation. Extraneous solutions are not solutions of an equation.

Example 2 **Solve each rational equation algebraically.**

(A) $\dfrac{3x+7}{x-5} = \dfrac{5x+17}{2x-10}$

Identify any excluded values. $\quad x-5=0 \qquad 2x-10=0$

The excluded value is 5. $\quad x=5 \qquad x=5$

Identify the LCD. $\quad 2x-10=2(x-5)$

The different factors are 2 and $x-5$. $\quad x-5=x-5$

The LCD is $2(x-5)$.

Multiply each term by the LCD. $\quad \dfrac{3x+7}{x-5}\cdot 2(x-5) = \dfrac{5x+17}{2(x-5)}\cdot 2(x-5)$

Divide out common factors. $\quad \dfrac{3x+7}{\cancel{x-5}}\cdot 2\cancel{(x-5)} = \dfrac{5x+17}{\cancel{2}\cancel{(x-5)}}\cdot \cancel{2}\cancel{(x-5)}$

Simplify. $\quad (3x+7)2 = 5x+17$

Use the Distributive Property. $\quad 6x+14=5x+17$

Solve for x. $\quad x+14=17$

$x=3$

The solution $x=3$ is not an excluded value. So, $x=3$ is the solution of the equation.

 $\dfrac{2x-9}{x-7}+\dfrac{x}{2}=\dfrac{5}{x-7}$

Identify any excluded values.

$x-7=0$

$x=\square$

The excluded value is ______.

Identify the LCD.

The different factors are ______.

The LCD is ______.

Multiply each term by the LCD. $\dfrac{2x-9}{x-7}\cdot\square+\dfrac{x}{2}\cdot\square=\dfrac{5}{x-7}\cdot\square$

Divide out common factors. $\dfrac{2x-9}{\cancel{x-7}}\cdot\square+\dfrac{x}{\cancel{2}}\cdot\square=\dfrac{5}{\cancel{x-7}}\cdot\square$

Simplify. $\square(2x-9)+x(\square)=5(\square)$

Use the Distributive Property. $\square+x^2-7x=\square$

Write in standard form. $\square=0$

Factor. $(\square)(\square)=0$

Use the Zero Product Property. $x-7=0$ or $\square=0$

Solve for x. $x=7$ or $x=\square$

The solution $x=\square$ is extraneous because it is an excluded value. The only solution is $x=\square$.

Your Turn

Solve each rational equation algebraically.

5. $\dfrac{8}{x+3}=\dfrac{x+1}{x+6}$

Explain 3 Solving a Real-world Problem with a Rational Equation

Rational equations are used to model real-world situations. These equations can be solved algebraically.

Ⓐ Kelsey is kayaking on a river. She travels 5 miles upstream and 5 miles downstream in a total of 6 hours. In still water, Kelsey can travel at an average speed of 3 miles per hour. What is the average speed of the river's current?

Analyze Information

Identify the important information:

- The answer will be the average speed of ____________.
- Kelsey spent ____________ kayaking.
- She traveled ____________ upstream and ____________ downstream.
- Her average speed in still water is ____________.

Formulate a Plan

Let c represent the speed of the current in miles per hour. When Kelsey is going upstream, her speed is equal to her speed in still water ________ c. When Kelsey is going downstream, her speed is equal to her speed in still water ________ c.

The variable c is restricted to ____________.

Complete the table.

	Distance (mi)	Average speed (mi/h)	Time (h)
Upstream	5		
Downsteam	5		

Use the results from the table to write an equation.
total time = time upstream + time downstream

$6 = \square + \square$

Solve

$3 - c = 0 \qquad 3 + c = 0$

$\square = c \qquad c = \square$

Excluded values: ________

LCD: ________

Multiply by the LCD. $6 \cdot \square = \frac{5}{3-c} \cdot \square + \frac{5}{3+c} \cdot \square$

Divide out common factors. $6 \cdot \square = \frac{5}{\cancel{3-c}} \cdot \square + \frac{5}{\cancel{3+c}} \cdot \square$

Simplify. $6 \cdot \square = 5 \cdot \square + 5 \cdot \square$

Use the Distributive Property. $\square = 15 + 5c + \square$

Write in standard form. $0 = \square$

Factor. $0 = 6(c+2)(\square)$

Use the Zero Product Property. $c + 2 = 0$ or $\square = 0$

Solve for c. $c = \square$ or $c = \square$

There ________ extraneous solutions. The solutions are ________.

Justify and Evaluate

The solution $c = \square$ is unreasonable because the speed of the current cannot be ________ but the solution $c = \square$ is reasonable because the speed of the current can be ________. If the speed of the current is ________, it would take Kelsey ______ hour(s) to go upstream and ______ hour(s) to go downstream, which is a total of ______ hours.

Reflect

6. Why does the domain of the variable have to be restricted in real-world problems that can be modeled with a rational equation?

____

____

Your Turn

7. Kevin can clean a large aquarium tank in about 7 hours. When Kevin and Lara work together, they can clean the tank in 4 hours. Write and solve a rational equation to determine how long, to the nearest tenth of an hour, it would take Lara to clean the tank if she works by herself. Explain whether the answer is reasonable.

Elaborate

8. Why is it important to check solutions to rational equations?

9. Why can extraneous solutions to rational equations exist?

10. **Essential Question Check In** How can you solve a rational equation without graphing?

Evaluate: Homework and Practice

- Online Homework
- Hints and Help
- Extra Practice

Solve each rational equation by graphing using a table of values.

1. $\frac{x}{x+4} = -3$

x	y	(x, y)
−8		
−6		
−5		
−3.5		
−2		
0		

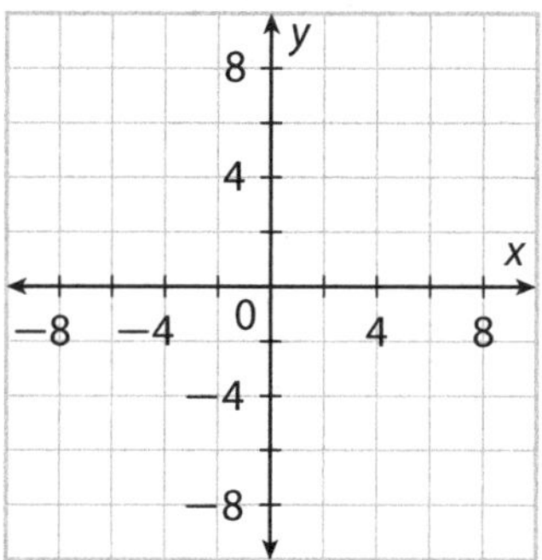

2. $\frac{x}{2x - 10} = 3$

x	y	(x, y)
0		
3		
4		
5.5		
7		
10		

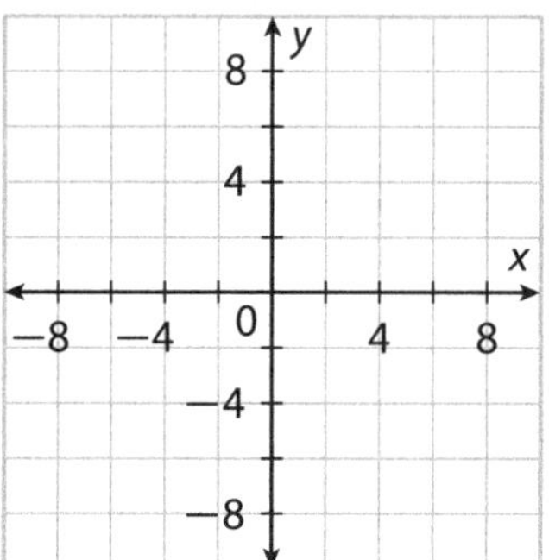

Find the LCD for each set of rational expressions.

3. $\frac{x}{2x + 16}$ and $\frac{-4x}{3x - 27}$

4. $\frac{x^2 - 4}{5x - 30}$ and $\frac{5x + 13}{7x - 42}$

5. $\dfrac{4x+12}{x^2+5x+6}$ and $\dfrac{5x+15}{10x+20}$

6. $\dfrac{-11}{x^2-3x-28}$ and $\dfrac{2}{x^2-2x-24}$

7. $\dfrac{12}{3x^2-21x-54}$ and $\dfrac{-1}{21x^2-84}$

8. $\dfrac{3x}{5x^2-40x-60}$ and $\dfrac{17}{-7x^2+56x+84}$

Solve each rational equation algebraically.

9. $\dfrac{9}{4x}-\dfrac{5}{6}=-\dfrac{13}{12x}$

10. $\dfrac{3}{x+1}+\dfrac{2}{7}=2$

11. $\dfrac{56}{x^2 - 2x - 15} - \dfrac{6}{x + 3} = \dfrac{7}{x - 5}$

12. $\dfrac{x^2 - 29}{x^2 - 10x + 21} = \dfrac{6}{x - 7} + \dfrac{5}{x - 3}$

13. $\dfrac{5}{2x+6} - \dfrac{1}{6} = \dfrac{2}{x+4}$

14. $\dfrac{5}{x^2 - 3x + 2} - \dfrac{1}{x-2} = \dfrac{x+6}{3x-3}$

For 15 and 16, write a rational equation for each real-world application. Do not solve.

15. A save percentage in lacrosse is found by dividing the number of saves by the number of shots faced. A lacrosse goalie saved 9 of 12 shots. How many additional consecutive saves s must the goalie make to raise his save percentage to 0.850?

16. Jake can mulch a garden in 30 minutes. Together, Jake and Ross can mulch the same garden in 16 minutes. How much time t, in minutes, will it take Ross to mulch the garden when working alone?

17. **Geometry** A new ice skating rink will be approximately rectangular in shape and will have an area of 18,000 square feet. Using an equation for the perimeter P, of the skating rink in terms of its width W, what are the dimensions of the skating rink if the perimeter is 580 feet?

18. Water flowing through both a small pipe and a large pipe can fill a water tank in 9 hours. Water flowing through the large pipe alone can fill the tank in 17 hours. Write an equation that can be used to find the amount of time t, in hours, it would take to fill the tank using only the small pipe.

19. A riverboat travels at an average of 14 km per hour in still water. The riverboat travels 110 km east up the Ohio River and 110 km west down the same river in a total of 17.5 hours. To the nearest tenth of a kilometer per hour, what was the speed of the current of the river?

20. A baseball player's batting average is equal to the number of hits divided by the number of at bats. A professional player had 139 hits in 515 at bats in 2012 and 167 hits in 584 at bats in 2013. Write and solve an equation to find how many additional consecutive hits h the batter would have needed to raise his batting average in 2012 to be at least equal to his average in 2013.

21. The time required to deliver and install a computer network at a customer's location is $t = 5 + \frac{2d}{r}$, where t is time in hours, d is in the distance (in miles) from the warehouse to the customer's location, and r is the average speed of the delivery truck. If it takes 8.2 hours for an employee to deliver and install a network for a customer located 80 miles from the warehouse, what is the average speed of the delivery truck?

22. **Art** A glassblower can produce several sets of simple glasses in about 3 hours. When the glassblower works with an apprentice, the job takes about 2 hours. How long would it take the apprentice to make the same number of sets of glasses when working alone?

23. Which of the following equations have at least two excluded values? Select all that apply.

A. $\frac{3}{x} + \frac{1}{5x} = 1$

B. $\frac{x-4}{x-2} + \frac{3}{x} = \frac{5}{6}$

C. $\frac{x}{x-6} + 1 = \frac{5}{2x-12}$

D. $\frac{2x-3}{x^2-10x+25} + \frac{3}{7} = \frac{1}{x-5}$

E. $\frac{7}{x+2} + \frac{3x-4}{x^2+5x+6} = 9$

H.O.T. Focus on Higher Order Thinking

24. **Critical Thinking** An equation has the form $\frac{a}{x} + \frac{x}{b} = c$, where a, b, and c are constants and $b \neq 0$. How many solutions could this equation have? Explain.

25. **Multiple Representations** Write an equation whose graph is a straight line, but with an open circle at $x = 4$.

26. **Justify Reasoning** Explain why the excluded values do not change when multiplying by the LCD to add or subtract rational expressions.

27. **Critical Thinking** Describe how you would find the inverse of the rational function $f(x) = \frac{x-1}{x-2}$, $x \neq 2$. Then find the inverse.

Lesson Performance Task

Kasey creates comedy sketch videos and posts them on a popular video website and is selling an exclusive series of sketches on DVD. The total cost to make the series of sketches is $989. The materials cost $1.40 per DVD and the shipping costs $2.00 per DVD. Kasey plans to sell the DVDs for $12 each.

a. Let d be the number of DVDs Kasey sells. Create a profit-per-item model from the given information by writing a rule for $C(d)$, the total costs in dollars, $S(d)$, the total sales income in dollars, $P(d)$, the profit in dollars, and $P_{PI}(d)$, the profit per item sold in dollars.

b. What is the profit per DVD if Kasey sells 80 DVDs? Does this value make sense in the context of the problem?

c. Use the function $P_{PI}(d)$ from part a to find how many DVDs Kasey would have to sell to break even. Identify all excluded values.

STUDY GUIDE REVIEW

Rational Expressions and Equations

MODULE 10

Essential Question: How can you use rational expressions and equations to solve real-world problems?

Key Vocabulary

closure *(cerradura)*
extraneous solution *(solución extraña)*
rational expression *(expresión racional)*
reciprocal *(recíproco)*

KEY EXAMPLE *(Lesson 10.1)*

Add $\frac{1}{3+x}$ and $\frac{3-x}{x}$, simplify the result, and note the excluded values.

$$\frac{1}{3+x}+\frac{3-x}{x}=\frac{1x}{(3+x)x}+\frac{(3-x)(3+x)}{x(3+x)}$$ Write with like denominators.

$$=\frac{x+(9-x^2)}{x(x+3)}$$ Add.

$$=\frac{-x^2+x+9}{x(x+3)},\ x\neq -3, 0$$ Simplify.

KEY EXAMPLE *(Lesson 10.2)*

Find the quotient $\frac{x+3}{x+2}\div\frac{x^2-9}{2x-4}$ and note any excluded values.

$$\frac{x+3}{x+2}\div\frac{x^2-9}{2x-4}=\frac{x+3}{x+2}\cdot\frac{2x-4}{x^2-9}$$ Multiply by the reciprocal.

$$=\frac{x+3}{x+2}\cdot\frac{2(x-2)}{(x+3)(x-3)}$$ Factor the numerators and denominators.

$$=\frac{2\cancel{(x+3)}(x-2)}{(x+2)\cancel{(x+3)}(x-3)}$$ Multiply and cancel the common factors.

$$=\frac{2(x-2)}{(x+2)(x-3)};\ x\neq \pm 2, \pm 3$$ Simplify.

KEY EXAMPLE *(Lesson 10.3)*

Solve the rational equation algebraically.

$$\frac{x}{x-3}+\frac{x}{2}=\frac{6x}{2x-6}$$

$$2\cancel{(x-3)}\frac{x}{\cancel{x-3}}+\cancel{2}(x-3)\frac{x}{\cancel{2}}=\cancel{2}\cancel{(x-3)}\frac{6x}{\cancel{2x-6}}$$ Multiply each term by the LCD and divide out common factors.

$$2x+x(x-3)=6x$$ Simplify.

$$x^2-7x=0$$ Write in standard form.

$$x(x-7)=0$$ Factor.

$$x=0 \text{ or } x=7$$ Solve for *x*.

EXERCISES

Add or subtract the given expressions, simplify the result, and note the excluded values. *(Lesson 10.1)*

1. $\frac{6x + 12}{x^2 - 9} + \frac{-3x - 3}{x^2 - 9}$

2. $\frac{4}{x^2 - 1} - \frac{x + 2}{x - 1}$

Multiply or divide the given expressions, simplify the result, and note the excluded values. *(Lesson 10.2)*

3. $\frac{x^2 - 4x - 5}{3x - 15} \cdot \frac{4}{x^2 - 2x - 3}$

4. $\frac{x + 2}{x - 4} \div \frac{x}{3x - 12}$

Solve each rational equation algebraically. *(Lesson 10.3)*

5. $x - \frac{10}{x} = 3$

6. $\frac{5}{x + 1} = \frac{2}{x + 4}$

MODULE PERFORMANCE TASK

Robots and Resistors

An engineer is designing part of a circuit that will control a robot. The circuit must have a certain total resistance to function properly. The engineer plans to use several resistors in *parallel*, which means each resistor is on its own branch of the circuit. The resistors available for this project are 20, 50, 80, and 200-ohm.

How can the engineer design a parallel circuit with a total resistance of 10 ohms using a maximum of 5 resistors, at least two of which must be different values? Find at least two possible circuit configurations that meet these criteria.

For another part of the circuit, the engineer wants to use resistors in parallel to create a total resistance of 6 ohms. Can she do it using the available resistor values? If so, how? If not, explain why not.

Begin by listing in the space below all of the information you will need to solve the problem. Then use your own paper to complete the task. Be sure to write down all your data and assumptions. Then use graphs, numbers, words, or algebra to explain how you reached your conclusion.

Ready to Go On?

10.1–10.3 Rational Expressions and Equations

- Online Homework
- Hints and Help
- Extra Practice

Perform the indicated operations, simplify the result and note any excluded values. *(Lessons 10.1, 10.2)*

1. $\frac{4}{x+5} + \frac{2x}{x^2 - 25}$

2. $\frac{3x+2}{x-2} - \frac{x+5}{x-2}$

3. $\frac{x+3}{x+2} \cdot \frac{2x-4}{x^2-9}$

4. $\frac{x-3}{x-4} \div \frac{x-2}{x^2-16}$

Solve each rational equation. *(Lesson 10.3)*

5. $\frac{3}{x+2} + \frac{3}{2x+4} = \frac{x}{2x+4}$

6. $\frac{x}{x-8} = \frac{24-2x}{x-8}$

7. $\frac{8x}{x^2-4} - \frac{4}{x+2} = \frac{8}{x^2-4}$

8. $\frac{3x}{x+1} + \frac{6}{2x} = \frac{7}{x}$

ESSENTIAL QUESTION

9. How do you add or subtract rational expressions and identify any excluded values?

MODULE 10
MIXED REVIEW

Assessment Readiness

1. A hiker averages 0.6 mile per hour walking up a mountain trail and 1.3 miles per hour walking down the trail. Find the total time in terms of d. Use the formula $t = \frac{d}{r}$.

A. $\frac{13}{60}d$

B. $\frac{7}{10}d$

C. $\frac{19}{20}d$

D. $\frac{95}{39}d$

2. If each of the following expressions is defined, which is equivalent to $x - 3$?

A. $\frac{(x-3)(x+5)}{x+3} \cdot \frac{x+3}{x+5}$

B. $\frac{(x+3)(x+5)}{x-5} \div \frac{x+5}{x-5}$

C. $\frac{x+3}{x+5} + \frac{x-3}{x+5}$

D. $\frac{5x-5}{x-5} - \frac{x-3}{x-5}$

3. Which equation does **not** have real roots?

A. $x^2 - 12 = 0$

B. $x^2 + 25 = 0$

C. $x^2 - 4x = -3$

D. $-8x^2 + 20 = 0$

4. What are the x-intercepts of the graph of $y = x^4 - 256$?

A. -256

B. $16, -16$

C. $4, -4$

D. $16, 4, -4, -16$

5. A restaurant has two pastry ovens. When both ovens are used, it takes about 3 hours to bake the bread needed for one day. When only the large oven is used, it takes about 4 hours to bake the bread for one day. About how long would it take to bake the bread for one day if only the small oven were used? Explain how you got your answer.

UNIT 4
MIXED REVIEW

Assessment Readiness

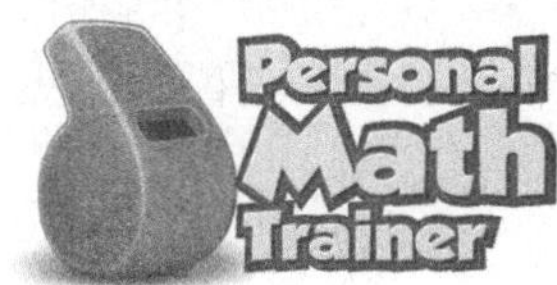

- Online Homework
- Hints and Help
- Extra Practice

1. What are the asymptotes of $y = \frac{3x^3}{x-5}$? Select the correct answer.

A. $x = 0, x = 5$

B. $x = 5$

C. $x = -5$

D. $y = 3$

2. How is the graph of $g(x) = 1 + \frac{1}{x+5}$ related to the graph of $f(x) = \frac{1}{x}$? Select the correct statement.

A. g is f translated 1 unit up and 5 units left

B. g is f translated 1 unit up and 5 units right

C. g is f translated 1 unit down and 5 units left

D. g is f translated 1 unit down and 5 units right

3. Which equation has the solutions $x = -6$ or $x = 4$? Select the correct answer.

A. $x + 2 = \frac{24}{x+6}$

B. $\frac{x+2}{x} = 24$

C. $x - 4 = \frac{24}{x}$

D. $x + 2 = \frac{24}{x}$

4. Which of the following expressions has the excluded values $x = 0$ and $x = 2$? Select the correct answer.

A. $\frac{1}{x+2} + \frac{4+x}{x}$

B. $\frac{1}{x-2} + \frac{4+x}{x}$

C. $\frac{1}{x^2-2} + \frac{4+x}{x^2+2}$

D. $\frac{x}{x+4} + \frac{x-2}{2}$

5. Which of the following operations simplifies to $\frac{(x+1)(x+3)}{(x+4)}$? Select the correct answer.

A. $\frac{x+1}{x-3} \div \frac{4}{x^2+9}$

B. $\frac{x+1}{x+4} \div \frac{x+4}{x+1}$

C. $\frac{x+1}{x-3} \div \frac{x+4}{x^2-9}$

D. $\frac{x+4}{x^2-9} \div \frac{x+1}{x-3}$

6. The time t it takes Sam to drive to his grandmother's house is inversely proportional to the speed v at which he drives. Write an equation for the one-way travel time. If it takes Sam 5 hours driving 50 miles per hour, how long would it take him if he drove at 65 miles per hour? *(Lesson 9.3)*

7. A town has two trucks to collect garbage. When both trucks are in use, they take 6 hours to collect all the garbage. When only the small truck is in use, it takes 24 hours to collect the garbage. How long would it take to collect the garbage if only the large truck is in service? *(Lesson 10.3)*

Performance Tasks

★ **8.** For a car moving with initial speed v_0 and acceleration a, the distance d that the car travels in time t is given by $d = v_0 t + \frac{1}{2}at^2$.

A. Write a rational expression in terms of t for the average speed of the car during a period of acceleration. Simplify the expression.

B. During a race, a driver accelerates for 3 s at a rate of 10 ft/s^2 in order to pass another car. The driver's initial speed was 264 ft/s. What was the driver's average speed during the acceleration?

★★ **9.** The average speed for the winner of the 2002 Indy 500 was 25 mi/h greater than the average speed for the 2001 winner. In addition, the 2002 winner completed the 500 mi race 32 min faster than the 2001 winner.

A. Let s represent the average speed of the 2001 winner in miles per hour. Write expressions in terms of s for the time in hours that it took the 2001 and 2002 winners to complete the race.

B. Write a rational equation that can be used to determine s. Solve your equation to find the average speed of the 2001 winner to the nearest mile per hour.

★★★10. **Architecture** The Renaissance architect Andrea Palladio preferred that the length and width of rectangular rooms be limited to certain ratios. These ratios are listed in the table. Palladio also believed that the height of a room with vaulted ceilings should be the harmonic mean of the length.

Rooms with a Width of 30 ft		
Length-to-Width Ratio	**Length (ft)**	**Height (ft)**
2:1		
3:2		
4:3		
5:3		
$\sqrt{2}$:1		

A. The harmonic mean of two positive numbers a and b is equal to $\frac{2}{\frac{1}{a}+\frac{1}{b}}$. Simplify this expression.

B. Complete the table for a rectangular room with a width of 30 feet that meets Palladio's requirements for its length and height. If necessary, round to the nearest tenth.

C. A Palladian room has a length-to-width ratio of 4:3. If the length of this room is doubled, what effect should this change have on the room's width and height, according to Palladio's principles?

MATH IN CAREERS

Chemist A chemist mixes 5 mL of an acid with 15 mL of water. The concentration of acid in the acid-and-water mix is $\frac{5}{5+15} = \frac{5}{20} = 25\%$. If the chemist adds more acid to the mix, then the concentration C becomes a function of the additional amount a of acid added to the mix.

a. Write a rule for the function $C(a)$.

b. What is a reasonable domain for this function? Explain.

c. What concentration of acid does pure water have? What concentration of acid does pure acid have? So, what are the possible values of $C(a)$?

d. Graph the function on the grid below.

e. Analyze the function's rule to determine the vertical asymptote of the function's graph. Why is the asymptote irrelevant in this situation?

f. Analyze the function's rule to determine the horizontal asymptote of the function's graph. What is the relevance of the asymptote in this situation?

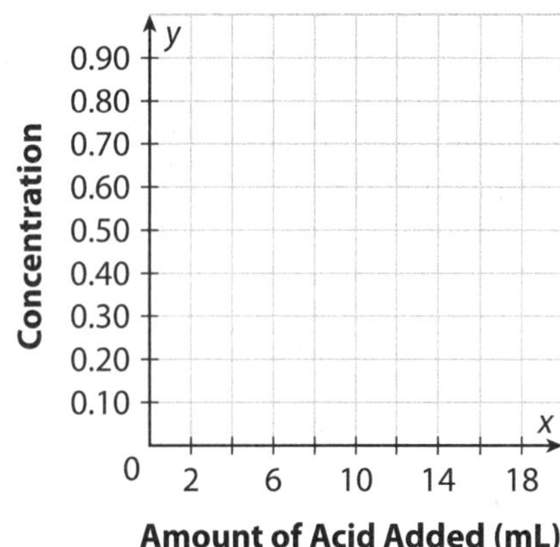

UNIT 5

Radical Functions, Expressions, and Equations

MODULE 11

Radical Functions

TEKS A2.2.A, A2.2.B, A2.4.E

MODULE 12

Radical Expressions and Equations

TEKS A2.4.F, A2.7.G

MATH IN CAREERS

Nutritionist Nutritionists provide services to individuals and institutions, such as schools and hospitals. Nutritionists must be able to calculate the amounts of different substances in a person's diet, including calories, fat, vitamins, and minerals. They must also calculate measures of fitness, such as body mass index. Nutritionists must use statistics when reviewing nutritional studies in scientific journals.

If you are interested in a career as a nutritionist, you should study these mathematical subjects:

- Algebra
- Statistics
- Business Math

Research other careers that require proficiency in understanding statistics in scientific articles. Check out the career activity at the end of the unit to find out how **Nutritionists** use math.

Reading Start-Up

Visualize Vocabulary

Use the ✓ words to complete the graphic. Put just one word in each section of the square.

The function that results from exchanging the input and output values of a function.	A function where each element of the range may correspond to more than one element of the domain.
Function Types	
A function where each element of the range corresponds to only one element of the domain.	A series of two functions in which the output of one function is used as the input for the other.

Vocabulary

Review Words

✔ composition of functions (composición de funciones)
extraneous solution (solución extraña)
✔ inverse function (función inversa)
✔ many-to-one function (función muchos a uno)
✔ one-to-one function (función uno a uno)
radical expression (expresión radical)

Preview Words

cube root function (función de raíz cúbica)
index (índice)
square root function (función de raíz cuadrada)

Understand Vocabulary

To become familiar with some of the vocabulary terms in the module, consider the following. You may refer to the module, the glossary, or a dictionary.

1. A function whose rule contains a variable under a square-root sign is a ______________.

2. A function whose rule contains a variable under a cube-root sign is a ______________.

3. In the radical expression $\sqrt[n]{x}$, n is the ________.

Active Reading

Pyramid Fold Before beginning a module, create a pyramid fold to help you take notes from each lesson in the module. The three sides of the pyramid can summarize information about function families, their graphs, and their characteristics.

MODULE 11

Radical Functions

Essential Question: How can you use radical functions to solve real-world problems?

TEKS

REAL WORLD VIDEO
A rocket must generate enough thrust to achieve escape velocity from Earth's gravitational field. Check out some of the calculations and preparations that go into a successful launch.

MODULE PERFORMANCE TASK PREVIEW

We Have Liftoff!

If you throw a ball straight up, it will eventually come back down. But if you could throw it with enough initial velocity, it would escape Earth's surface and go into orbit. If you could throw it even faster, it might even escape the solar system. What is the escape velocity for Earth, the minimum velocity for an object to leave Earth's surface and not return? What about the velocity necessary to escape other planets? Let's take off and find out!

Are YOU Ready?

Complete these exercises to review skills you will need for this module.

- Online Homework
- Hints and Help
- Extra Practice

Exponents

Example 1 Simplify .

$$x^2 \cdot x^3 - 2x^4 \cdot x = x^{2+3} - 2x^{4+1}$$
$$= x^5 - 2x^5$$
$$= -x^5$$

Simplify each expression.

1. $5x^3 \cdot 2x$

2. $-x^4 \cdot x^3$

3. $4x^2\left(2xy - x^2\right)$

Inverse Linear Functions

Example 2 Write the inverse function of $y = x + 9$.

$y - 9 = x + 9 - 9$ Subtract.

$y - 9 = x$ Simplify.

$x - 9 = y$ Switch x and y.

The inverse function of $y = x + 9$ is $y = x - 9$.

Example 3 Write the inverse function of $y = \frac{x}{-22}$.

$(-22)y = -\frac{x}{22}(-22)$ Multiply.

$-22y = x$ Simplify.

$-22x = y$ Switch x and y.

The inverse function of $y = \frac{x}{-22}$ is $y = -22x$.

Write the inverse of each function.

4. $y = x - 6$

5. $y = 7x$

6. $y = \frac{1}{2}x$

7. $y = x + 11$

8. $y = -18x$

9. $y = 21 + x$

Name________________ Class__________ Date________

11.1 Inverses of Simple Quadratic and Cubic Functions

Resource Locker

Essential Question: What functions are the inverses of quadratic functions and cubic functions, and how can you find them?

A2.2.B Graph and write the inverse of a function using notation such as $f^{-1}(x)$. Also A2.2.A, A2.2.C, A2.7.I

Explore Finding the Inverse of a Many-to-One Function

The function $f(x)$ is defined by the following ordered pairs: $(-2, 4)$, $(-1, 2)$, $(0, 0)$, $(1, 2)$, and $(2, 4)$.

(A) Find the inverse function of $f(x)$, $f^{-1}(x)$, by reversing the coordinates in the ordered pairs.

____

(B) Is the inverse also a function? Explain.

____

(C) If necessary, restrict the domain of $f(x)$ such that the inverse, $f^{-1}(x)$, is a function.

To restrict the domain of $f(x)$ so that its inverse is a function, you can restrict it to

$\{x \mid x \geq \square\}$

(D) With the restricted domain of $f(x)$, what ordered pairs define the inverse function $f^{-1}(x)$?

____

Reflect

1. **Discussion** Look again at the ordered pairs that define $f(x)$. Without interchanging the coordinates, how could you have known that the inverse of $f(x)$ would not be a function?

____

2. How will restricting the domain of $f(x)$ affect the range of its inverse?

____

Explain 1 Finding and Graphing the Inverse of a Simple Quadratic Function

The function $f(x) = x^2$ is a many-to-one function, so its domain must be restricted in order to find its inverse function. If the domain is restricted to $x \geq 0$, then the inverse function is $f^{-1}(x) = \sqrt{x}$; if the domain is restricted to $x \leq 0$, then the inverse function is $f^{-1}(x) = -\sqrt{x}$.

The inverse of a quadratic function is a **square root function**, which is a function whose rule involves $\sqrt{x}$. **The parent square root function** is $g(x) = \sqrt{x}$. A square root function is defined only for values of x that make the expression under the radical sign nonnegative.

Example 1 **Restrict the domain of each quadratic function and find its inverse. Confirm the inverse relationship using composition. Graph the function and its inverse.**

 $f(x) = 0.5$

Restrict the domain. $\left\{x \mid x \geq 0\right\}$

Find the inverse.

Replace $f(x)$ with y. $\qquad y = 0.5x^2$

Multiply both sides by 2. $\qquad 2y = x^2$

Use the definition of positive square root. $\qquad \sqrt{2y} = x$

Switch x and y to write the inverse. $\qquad \sqrt{2x} = y$

Replace y with $f^{-1}(x)$. $\qquad f^{-1}(x) = \sqrt{2x}$

Confirm the inverse relationship using composition.

$$\begin{aligned} f^{-1}\left(f(x)\right) &= f^{-1}\left(0.5x^2\right) \\ &= \sqrt{2\left(0.5x^2\right)} \\ &= \sqrt{x^2} \\ &= x \text{ for } x \geq 0 \end{aligned}$$

$$\begin{aligned} f(f^{-1}(x)) &= 0.5\,(\sqrt{2x})^2 \\ &= 0.5(2x) \\ &= x \text{ for } x \geq 0 \end{aligned}$$

Since $f^{-1}\left(f(x)\right) = x$ for $x \geq 0$ and $f(f^{-1}(x)) = x$ for $x \geq 0$, it has been confirmed that $f^{-1}(x) = \sqrt{2x}$ for $x \geq 0$ is the inverse function of $f(x) = 0.5x^2$ for $x \geq 0$.

Graph $f^{-1}(x)$ by graphing $f(x)$ over the restricted domain and reflecting the graph over the line $y = x$.

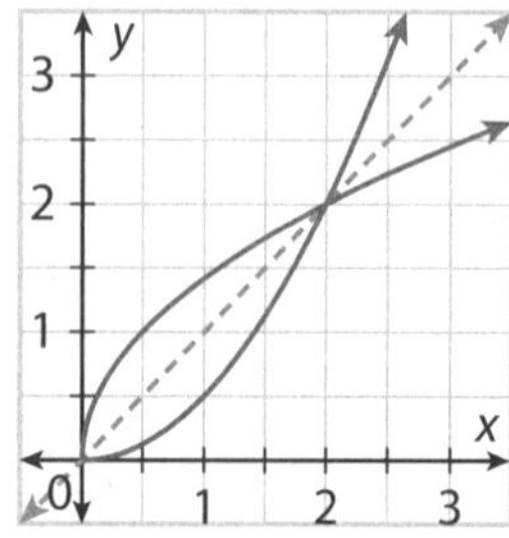

(B) $f(x) = x^2 - 7$

Restrict the domain. $\{x|\ x \geq \square\}$

Find the inverse.

Replace $f(x)$ with y. $\square = x^2 - 7$

Add 7 to both sides. $\square = x^2$

Use the definition of positive square root. $\square = x$

Switch x and y to write the inverse. ______________

Replace y with $f^{-1}(x)$. ______________

Identify the domain of $f^{-1}(x)$. ______________

Confirm the inverse relationship using composition.

$f^{-1}(f(x)) = f^{-1}(\square)$ $\qquad$ $f(f^{-1}(x)) = f(\square)$

$= \square$ $\qquad$ $= \square$

$= \square$ $\qquad$ $= \square$

$= \square$ $\qquad$ $= \square$

Since $f^{-1}(f(x)) = \square$ for ______ and $f(\square) = x$ for $x \geq \square$, it has been confirmed

that $f^{-1}(x) = \square$ for $\square$ is the inverse function of $f(x) = x^2 - 7$ for ______.

Graph $f^{-1}(x)$ by graphing $f(x)$ over the restricted domain and reflecting the graph over the line $y = x$.

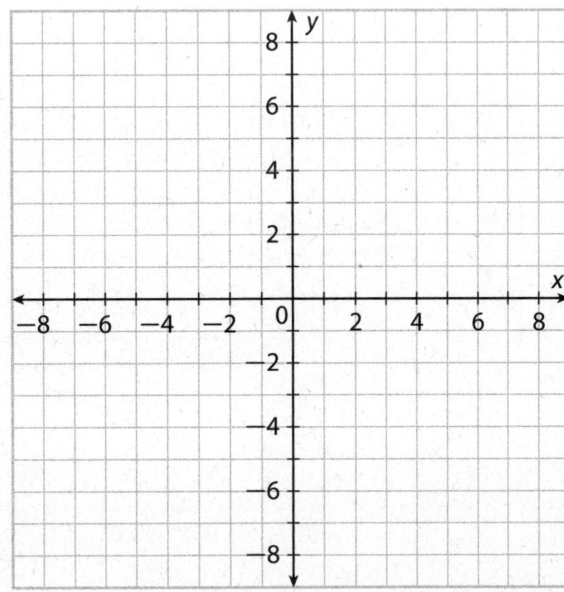

Your Turn

Restrict the domain of each quadratic function and find its inverse. Confirm the inverse relationship using composition. Graph the function and its inverse.

3. $f(x) = 3x^2$

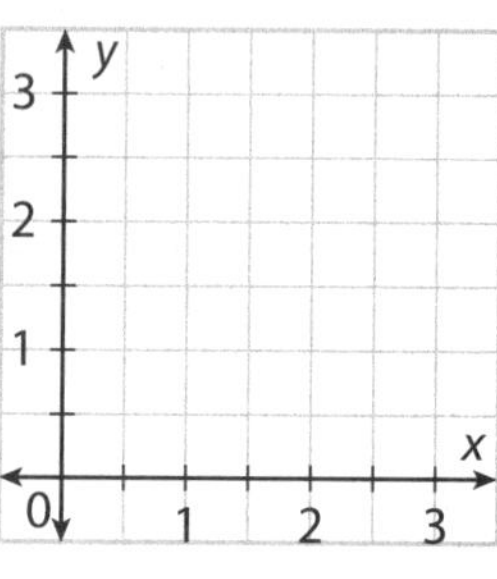

4. $f(x) = 5x^2$

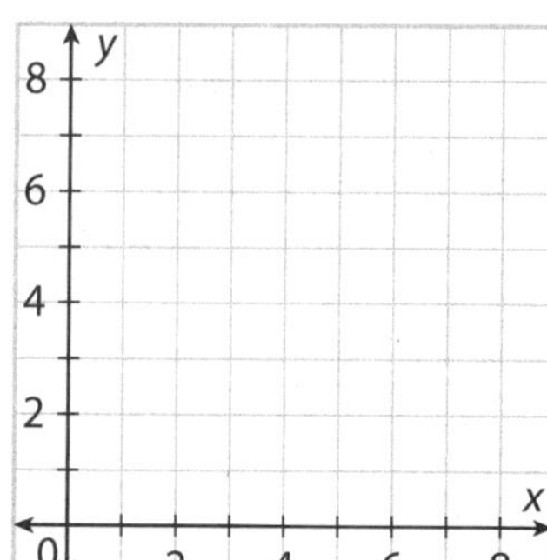

Explain 2 Finding the Inverse of a Quadratic Model

In many instances, quadratic functions are used to model real-world applications. It is often useful to find and interpret the inverse of a quadratic model. Note that when working with real-world applications, it is more useful to use the notation $x(y)$ for the inverse of $y(x)$ instead of the notation $y^{-1}(x)$.

Example 2 **Find the inverse of each of the quadratic functions. Use the inverse to solve the application.**

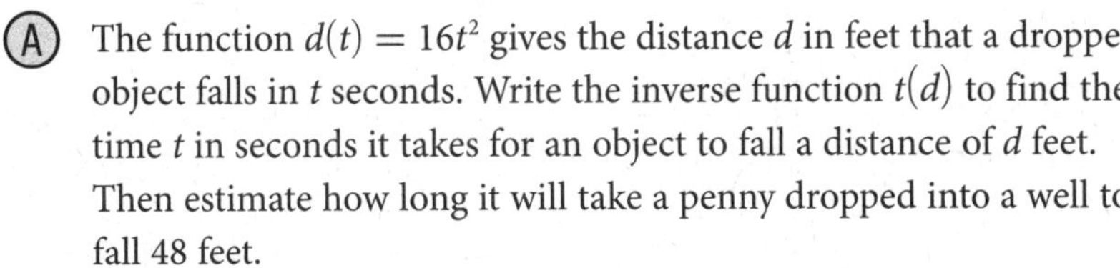

Ⓐ The function $d(t) = 16t^2$ gives the distance d in feet that a dropped object falls in t seconds. Write the inverse function $t(d)$ to find the time t in seconds it takes for an object to fall a distance of d feet. Then estimate how long it will take a penny dropped into a well to fall 48 feet.

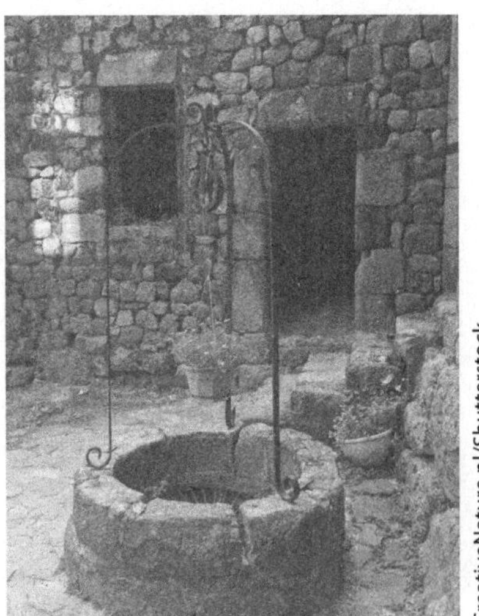

©CreativeNature.nl/Shutterstock

The original function $d(t) = 16t^2$ is a quadratic function with a domain restricted to $t \geq 0$.

Find the inverse function.

Write $d(t)$ as d. $\quad d = 16t^2$

Divide both sides by 16. $\quad \frac{d}{16} = t^2$

Use the definition of positive square root. $\quad \sqrt{\frac{d}{16}} = t$

Write t as $t(d)$. $\quad \sqrt{\frac{d}{16}} = t(d)$

The inverse function is $t(d) = \sqrt{\frac{d}{16}}$ for $d \geq 0$.

Use the inverse function to estimate how long it will take a penny dropped into a well to fall 48 feet. Substitute $d = 48$ into the inverse function.

Write the function. $\quad t(d) = \sqrt{\frac{d}{16}}$

Substitute 48 for d. $\quad t(48) = \sqrt{\frac{48}{16}}$

Simplify. $\quad t(48) = \sqrt{3}$

Use a calculator to estimate. $\quad t(48) \approx 1.7$

So, it will take about 1.7 seconds for a penny to fall 48 feet into the well.

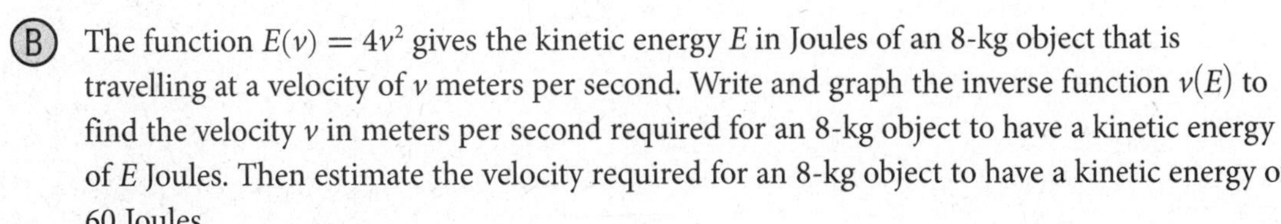

Ⓑ The function $E(v) = 4v^2$ gives the kinetic energy E in Joules of an 8-kg object that is travelling at a velocity of v meters per second. Write and graph the inverse function $v(E)$ to find the velocity v in meters per second required for an 8-kg object to have a kinetic energy of E Joules. Then estimate the velocity required for an 8-kg object to have a kinetic energy of 60 Joules.

The original function $E(v) = 4v^2$ is a ______________ function with a domain restricted to v ______________.

Find the inverse function.

Write $E(v)$ as E. $\square = 4v^2$

Divide both sides by 4. $\square = v^2$

Use the definition of positive square root. $\square = v$

Write v as $v(E)$. ______

The inverse function is $v(E) =$ ______ for E ______.

Use the inverse function to estimate the velocity required for an 8-kg object to have a kinetic energy of 60 Joules.

Substitute $E = 60$ into the inverse function.

Write the function. $v(E) = \square$

Substitute 60 for E. $v\left(\square\right) = \square$

Simplify. ______

Use a calculator to estimate. ______

So, an 8-kg object with kinetic energy of 60 Joules is traveling at a velocity of ______ meters per second.

Your Turn

Find the inverse of the quadratic function. Use the inverse to solve the application.

5. The function $A(r) = \pi r^2$ gives the area of a circular object with respect to its radius r. Write the inverse function $r(A)$ to find the radius r required for area of A. Then estimate the radius of a circular object that has an area of 40 cm^2.

Explain 3 Finding and Graphing the Inverse of a Simple Cubic Function

Note that the function $f(x) = x^3$ is a one-to-one function, so its domain does not need to be restricted in order to find its inverse function. The inverse of $f(x) = x^3$ is $f^{-1}(x) = \sqrt[3]{x}$.

The inverse of a cubic function is a **cube root function**, which is a function whose rule involves $\sqrt[3]{x}$. The **parent cube root function** is $g(x) = \sqrt[3]{x}$.

Example 3 **Find the inverse of each cubic function. Confirm the inverse relationship using composition. Graph the function and its inverse.**

(A) $f(x) = 0.5x^3$

Find each inverse. Graph the function and its inverse.

Replace $f(x)$ with y.	$y = 0.5x^3$
Multiply both sides by 2.	$2y = x^3$
Use the definition of cube root.	$\sqrt[3]{2y} = x$
Switch x and y to write the inverse.	$\sqrt[3]{2x} = y$
Replace y with $f^{-1}(x)$.	$\sqrt[3]{2x} = f^{-1}(x)$

Confirm the inverse relationship using composition.

$$\begin{aligned} f^{-1}(f(x)) &= f^{-1}(0.5x^3) \\ &= \sqrt[3]{2(0.5x^3)} \\ &= \sqrt[3]{x^3} \\ &= x \end{aligned}$$

$$\begin{aligned} f(f^{-1}(x)) &= f(\sqrt[3]{2x}) \\ &= 0.5(\sqrt[3]{2x})^3 \\ &= 0.5(2x) \\ &= x \end{aligned}$$

Since $f^{-1}(f(x)) = x$, and $f(f^{-1}(x)) = x$, it has been confirmed that $f^{-1}(x) = \sqrt[3]{2x}$ is the inverse function of $f(x) = 0.5x^3$.

Graph $f^{-1}(x)$ by graphing $f(x)$ and reflecting the graph over the line $y = x$.

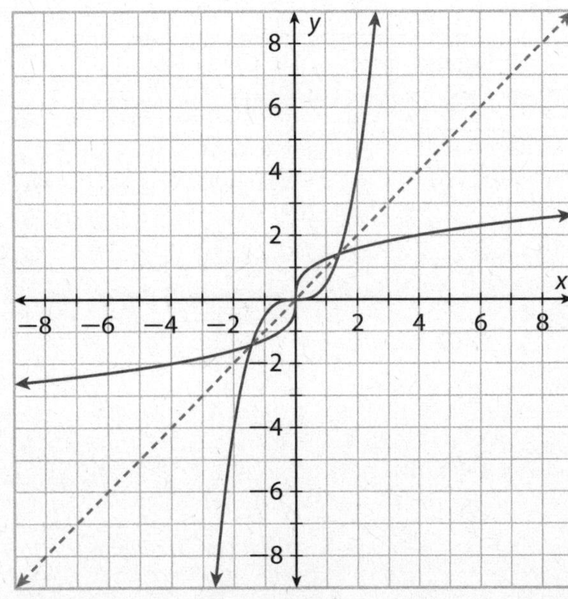

(B) $f(x) = x^3 - 9$

Find the inverse.

Replace $f(x)$ with y. $\square = x^3 - 9$

Add 9 to both sides. $\square = x^3$

Use the definition of cube root. $\square = x$

Switch x and y to write the inverse. ______

Replace y with $f^{-1}(x)$. ______

Confirm the inverse relationship using composition.

$f^{-1}(f(x)) = f^{-1}(\square)$

$= \square$

$= \square$

$= \square$

$f(f^{-1}(x)) = f(\square)$

$= \square$

$= \square$

$= \square$

Since $f^{-1}(f(x)) = \square$ and $f(f^{-1}(x)) = x$, it has been confirmed that $f^{-1}(x) = \square$ is the inverse function of $f(x) = x^3 - 9$.

Graph $f^{-1}(x)$ by graphing $f(x)$ and reflecting the graph over the line $y = x$.

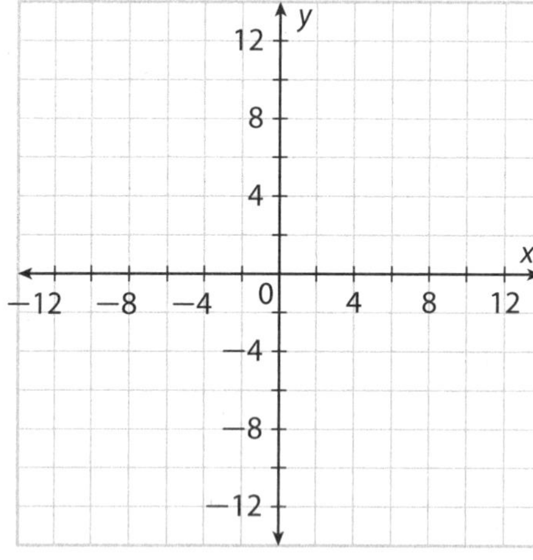

Your Turn

Find each inverse. Graph the function and its inverse.

6. $f(x) = 2x^3$

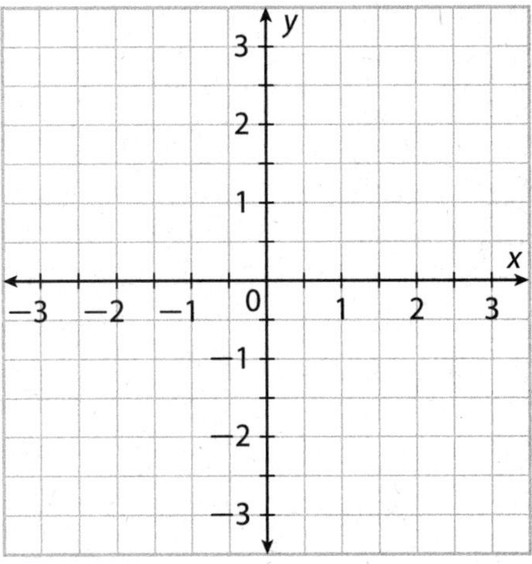

7. $f(x) = 6x^3$

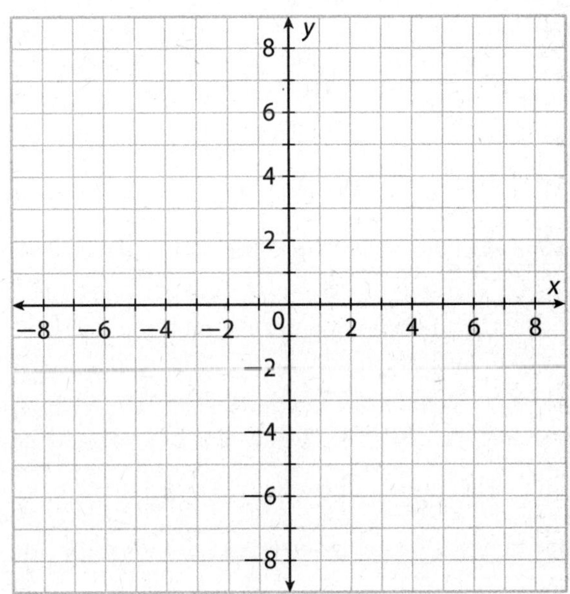

Explain 4 Finding the Inverse of a Cubic Model

In many instances, cubic functions are used to model real-world applications. It is often useful to find and interpret the inverse of cubic models. As with quadratic real-world applications, it is more useful to use the notation $x(y)$ for the inverse of $y(x)$ instead of the notation $y^{-1}(x)$.

Example 4 **Find the inverse of each of the following cubic functions.**

(A) The function $m(L) = 0.00001L^3$ gives the mass m in kilograms of a red snapper of length L centimeters. Find the inverse function $L(m)$ to find the length L in centimeters of a red snapper that has a mass of m kilograms.

The original function $m(L) = 0.00001L^3$ is a cubic function.

Find the inverse function.

Write $m(L)$ as m. $m = 0.00001L^3$

Multiply both sides by 100,000. $100{,}000m = L^3$

Use the definition of cube root. $\sqrt[3]{100{,}000m} = L$

Write L as $L(m)$. $\sqrt[3]{100{,}000m} = L(m)$

The inverse function is $L(m) = \sqrt[3]{100,\ 000m}$.

(B) The function $A(r) = \frac{4}{3}\pi r^3$ gives the surface area A of a sphere with radius r. Find the inverse function $r(A)$ to find the radius r of a sphere with surface area A.

The original function $A(r) = \frac{4}{3}\pi r^3$ is a __________ function.

Find the inverse function.

Write $A(r)$ as A. $\square = \frac{4}{3}\pi r^3$

Divide both sides by $\frac{4}{3}\pi$. $\square = r^3$

Use the definition of cube root. $\square = r$

Write r as $r(A)$. __________

The inverse function is $r(A) =$ __________.

Your Turn

8. The function $m(r) = \frac{44}{3}\pi r^3$ gives the mass in grams of a spherical lead ball with a radius of r centimeters.

9. Find the inverse function $r(m)$ to find the radius r of a lead sphere with mass m.

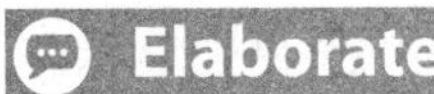

Elaborate

10. What is the general form of the inverse function for the function $f(x) = ax^2$? State any restrictions on the domains.

11. What is the general form of the inverse function for the function $f(x) = ax^3$? State any restrictions on the domains.

12. **Essential Question Check-In** Why must the domain be restricted when finding the inverse of a quadratic function, but not when finding the inverse of a cubic function?

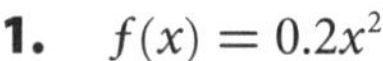

Evaluate: Homework and Practice

- Online Homework
- Hints and Help
- Extra Practice

Restrict the domain of the quadratic function and find its inverse. Confirm the inverse relationship using composition. Graph the function and its inverse.

1. $f(x) = 0.2x^2$

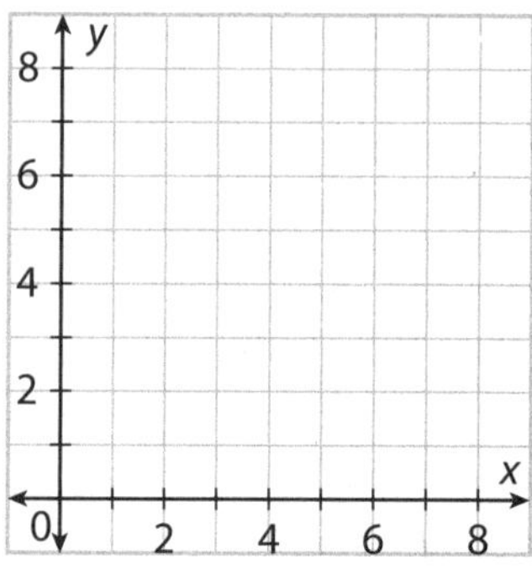

2. $f(x) = 8x^2$

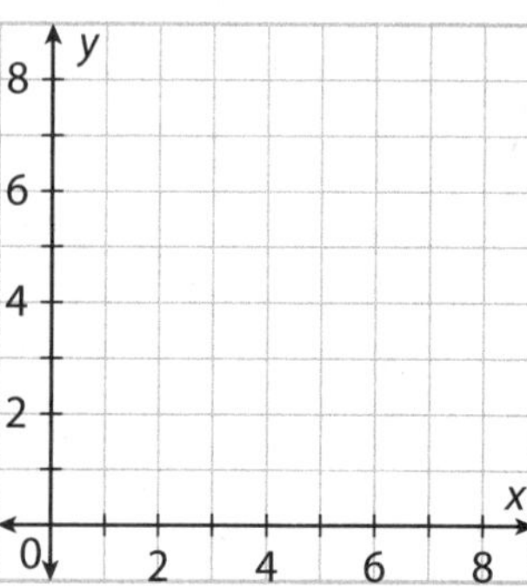

3. $f(x) = x^2 + 10$

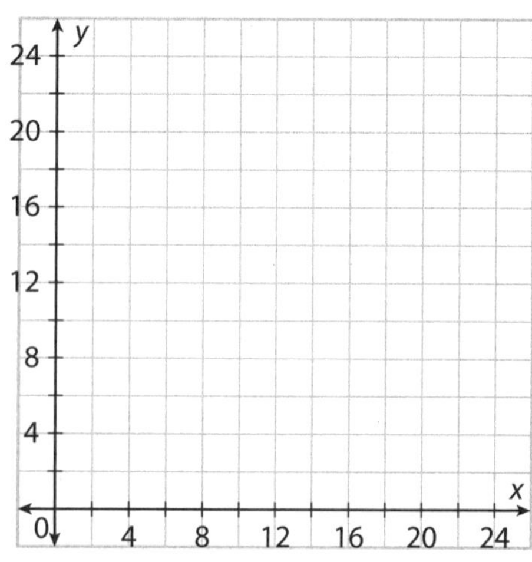

Restrict the domain of the quadratic function and find its inverse. Confirm the inverse relationship using composition.

4. $f(x) = 15x^2$

5. $f(x) = x^2 - \frac{3}{4}$

6. $f(x) = 0.7x^2$

7. The function $d(s) = \frac{1}{14.9}s^2$ models the average depth d in feet of the water over which a tsunami travels, where s is the speed in miles per hour. Write the inverse function $s(d)$ to find the speed required for a depth of d feet. Then estimate the speed of a tsunami over water with an average depth of 1500 feet.

8. The period of a pendulum is the time it takes the pendulum to complete one back-and-forth swing.

The function $x(T) = 9.8\left(\frac{T}{2\pi}\right)^2$ gives the length x in meters for a pendulum that has a period of T seconds. Write the inverse function to find the period of a pendulum that has length x. Then find the period of a pendulum with a length of 5 meters.

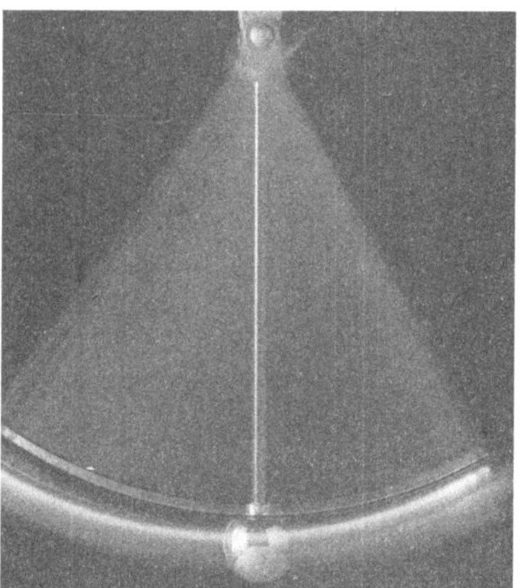

Find the inverse of each cubic function. Confirm the inverse relationship using composition. Graph the function and its inverse.

9. $f(x) = 0.25x^3$

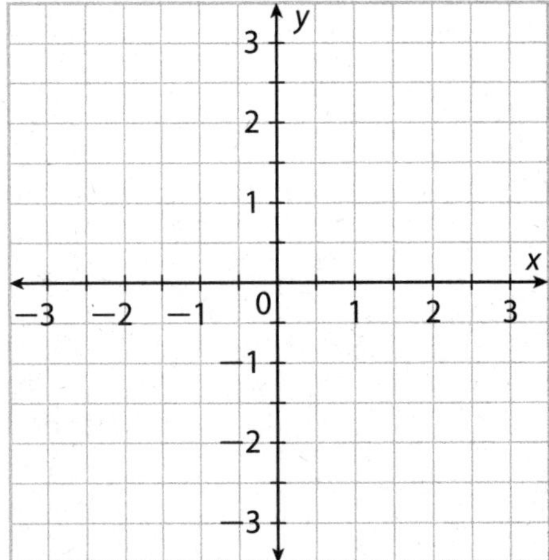

10. $f(x) = -12x^3$

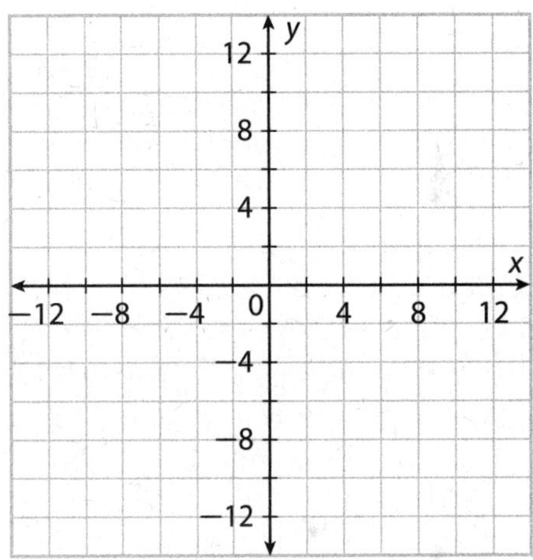

Find the inverse of the cubic function. Confirm the inverse relationship using composition.

11. $f(x) = x^3 - \frac{5}{6}$

12. $f(x) = x^3 + 9$

13. The function $m(r) = 31r^3$ models the mass in grams of a spherical zinc ball as a function of the ball's radius in centimeters. Write the inverse model to represent the radius r in cm of a spherical zinc ball as a function of the ball's mass m in g.

14. The function $m(r) = 21r^3$ models the mass in grams of a spherical titanium ball as a function of the ball's radius in centimeters. Write the inverse model to represent the radius r in centimeters of a spherical titanium ball as a function of the ball's mass m in grams.

15. The weight w in pounds that a shelf can support can be modeled by $w(d) = 82.9d^3$ where d is the distance, in inches, between the supports for the shelf. Write the inverse model to represent the distance d in inches between the supports of a shelf as a function of the weight w in pounds that the shelf can support.

H.O.T. Focus on Higher Order Thinking

16. Explain the Error A student was asked to find the inverse of the function $f(x) = \left(\frac{x}{2}\right)^3 + 9$. What did the student do wrong? Find the correct inverse.

$$f(x) = \left(\frac{x}{2}\right)^3 + 9$$

$$y = \left(\frac{x}{2}\right)^3 + 9$$

$$y - 9 = \left(\frac{x}{2}\right)^3$$

$$2y - 18 = x^3$$

$$\sqrt[3]{2y - 18} = x$$

$$y = \sqrt[3]{2x - 18}$$

$$f^{-1}(x) = \sqrt[3]{2x - 18}$$

17. Make a Conjecture The function $f(x) = x^2$ must have its domain restricted to have its inverse be a function. The function $f(x) = x^3$ does not need to have its domain restricted to have its inverse be a function. Make a conjecture about which power functions need to have their domains restricted to have their inverses be functions and which do not.

18. Multi-Step A framing store uses the function $\left(\frac{c - 0.2}{0.5}\right)^2 = a$ to determine the total area of a piece of glass with respect to the cost before installation of the glass. Write the inverse function for the cost c in dollars of glass for a picture with an area of a in square inches. Then write a new function to represent the total cost C the store charges if it costs \$6.00 for installation. Use the total cost function to estimate the cost if the area of the glass is 192 cm².

Lesson Performance Task

One method used to irrigate crops is the center-pivot irrigation system. In this method, sprinklers rotate in a circle to water crops. The challenge for the farmer is to determine where to place the pivot in order to water the desired number of acres. The farmer knows the area but needs to find the radius of the circle necessary to define that area. How can the farmer determine this from the formula for the area of a circle $A = \pi r^2$? Find the formula the farmer could use to determine the radius necessary to irrigate a given number of acres, A. (Hint: One acre is 43,560 square feet.) What would be the radius necessary for the sprinklers to irrigate an area of 133 acres?

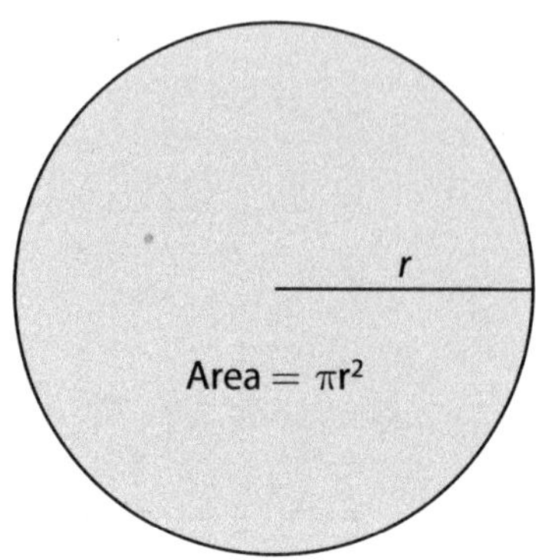

Name________________ Class__________ Date________

11.2 Graphing Square Root Functions

Resource Locker

Essential Question: How can you use transformations of a parent square root function to graph functions of the form $g(x) = a\sqrt{(x-h)} + k$ or $g(x) = \sqrt{\frac{1}{b}(x-h)} + k$?

A2.4.C Determine the effect on the graphs of $f(x) = \sqrt{x}$ when $f(x)$ is replaced by $af(x)$, $f(x) + d$, $f(bx)$, and $f(x - c)$ for specific positive and negative values of a, b, c, and d. Also A2.2.A, A2.7.I

Explore Graphing and Analyzing the Parent Square Root Function

Although you have seen how to use imaginary numbers to evaluate square roots of negative numbers, graphing complex numbers and complex valued functions is beyond the scope of this course. For purposes of graphing functions based on the square roots (and in most cases where a square root function is used in a real-world example), the domain and range should both be limited to real numbers.

The square root function is the inverse of a quadratic function with a domain limited to positive real numbers. The quadratic function must be a one-to-one function in order to have an inverse, so the domain is limited to one side of the vertex. The square root function is also a one-to-one function as all inverse functions are.

(A) The domain of the square root function (limited to real numbers) is given by $\{x \mid x \geq \square\}$

(B) Fill in the table.

x	$f(x) = \sqrt{x}$
0	
1	
4	
9	

(C) Plot the points on the graph, and connect them with a smooth curve.

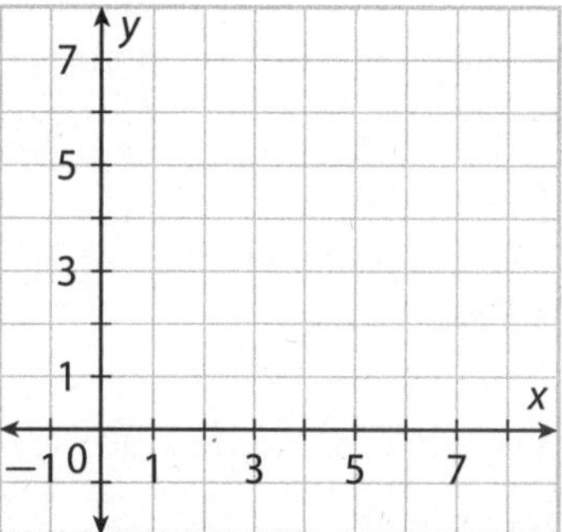

(D) Recall that this function is the inverse of the parent quadratic $\left(f(x) = x^2\right)$ with a domain limited to the nonnegative real numbers. Write the range of this square root function:

$\{y \mid y \geq \square\}$.

(E) The graph appears to be getting flatter as x increases, indicating that the rate of change __________ as x increases.

(F) Describe the end behavior of the square root function, $f(x) = \sqrt{x}$. $f(x) \to \square$ as $x \to \square$

Reflect

1. **Discussion** Why does the end behavior of the square root function only need to be described at one end?

2. The solution to the equation $x^2 = 4$ is sometimes written as $x = \pm 2$. Explain why the inverse of $f(x) = x^2$ cannot similarly be written as $g(x) = \pm\sqrt{x}$ in order to use all reals as the domain of $f(x)$.

Explore 2 Predicting the Effects of Parameters on the Graphs of Square Root Functions

You have learned how to transform the graph of a function using reflections across the x- and y-axes, vertical and horizontal stretches and compressions, and translations. Here, you will apply those transformations to the graph of the square root function $f(x) = \sqrt{x}$.

When transforming the parent function $f(x) = \sqrt{x}$, you can get functions of the form

$g(x) = a\sqrt{(x - h)} + k$ or $g(x) = \sqrt{\frac{1}{b}(x - h)} + k$.

For each parameter, predict the effect on the graph of the parent function, and then confirm your prediction with a graphing calculator.

(A) Predict the effect of the parameter, h, on the graph of $g(x) = \sqrt{x - h}$ for each function.

a. $g(x) = \sqrt{x - 2}$: The graph is a ________ of the graph of $f(x)$ [right/left/up/down] 2 units.

b. $g(x) = \sqrt{x + 2}$: The graph is a ________ of the graph of $f(x)$ [right/left/up/down] 2 units.

Check your answers using a graphing calculator.

(B) Predict the effect of the parameter k on the graph of $g(x) = \sqrt{x} + k$ for each function.

a. $g(x) = \sqrt{x} + 2$: The graph is a ________ of the graph of $f(x)$ [right/up/left/down] 2 units.

b. $g(x) = \sqrt{x} - 2$: The graph is a ________ of the graph of $f(x)$ [right/up/left/down] 2 units.

Check your answers using a graphing calculator.

Ⓒ Predict the effect of the parameter a on the graph of $g(x) = a\sqrt{x}$ for each function.

a. $g(x) = 2\sqrt{x}$: The graph is a ____________ stretch of the graph of $f(x)$ by a factor of ____.

b. $g(x) = \frac{1}{2}\sqrt{x}$: The graph is a ____________ compression of the graph of $f(x)$ by a factor of ____.

c. $g(x) = -\frac{1}{2}\sqrt{x}$: The graph is a ____________ compression of the graph of $f(x)$ by a factor of ____ as well as a ____________ across the ____________.

d. $g(x) = -2\sqrt{x}$: The graph is a ____________ stretch of the graph of $f(x)$ by a factor of ____ as well as a ____________ across the ____________.

Check your answers using a graphing calculator.

Ⓓ Predict the effect of the parameter b on the graph of $g(x) = \sqrt{\frac{1}{b}x}$ for each function.

a. $g(x) = \sqrt{\frac{1}{2}x}$: The graph is a ____________ stretch of the graph of $f(x)$ by a factor of ____.

b. $g(x) = \sqrt{2x}$: The graph is a ____________ compression of the graph of $f(x)$ by a factor of ____.

c. $g(x) = \sqrt{-\frac{1}{2}x}$: The graph is a ____________ stretch of the graph of $f(x)$ by a factor of ____ as well as a ____________ across the ____________.

d. $g(x) = \sqrt{-2x}$: The graph is a ____________ compression of the graph of $f(x)$ by a factor of ____ as well as a ____________ across the ____________.

Check your answers using a graphing calculator.

Reflect

3. **Discussion** Describe what the effect of each of the transformation parameters is on the domain and range of the transformed function.

Explain 1 Graphing Square Root Functions

When graphing transformations of the square root function, it is useful to consider the effect of the transformation on two reference points, $(0, 0)$ and $(1, 1)$, that lie on the parent function, and where they map to on the transformed function, $g(x)$.

$f(x) = \sqrt{x}$		$g(x) = a\sqrt{x-h} + k$		$g(x) = \sqrt{\frac{1}{b}(x-h)} + k$	
x	y	x	y	x	y
0	0	h	k	h	k
1	1	$h+1$	$k+a$	$h+b$	$k+1$

The transformed reference points can be found by recognizing that the initial point of the graph is translated from $(0, 0)$ to (h, k).

When $g(x)$ involves the parameters a, h, and k, the second transformed reference point is 1 unit to the right of (h, k) and $|a|$ units up or down from (h, k), depending on the sign of a. When $g(x)$ involves the parameters b, h, and k, the second transformed reference point is $|b|$ units left or right from (h, k), depending on the sign of b, and 1 unit above (h, k).

Transformations of the square root function also affect the domain and range. In order to work with real valued inputs and outputs, the domain of the square root function cannot include values of x that result in a negative-valued expression. Negative values of x can be in the domain, as long as they result in nonnegative values of the expression that is inside the square root. Similarly, the value of the square root function is positive by definition, but multiplying the square root function by a negative number, or adding a constant to it changes the range and can result in negative values of the transformed function.

Example 1 **For each of the transformed square root functions, find the transformed reference points and use them to plot the transformed function on the same graph with the parent function. Describe the domain and range using set notation.**

 $g(x) = 2\sqrt{x-3} - 2$

To find the domain:

Square root input must be nonnegative. $x - 3 \geq 0$

Solve the inequality for x. $x \geq 3$

The domain is $\{x \mid x \geq 3\}$.

To find the range:

The square root function is nonnegative. $\sqrt{x-3} \geq 0$

Multiply by 2 $2\sqrt{x-3} \geq 0$

Subtract 2. $2\sqrt{x-3} - 2 \geq -2$

The expression on the left is $g(x)$. $g(x) \geq -2$

Since $g(x)$ is greater than or equal to -2 for all x in the domain, the range is $\{y \mid y \geq -2\}$.

$(0, 0) \rightarrow (3, -2)$

$(1, 1) \rightarrow (4, 0)$

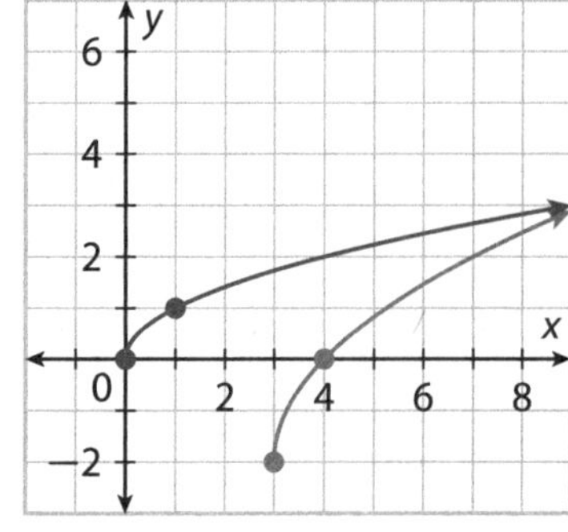

Ⓑ $g(x) = \sqrt{-\frac{1}{2}(x - 2)} + 1$

To find the domain:

Square root input must be nonnegative. $-\frac{1}{2}(x - 2) \geq$ ☐

Multiply both sides by −2. $x - 2$ ☐ 0

Add 2 to both sides. ☐ ≤ 2

Expressed in set notation, the domain is $\{x \mid$ ☐ $\}$.

To find the range:

The square root function is nonnegative. $\sqrt{-\frac{1}{2}(x - 2)}$ ☐ 0

Add 1 both sides $\sqrt{-\frac{1}{2}(x - 2)} + 1 \geq$ ☐

The expression on the left is ☐. $g(x) \geq 1$

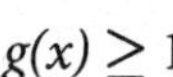

Since $g(x)$ is greater than 1 for all x in the domain,

the range (in set notation) is $\{y \mid$ ☐ $\}$.

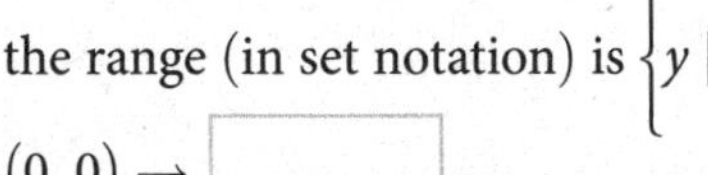

$(0, 0) \rightarrow$ ☐

$(1, 1) \rightarrow$ ☐

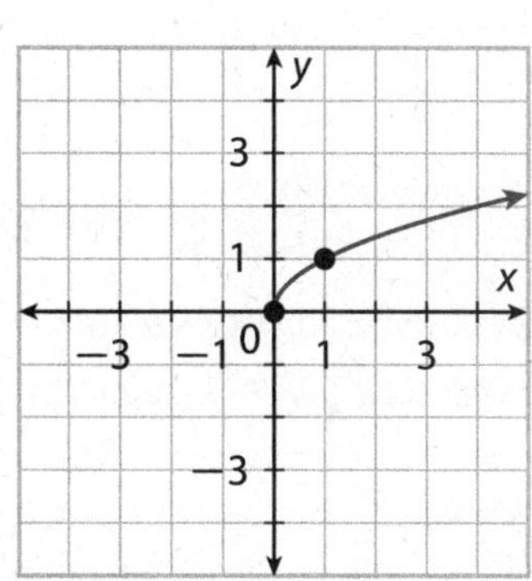

Your Turn

For each of the transformed square root functions, find the transformed reference points and use them to plot the transformed function on the same graph with the parent function. Describe the domain and range using set notation.

4. $g(x) = -3\sqrt{x - 2} + 3$

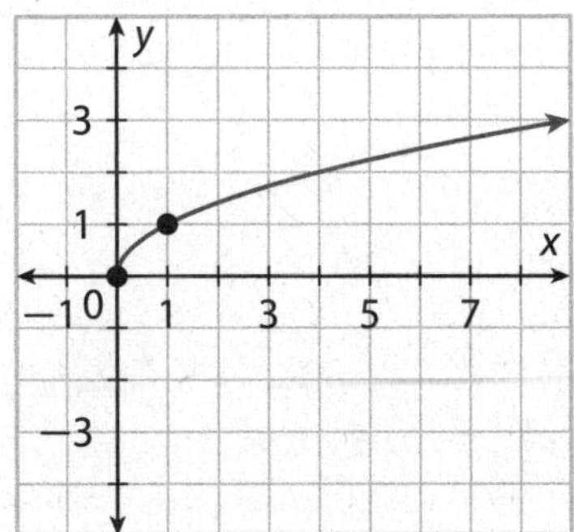

5. $g(x) = \sqrt{\frac{1}{3}(x + 2)} + 1$

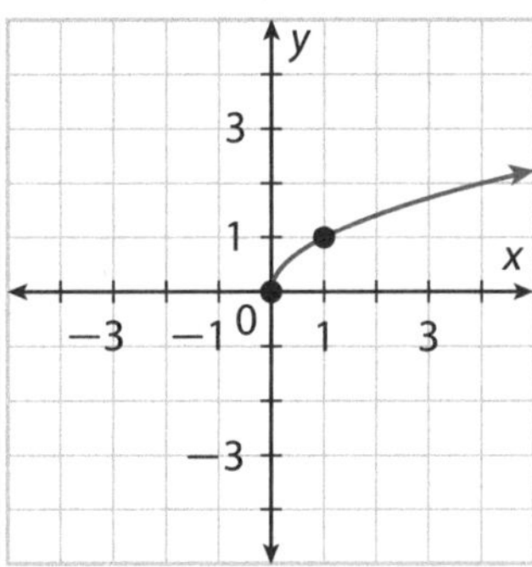

Explain 2 Writing Square Root Functions

Given the graph of a square root function and the form of the transformed function, either $g(x) = a\sqrt{x - h} + k$ or $g(x) = \sqrt{\frac{1}{b}(x - h)} = k$, the transformation parameters can be determined from the transformed reference points. In either case, the initial point will be at (h, k) and readily apparent. The parameter a can be determined by how far up or down the second point (found at $x = h + 1$) is from the initial point, or the parameter b can be determined by how far to the left or right the second point (found at $y = k + 1$) is from the initial point.

Example 2 **Write the function that matches the graph using the indicated transformation format.**

Ⓐ $g(x) = \sqrt{\frac{1}{b}(x - h)} + k$

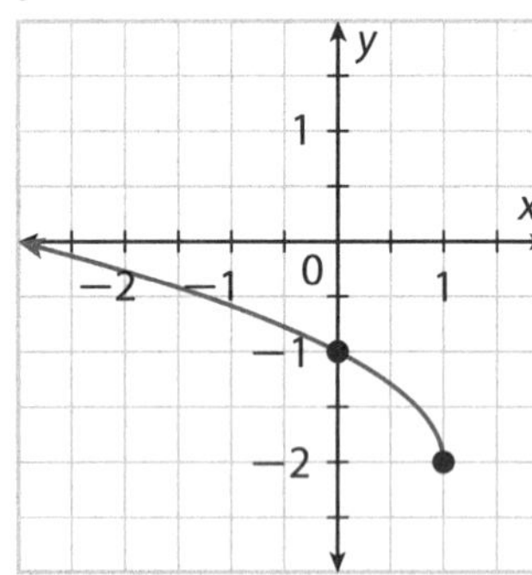

Initial point: $(h, k) = (1,-2)$

Second point:

$(h + b, k + 1) = (0, -1)$

$1 + b = 0$

$b = -1$

The function is $g(x) = \sqrt{-1(x - 1)} - 2$.

(B) $g(x) = a\sqrt{x - h} + k$

Initial point: $(h, k) = \left(\square, \square\right)$

Second point:

$\left(h + 1, k + \square\right) = \left(-1, \square\right)$

$\square + a = 2$

$a = \square$

The function is $g(x) = \square \sqrt{x \square} 2 - \square$.

Your Turn

Write the function that matches the graph using the indicated transformation format.

6. $g(x) = \sqrt{\frac{1}{b}(x - h)} + k$

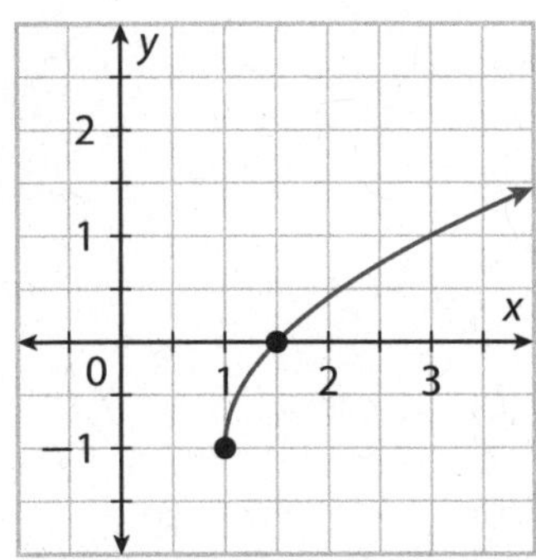

7. $g(x) = a\sqrt{(x - h)} + k$

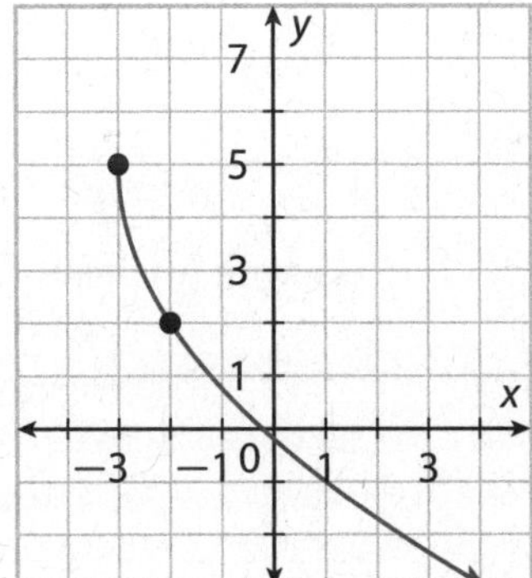

Explain 3 Modeling with Square Root Functions

Square root functions that model real-world situations can be used to investigate average rates of change.

Recall that the average rate of change of the function $f(x)$ over an interval from x_1 to x_2 is given by

$$\frac{f(x_2) - f(x_1)}{x_2 - x_1}.$$

Example 3 **Use a calculator to evaluate the model at the indicated points, and connect the points with a curve to complete the graph of the model. Calculate the average rates of change over the first and last intervals and explain what the rate of change represents.**

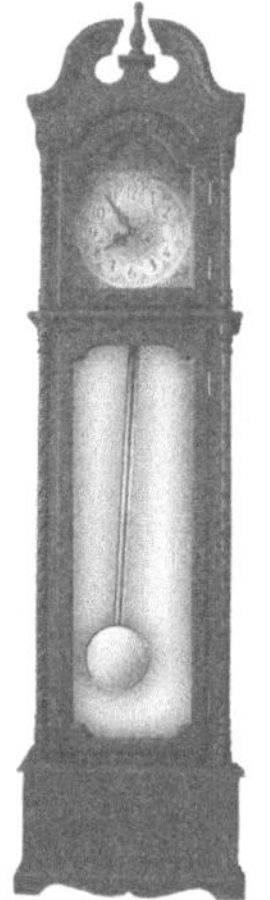

(A) The approximate period T of a pendulum (the time it takes a pendulum to complete one swing) is given in seconds by the formula $T = 0.32\sqrt{\ell}$, where ℓ is the length of the pendulum in inches. Use lengths of 2, 4, 6, 8, and 10 inches.

First find the points for the given x-values.

Length (inches)	Period (seconds)
2	0.45
4	0.64
6	0.78
8	0.91
10	1.01

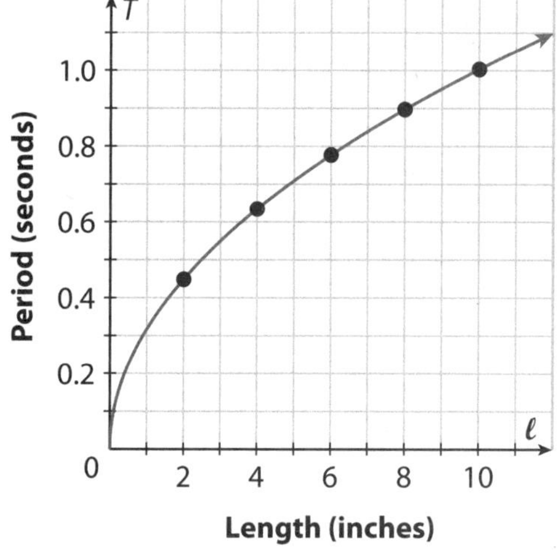

Plot the points and draw a smooth curve through them.

Find the average increase in period per inch increase in the pendulum length for the first interval and the last interval.

First interval:

$$\text{rate of change} = \frac{0.64 - 0.45}{4 - 2}$$
$$= 0.095$$

Last Interval:

$$\text{rate of change} = \frac{1.01 - 0.91}{10 - 8}$$
$$= 0.05$$

The average rate of change is less for the last interval. The average rate of change represents the increase in pendulum period with each additional inch of length. As the length of the pendulum increases, the increase in period time per inch of length becomes less.

(B) A car with good tires is on a dry road. The speed, in miles per hour, from which the car can stop in a given distance d, in feet, is given by $s(d) = \sqrt{96d}$. Use distances of 20, 40, 60, 80, and 100 feet.

First, find the points for the given x-values.

Distance	20	40	60	80	100
Speed					

Plot the points and draw a smooth curve through them.

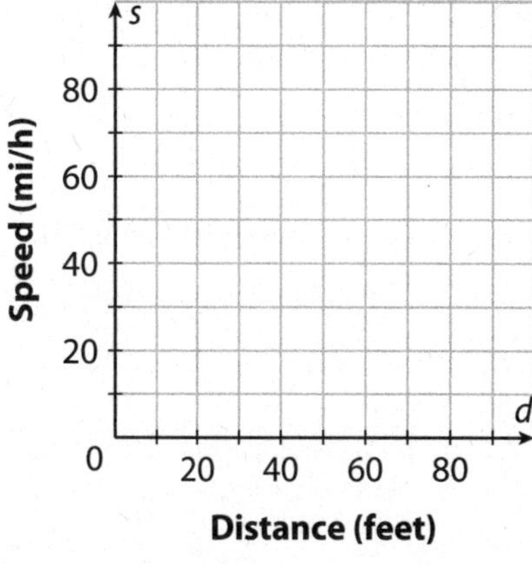

First interval:

$$\text{rate of change} = \frac{\square - \square}{40 - 20}$$

$$= \square$$

Last Interval:

$$\text{rate of change} = \frac{\square - \square}{100 - 80}$$

$$= \square$$

The average rate of change is ______ for the last interval. The average rate of change represents the increase in ________ with each additional ________________. As the available stopping distance increases, the additional increase in speed per foot of stopping distance __________.

Your Turn

Use a calculator to evaluate the model at the indicated points, and connect the points with a curve to complete the graph of the model. Calculate the average rates of change over the first and last intervals and explain what the rate of change represents.

8. The speed in miles per hour of a tsunami can be modeled by the function $s(d) = 3.86\sqrt{d}$, where d is the average depth in feet of the water over which the tsunami travels. Graph this function from depths of 1000 feet to 5000 feet and compare the change in speed with depth from the shallowest interval to the deepest. Use depths of 1000, 2000, 3000, 4000, and 5000 feet for the x-values.

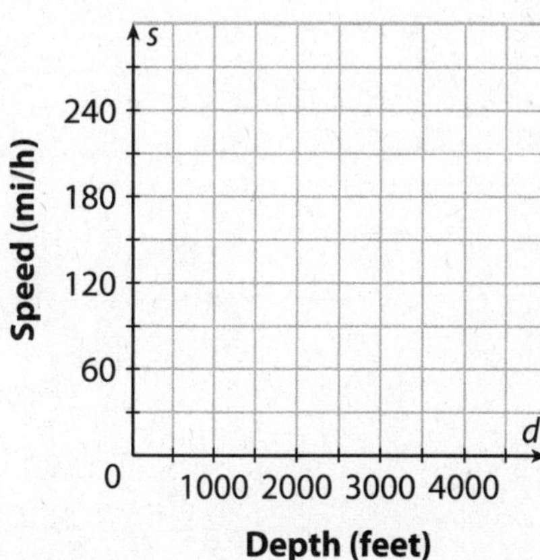

Elaborate

9. What is the difference between the parameters inside the radical (b and h) and the parameters outside the radical (a and k)?

10. Which transformations change the square root function's end behavior?

11. Which transformations change the square root function's initial point location?

12. Which transformations change the square root function's domain?

13. Which transformations change the square root function's range?

14. **Essential Question Check-In** Describe in your own words the steps you would take to graph a function of the form $g(x) = a\sqrt{x - h} + k$ or $g(x) = \sqrt{\frac{1}{b}(x - h)} + k$ if you were given the values of h and k and using either a or b.

Evaluate: Homework and Practice

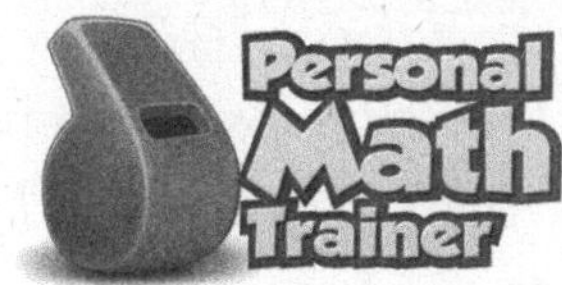

- Online Homework
- Hints and Help
- Extra Practice

1. Graph the functions $f(x) = \sqrt{x}$ and $g(x) = -\sqrt{x}$ on the same grid. Describe the domain, range and end behavior of each function. How are the functions related?

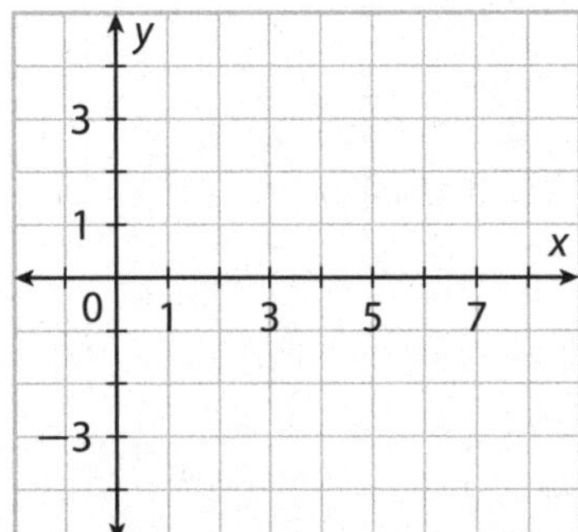

Describe the transformations of $g(x)$ from the parent function $f(x) = \sqrt{x}$.

2. $g(x) = \sqrt{\frac{1}{2}x + 1}$

3. $g(x) = -5\sqrt{x+1} - 3$

4. $g(x) = \frac{1}{4}\sqrt{x-5} - 2$

5. $g(x) = \sqrt{-7(x-7)}$

Describe the domain and range of each function using set notation.

6. $g(x) = \sqrt{\frac{1}{3}(x-1)}$

7. $g(x) = 3\sqrt{x+4} + 3$

8. $g(x) = \sqrt{-5(x+1)} + 2$

9. $g(x) = -7\sqrt{x-3} - 5$

Plot the transformed function $g(x)$ on the grid with the parent function, $f(x) = \sqrt{x}$. Describe the domain and range of each function using set notation.

10. $g(x) = -\sqrt{x} + 3$

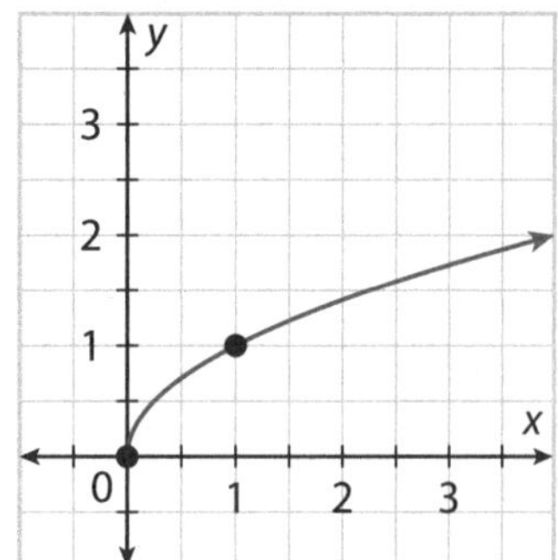

11. $g(x) = \sqrt{\frac{1}{3}(x + 4)} - 1$

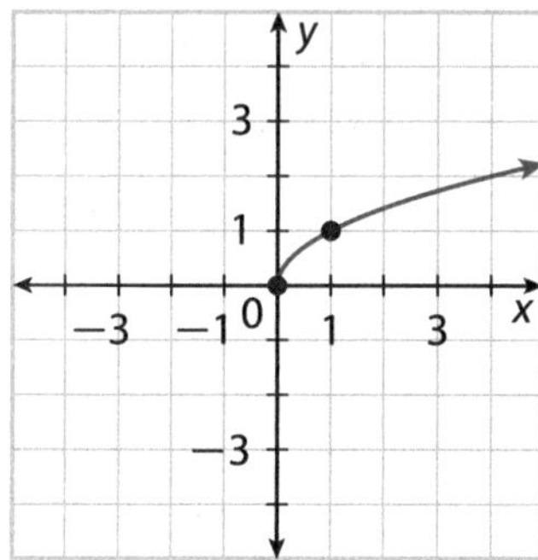

12. $g(x) = \sqrt{-\frac{2}{3}\left(x - \frac{1}{2}\right)} - 2$

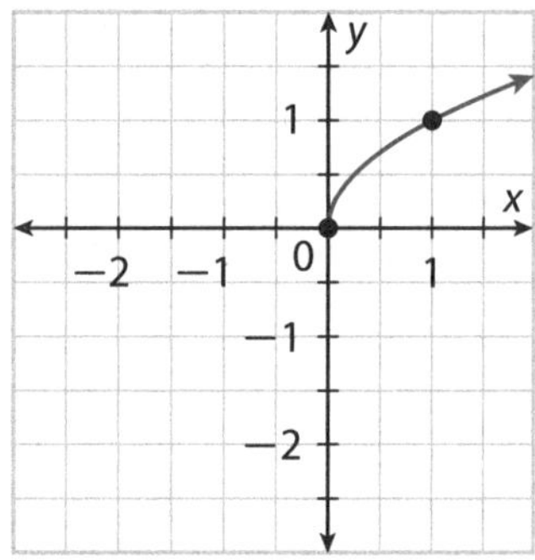

13. $g(x) = 4\sqrt{x + 3} - 4$

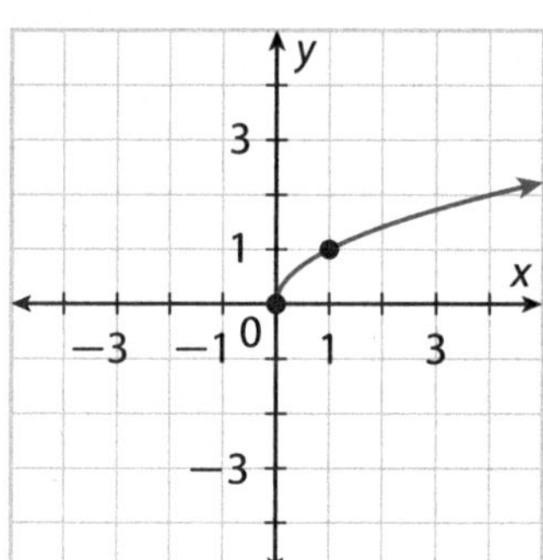

Write the function that matches the graph using the indicated transformation format.

14. $g(x) = \sqrt{\frac{1}{b}(x - h)} + k$

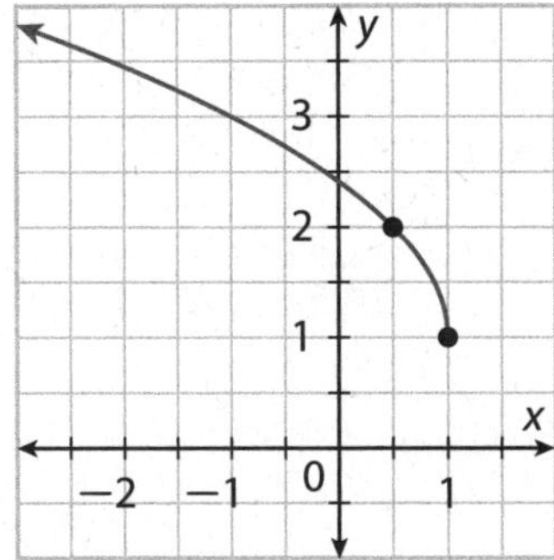

15. $g(x) = \sqrt{\frac{1}{b}(x - h)} + k$

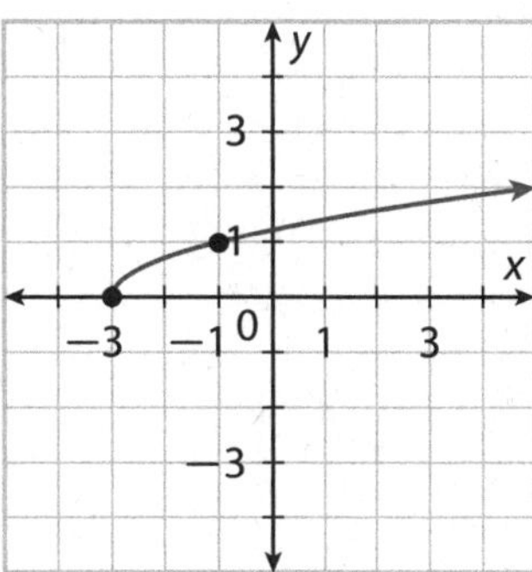

16. $g(x) = a\sqrt{x - h} + k$

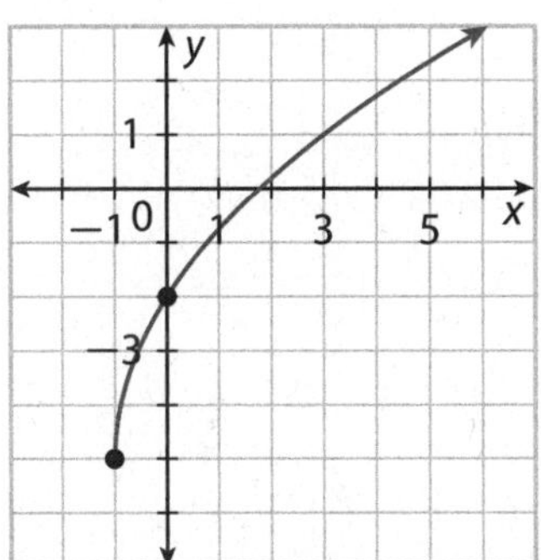

17. $g(x) = a\sqrt{x - h} + k$

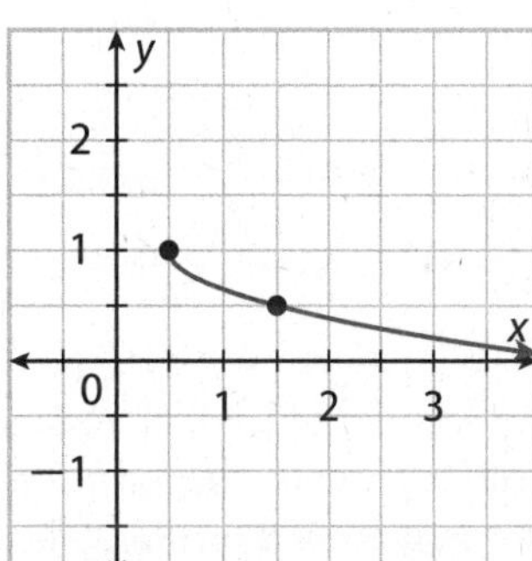

Use a calculator to evaluate the model at the indicated points, and connect the points with a curve to complete the graph of the model. Calculate the average rates of change over the first and last intervals and explain what the rate of change represents.

18. A farmer is trying to determine how much fencing to buy to make a square holding pen with a 6-foot gap for a gate. The length of fencing, f, in feet, required as a function of area, A, in square feet, is given by $f(A) = 4\sqrt{A} - 6$. Evaluate the function from 20 ft² to 100 ft² by calculating points every 20 ft².

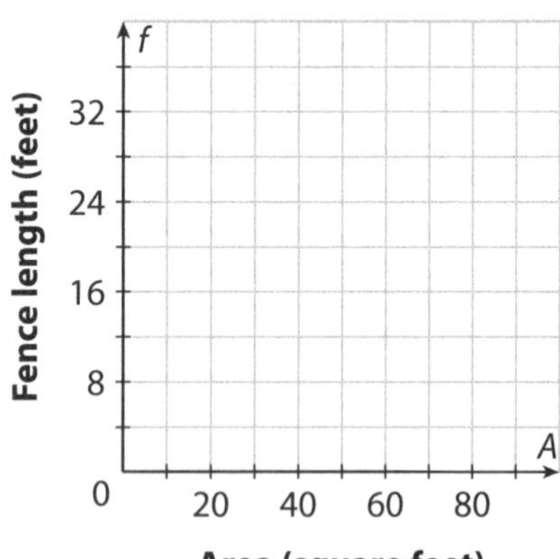

19. The speed, s, in feet per second, of an object dropped from a height, h, in feet, is given by the formula $s(h) = \sqrt{64h}$. Evaluate the function for heights of 0 feet to 25 feet by calculating points every 5 feet.

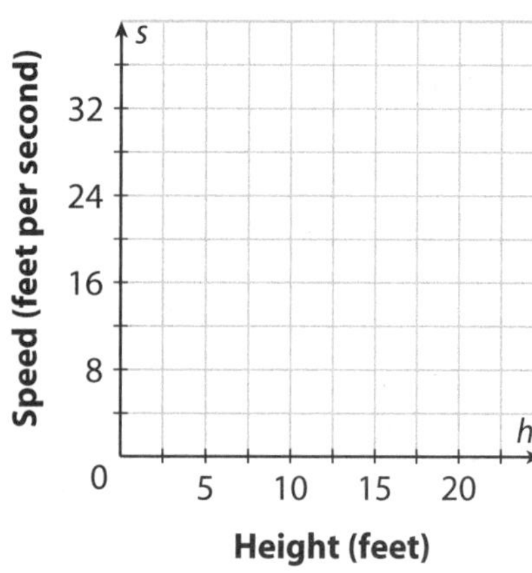

20. Water is draining from a tank at an average speed, s, in feet per second, characterized by the function $s(d) = 8\sqrt{d - 2}$, where d is the depth of the water in the tank in feet. Evaluate the function for depths of 2, 3, 4, and 5 feet.

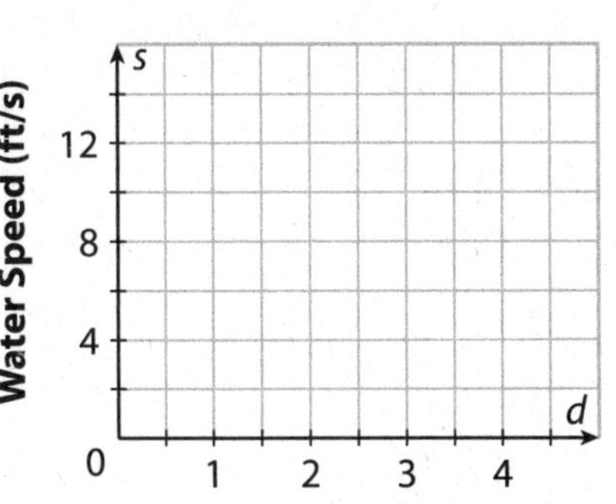

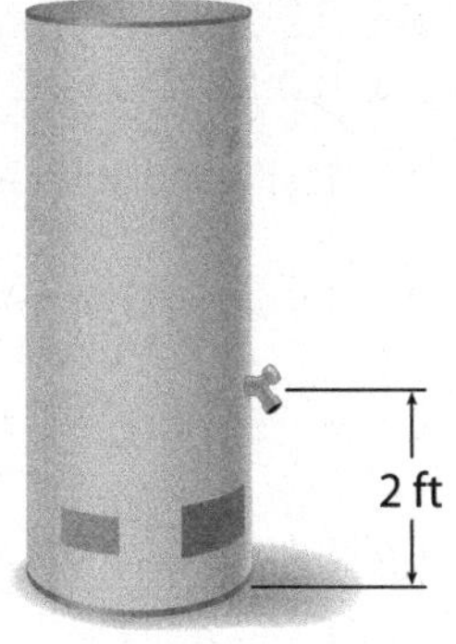

21. A research team studies the effects from an oil spill to develop new methods in oil clean-up. In the spill they are studying, the damaged oil tanker spilled oil into the ocean, forming a roughly circular spill pattern. The spill expanded out from the tanker, increasing the area at a rate of 100 square meters per hour. The radius of the circle is given by the function $r = \sqrt{\frac{100}{\pi}t}$, where t is the time (in hours) after the spill begins. Evaluate the function at hours 0, 1, 2, 3, and 4.

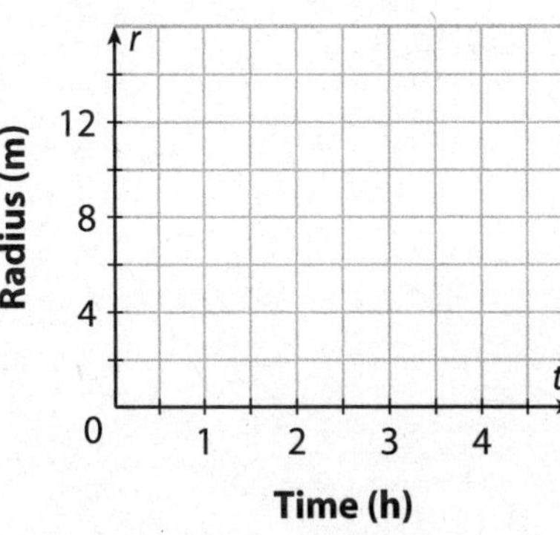

22. Give all of the transformations of the parent function $f(x) = \sqrt{x}$ that result in the function $g(x) = \sqrt{-2(x-3)} + 2$.

A. Horizontal stretch

B. Horizontal compression

C. Horizontal reflection

D. Horizontal translation

E. Vertical stretch

F. Vertical compression

G. Vertical reflection

H. Vertical translation

H.O.T. Focus on Higher Order Thinking

23. **Draw Conclusions** Describe the transformations to $f(x) = \sqrt{x}$ that result in the function $g(x) = \sqrt{-8x + 16} + 3$.

24. **Analyze Relationships** Show how a horizontally stretched square root function can sometimes be replaced by a vertical compression by equating the two forms of the transformed square root function.

$$g(x) = a\sqrt{x} = \sqrt{\frac{1}{b}x}$$

What must you assume about a and b for this replacement to result in the same function?

25. Multi-Step On a clear day, the view across the ocean is limited by the curvature of Earth. Objects appear to disappear below the horizon as they get farther from an observer. For an observer at height h above the water looking at an object with a height of H (both in feet), the approximate distance (d) in miles at which the object drops below the horizon is given by $d(h) = 1.21\sqrt{h + H}$.

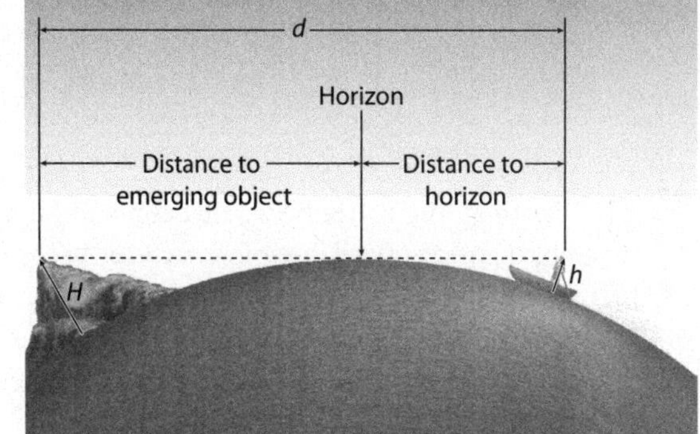

a. What is the effect of the object height, H, on the graph of $d(h)$?

b. What is the domain of the function $d(h)$? Explain your answer.

c. Plot two functions of distance required to see an object over the horizon versus observer height: one for seeing a 2-foot-tall buoy and one for seeing a 20-foot-tall sailboat. Calculate points every 10 feet from 0 to 40 feet.

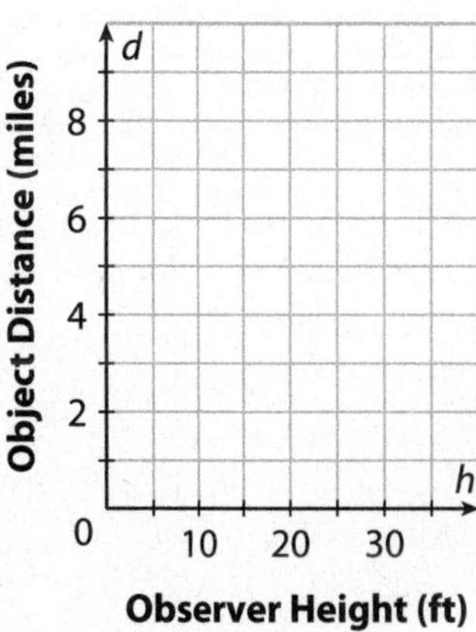

d. Where is the greatest increase in viewing distance with observer height?

Lesson Performance Task

With all the coffee beans that come in for processing, a coffee manufacturer cannot sample all of them. Suppose one manufacturer uses the function $s(x) = \sqrt{x} + 1$ to determine how many samples it must take from x containers in order to obtain a good representative sampling of beans. How does this function relate to the function $f(x) = \sqrt{x}$? Graph both functions. How many samples should be taken from a shipment of 45 containers of beans? Explain why this can only be a whole number answer.

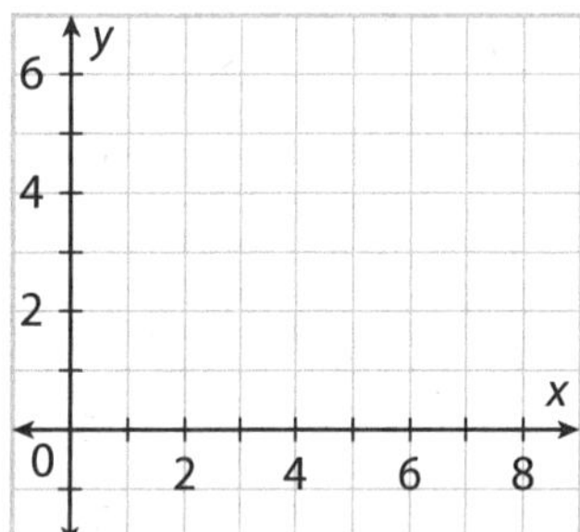

Name________________________ Class______________ Date__________

11.3 Fitting Square Root Functions to Data

Resource Locker

Essential Question: How can you find the equation of a square root function to model data?

A2.4.E Formulate…square root equations using technology given a table of data.

Explore Investigating the Inverse Relationship between General Quadratic Functions and Radical Functions

Investigate the relationship between the quadratic function $f(x) = 4x^2 - 16x + 21$ and its inverse.

(A) First, complete the square so that the function is in vertex form. Factor the first two terms so that the coefficient of x^2 is 1.

$f(x) = 4x^2 - 16x + 21$

$f(x) =$ ______

(B) Using the vertex form of $f(x) = 4x^2 - 16x + 21$, give the vertex of the function.

____

(C) The graph of $f(x)$ is the graph of $y = x^2$ ________ stretched by a factor of 4, translated to the right ____ unit(s), and translated ______ 5 units.

(D) The inverse of $f(x)$ will be a square root function. Does $f(x)$ pass the horizontal line test?

____

(E) Based on your answer to Step D, will the domain of $f(x)$ have to be restricted so that a positive square root function will be its inverse? If so, give the restricted domain.

____

(F) Use the vertex of $f(x)$ to predict the initial point of its inverse $f^{-1}(x)$.

The initial point of $f^{-1}(x)$ is ______.

(G) Find $f^{-1}(x)$ by only using the positive square root.

$f(x) = 4(x - 2)^2 + 5$

$f^{-1}(x) =$ ______

(H) The inverse of $f(x) = 4(x - 2)^2 + 5$ is $f^{-1}(x) = \sqrt{\frac{x-5}{4}} + 2$. What is the initial point of $f^{-1}(x)$?

____

(I) The graph of $f^{-1}(x)$ is the graph of $y = \sqrt{x}$ __________ stretched by a factor of 4, translated to the __________ 5 units, and translated up __________ unit(s).

Reflect

1. Find the range of $f(x)$ and the domain and range of $f^{-1}(x)$. How do the restricted domain and range of $f(x)$ compare to the domain and range of $f^{-1}(x)$?

____

____

____

Explain 1 Roughly Fitting a Square Root Function to Data

A square root function in the form $y = a\sqrt{x - h} + k$ can be fit to model a set of data that appear to follow a square root model by visually estimating the model's initial point, selecting one other point in the data, and solving for a.

Example 1 **Plot the data given in the table on a coordinate grid, estimate the initial point, and draw a curve similar to a square root function that appears to go through the data points. Then write an equation for the estimated square root function.**

(A) The data represents the number of bacteria in refrigerated food, where x is the number of bacteria and y is the temperature in degrees Celsius.

x	y
100	1
105	1.4
120	2
125	2.2
140	2.4
145	2.5
180	3

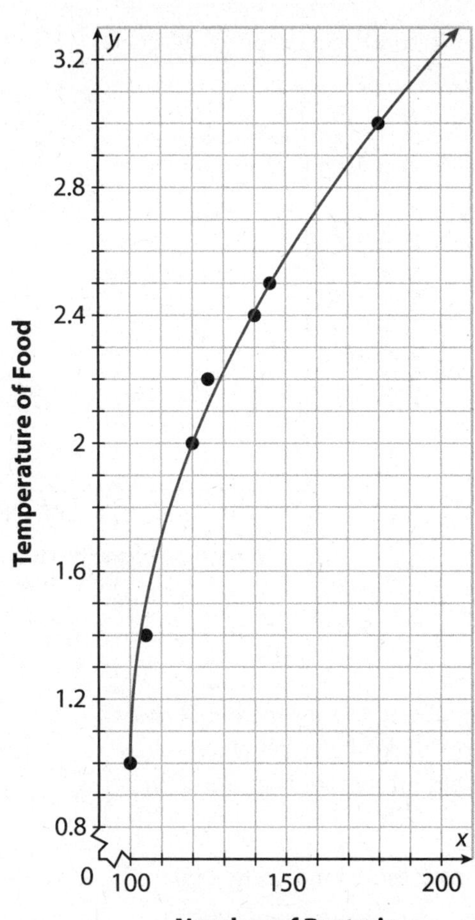

From the plot, it appears that the first data point makes a good estimate for the initial point of a curve that approximates all of the data points and is shaped like the graph of a square root function. Sketch a curve from this initial point that passes through or near the other data points.

Now determine the values of h and k for your model using your estimation of the initial point for the graph.

initial point: $= (100, 1)$

$h = 100$

$k = 1$

Substitute $h = 100$ and $k = 1$ into $y = a\sqrt{x - h} + k$.

$y = a\sqrt{x - 100} + 1$

To find the value of a, substitute the coordinates of any other point that appears to lie on or very near the curve into $y = a\sqrt{x - 100} + 1$ and solve for a. In this case, use $x = 120$ and $y = 2$.

$2 = a\sqrt{120 - 100} + 1$

$1 = a\sqrt{20}$

$0.224 \approx a$

An equation for a square root function that models the data is $y = 0.224\sqrt{x - 100} + 1$.

(B) The data represents the radar-detected speed, s, in inches per second, of a dropped object at the end of a fall as a function of its starting height, v, in inches.

Starting Height (v) (inches)	Speed at Sensor (s) (in./sec)
30	56
35	101
45	129
70	193
90	226
125	273
150	313
200	354

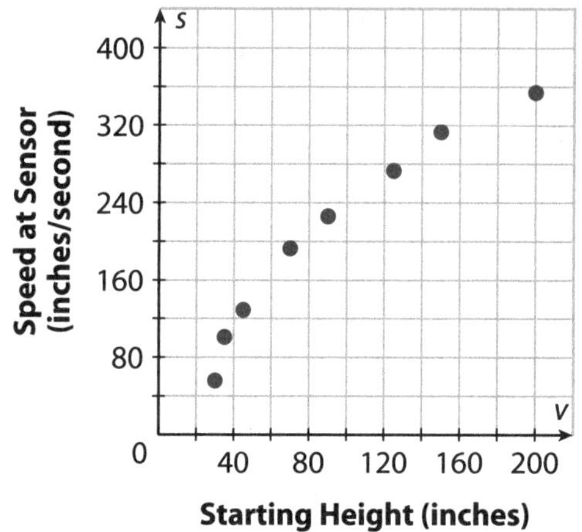

From the plot, it appears that the first data point makes a good estimate for the initial point of a curve that approximates all of the data points and is shaped like the graph of a square root function. Sketch a curve from this initial point that passes through or near the other data points.

Now determine the values of h and k for your model using your estimation of the initial point for the graph.

initial point: $(30, 56)$

$h = \square$

$k = \square$

$s(v) = a\sqrt{v - \square} + \square$

Find the value of a by using another point on the curve, substituting it in, and solving for a.

Reference Point $= (125, \square)$

$\square = a\sqrt{\square - h} + k$

$273 - \square = a\sqrt{125 - \square}$

$a = \dfrac{\square}{\sqrt{\square}}$

$\approx \square$

An equation for a square root function that models the data is

$s(v) =$ ____________________.

Reflect

2. **Discussion** How did you arrive at your estimate for the vertex? Can you think of another way to estimate the vertex? Discuss the merits or demerits of each approach.

Your Turn

Plot the data given in the table on a coordinate grid, estimate the initial point, and draw a curve shaped similar to a square root function that appears to go through the data points. Then write an equation for the estimated square root function.

3. Thomas opens a new boat dealership at the beginning of 2008 and sees his sales grow every year. The data represent the years and the number of boats sold each year.

Year	Sales
2008	37
2009	45
2010	50
2011	54
2012	54
2013	57
2014	57
2015	64

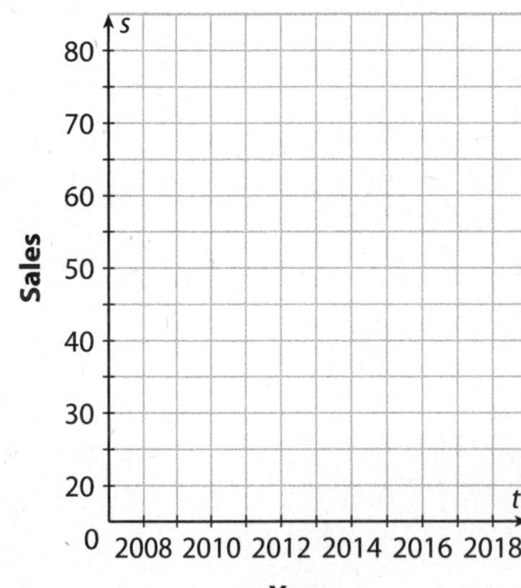

Explain 2 Fitting a Square Root Function to Data Using Technology

Graphing calculators can be used to fit a selection of function types to a data set, as you have seen in previous lessons involving linear and quadratic models. Regression to a square root model is not usually available on a graphing calculator but a quadratic regression is. The results of a quadratic regression can be used to find an inverse function following the method in the Explore.

There are two options for restricting the domain of a quadratic function so that it is a one-to-one function, and the correct choice is the one that includes the given data set. The data determine if the square root function will be positive or negative.

Example 2 **Using the same data from Example 1, find a square root model by performing a quadratic regression.**

 Use the data from Example 1A.

Using the statistics features on your calculator, enter the data, using List 1 for the temperature and List 2 for the number of bacteria. Entering the data this way will switch the variables x and y from Example 1A so that you can perform quadratic regression on the switched data. You can then rewrite the quadratic regression equation in vertex form and solve for y to find the equation of a square root function model.

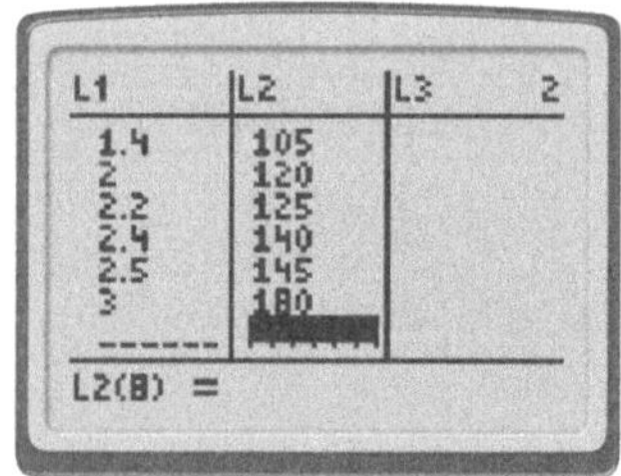

From the statistics calculations menu, select quadratic regression.

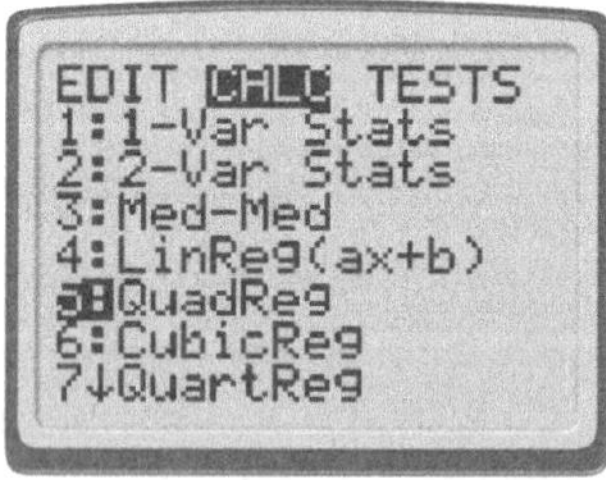

Push ENTER twice and you should see a result like this:

An approximate model is $y = 21x^2 - 44x + 124$.

Complete the square so that the function is in vertex form.

$$\begin{aligned} y &= 21x^2 - 44x + 124 \\ &= 21(x^2 - 2.1x) + 124 \\ &= 21(x^2 - 2.1x + 1.10) - 23.2 + 124 \\ &= 21(x - 1.05)^2 + 101 \end{aligned}$$

The restricted domain of y that includes the data points is given by $x \geq 1.05$.

You can find the inverse by using the positive square root function written in vertex form.

$$\begin{aligned} y &= \sqrt{\frac{1}{21}(x - 101)} + 1.05 \\ &= 0.218\sqrt{x - 101} + 1.05 \end{aligned}$$

Comparing this function to the function from Example 1A shows that they are very similar. Plotting this function and comparing its graph to the graph of the function in Example 1A, you can see that they are nearly indistinguishable.

(B) Use the data from Example 1B.

If the square root function is $s(v)$, then its inverse is a quadratic function, $\square$. To find the quadratic function in standard form, perform a quadratic regression on the data from Example 1B.

$$v(s) = \square\, s^2 - \square\, s + \square$$

Complete the square to convert $v(s)$ to vertex form.

$$\begin{aligned} v(s) &= 0.001657s^2 - 0.1194s + 31.48 \\ &= 0.001657\left(s^2 - \square\, s\right) + 31.48 \\ &= 0.001657\left(s^2 - 72.06s + \square\right) - \square + 31.48 \\ &= 0.001657\left(s^2 - 72.06s + \square\right) + \square \\ &= 0.001657\left(s - \square\right)^2 + \square \end{aligned}$$

The restricted domain of $v(s)$ that includes the data is $s \,\square\, 36.03$.

The inverse of $v(s)$ can be written in vertex form using a positive/negative square root as:

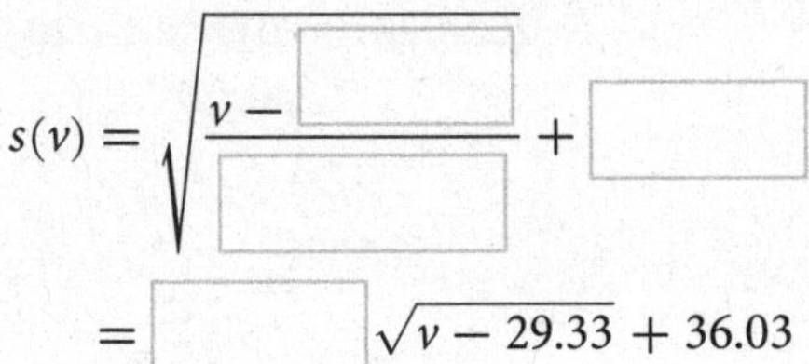

$$s(v) = \sqrt{\frac{v - \square}{\square}} + \square$$

$$= \square\sqrt{v - 29.33} + 36.03$$

Plot this function and your previous function from Example 1B on the same grid and compare them.

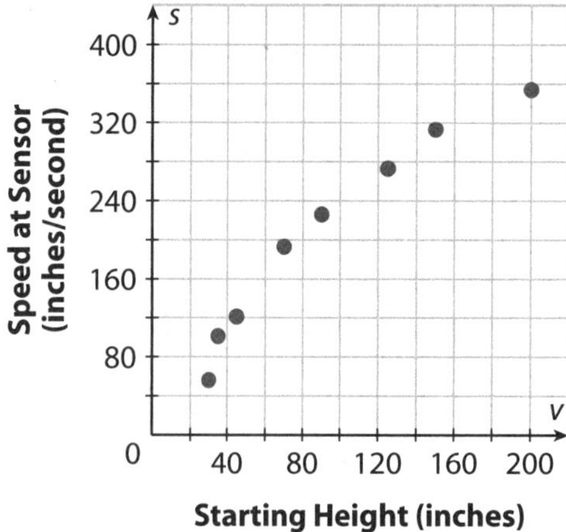

Your Turn

4. Use the data from Your Turn 3 and perform a quadratic regression to find a square root function using your graphing calculator. Plot this function and your previous function from Your Turn 3 on the same grid and compare them.

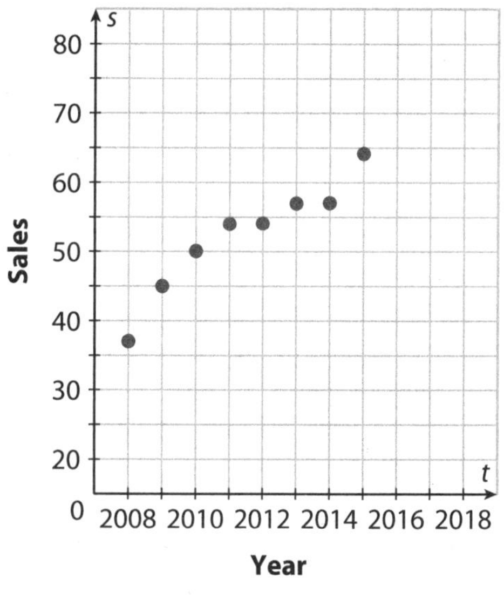

Explain 3 Solving a Real-World Problem

Models can be used to make predictions in the real world even at points along a curve that do not correspond to any existing data points.

Example 3 **Read the problem description and solve using the four-step problem solving process.**

Use the models from Example 1B and 2B to determine the answer to the problem.

You are tasked with testing the durability of a hard plastic shell on a new phone design and you need to test the material by checking for damage after repeated impacts at 360 inches per second to simulate the kind of accidental damage a typical user inflicts over the life of a phone. You decide that the easiest way to set this experiment up is to determine the height required to achieve a speed of 360 inches per second on impact. You tried a series of heights with a radar gun to measure the speed and collected the data shown in Example 1B.

Analyze

Looking over the data you collected, there are no data points with a speed of 360 inches per second, and continued attempts do not seem likely to randomly produce the correct speed. On the other hand, the pattern of the data appears to match a ______________ function.

Formulate a Plan

Since you cannot easily pick a height and hope it produces the desired speed, you need to use a ______________________ to determine an appropriate height.

You have two methods of fitting the data based on examples in this lesson, one based on estimation of parameters and one based on ______________.

Solve

Find the height by solving the equation: $s = a\sqrt{v - h} + k$ using $s = 360$ in./s and the model parameters you found in Examples 1B and 2B.

Estimate model: $\square = \square\sqrt{v - \square} + \square$

Regression model: $\square = \square\sqrt{v - \square} + \square$

In order to solve this equation, use a graphical approach. Using the curves you made in the previous examples, find the intersection of the two model curves with the constant function $s = \square$.

Estimate model: $v = \square$

Regression model: $v = \square$

Justify and Evaluate

First check that the graphically solved numbers produce the desired speed.

Estimate model: $s(\square) = \square$

Regression model: $s(\square) = \square$

The regression model suggests a slightly lower/higher height to reach the same desired speed.

Your Turn

5. Using the graphs of the two models from Your Turns 3 and 4, predict the expected year when boat sales will reach 65. Compare the results of the two models.

Elaborate

6. When selecting a second point in order to determine the value of a, is it better to pick a point near the initial point or farther from it? Explain.

7. Why is a quadratic model used to find the regression equation of a square root model?

8. **Discussion** What are the advantages and disadvantages of estimating a square root function compared to using quadratic regression?

9. **Essential Question Check-In** How can you use a calculator to find the equation of a square root function to model data?

Evaluate: Homework and Practice

- Online Homework
- Hints and Help
- Extra Practice

Find the inverse positive square root function for each quadratic function by first writing each quadratic function in vertex form. Compare the vertex of the quadratic function with the initial point of its inverse function. Then compare the domains and ranges, and the transformations from their respective parent functions.

1. $f(x) = x^2 + 4x + 1$

2. $f(x) = -x^2 - 10x - 21$

Estimate and then plot the best fitting square root function to the given data. Then give the domain and range of the function.

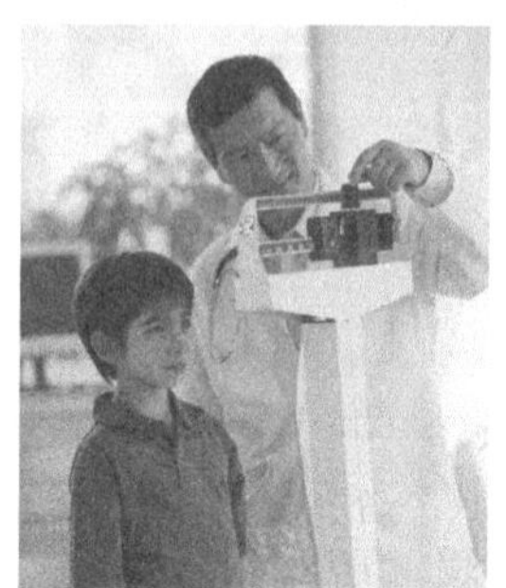

3. Body mass index (BMI) is a measure used to determine healthy body mass based on a person's height. BMI is calculated by dividing a person's mass in kilograms by the square of his or her height in meters. The median BMI measures for a group of boys are given in the chart.

Median BMI (B)	Age of Boys (A)
15.4	6
15.5	7
15.8	8
16.2	9
16.6	10

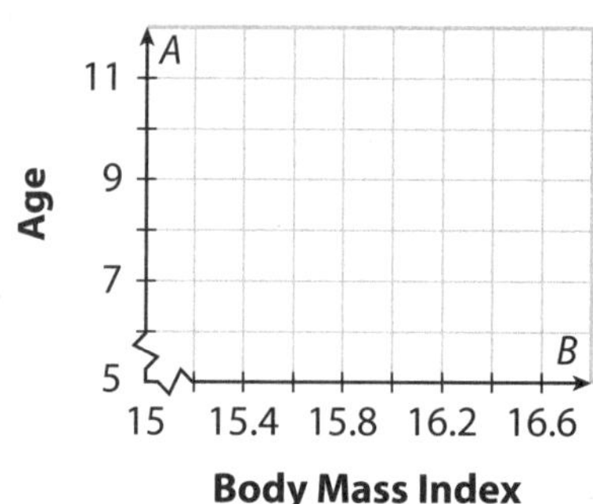

4. An object traveling at a certain speed has a kinetic energy of 20 kJ. Model the speed of the object based on its kinetic energy given in the table below.

Kinetic Energy (kJ)	Speed above Initial Speed (m/s)
20	0.00
80	5.06
185	10.05
330	14.94
525	20.02
1040	29.99
1365	35.01
1730	39.98
2140	44.96
2600	50.01
3095	54.97
3645	60.03

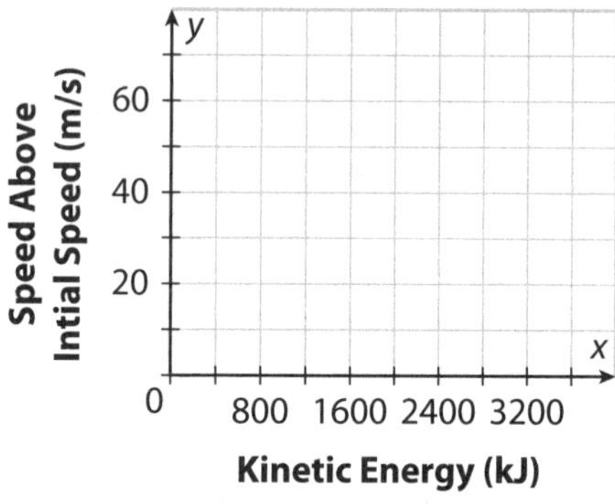

Use the data in the table and a graphing calculator to complete parts a–c.

5.

x	y
2	3.1
2.1	2.2
2.4	4.2
3	8.9
4	10.8
5	11
6	13.2
7	13.5

a. Use quadratic regression on a graphing calculator to find the quadratic regression function for the data pairs (y, x). Convert the function to vertex form and indicate the domain and range that contains the data.

b. Find a square root function by solving the quadratic regression equation for y. Then graph this square root function on the grid with the data points.

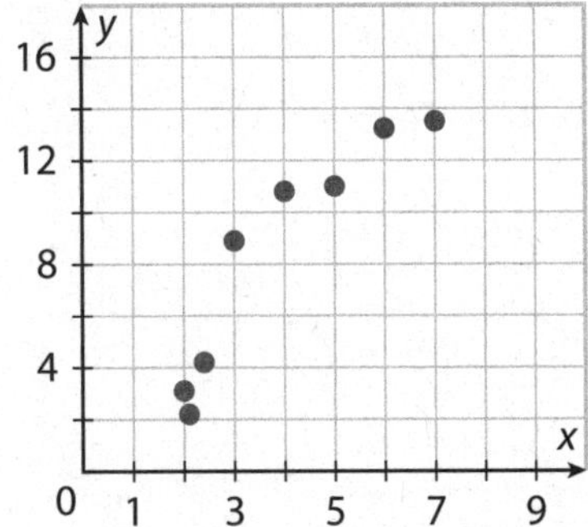

c. Use the square root function to predict the value of y to the nearest tenth when $x = 8$.

6.

x	y
−3	3.9
−2.5	2.8
−2	1.9
−1.5	0.5
0	−0.2
2	−0.9
4	−2.6
6	−2.4

a. Use quadratic regression on a graphing calculator to find the quadratic regression function for the data pairs (y, x). Convert the function to vertex form and indicate the domain and range that contains the data.

b. Find a square root function by solving the quadratic regression equation for y. Then graph this square root function on the grid with the data points.

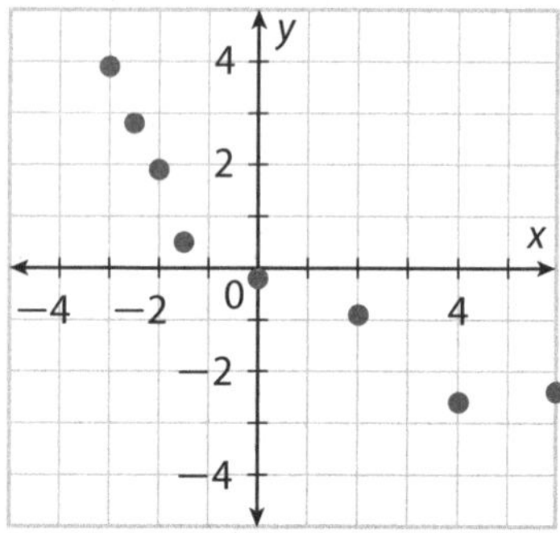

c. Use the square root function to predict the value of y to the nearest tenth when $x = 3$.

H.O.T. Focus on Higher Order Thinking

7. **Draw Conclusions** For the data shown, which are repeated from Evaluate 4, an informal model using the initial point (20, 0) and (3645, 60.03) as the second point gives the function $f(x) = 0.997\sqrt{x - 20}$. A model derived by first finding a quadratic regression equation and then finding its inverse is $r(x) = 1.067\sqrt{x - 2.322} - 4.386$. Complete the table to find the residuals for each model. Then plot the residuals and determine which model is a better fit for the data.

x	y_d	$y_f = f(x)$	$y_d - y_f$	$y_r = r(x)$	$y_d - y_r$
20	0.00				
80	5.06				
185	10.05				
330	14.94				
525	20.02				
1040	29.99				
1365	35.01				
1730	39.98				
2140	44.96				
2600	50.01				
3095	54.97				
3645	60.03				

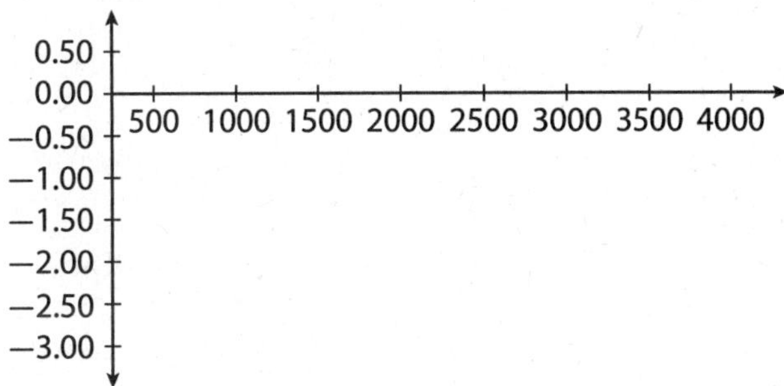

8. **Critique Reasoning** A student makes the following claim: When roughly fitting a square root function to data for which a model curve will open to the right, the initial point is always either the leftmost data point or will be to the left of all of the data. Is the student correct? Justify your answer or find a counterexample.

Lesson Performance Task

The speed of sound is typically given as 767 mi/h, but that value is only accurate if the temperature is 68°F. We can calculate the speed of sound, s, in air in miles per hour using the function $s = \sqrt{k(T + C)}$, where T is the temperature in degrees Fahrenheit, and k and C are constants. The table contains some values returned by this function.

Temperature (°F)	Speed of Sound in Air (mi/h)
−460	0
−100	634
0	717
68	768
212	866

a. Determine the values of the constants k and C of the function.

b. Suppose one day in Antarctica it is −110°F and in Texas it is 110°F. What is the difference in the speed of sound in these two locations?

c. How much longer would it take to hear a very loud sound from 5 miles away in Antarctica than in Texas?

Name ______________________ Class ____________ Date ________

11.4 Graphing Cube Root Functions

Resource Locker

Essential Question: How can you use transformations of the parent cube root function to graph functions of the form $f(x) = a\sqrt[3]{(x-h)} + k$ or $g(x) = \sqrt[3]{\frac{1}{b}(x-h)} + k$?

A2.6.A Analyze the effect on the graphs of ... $f(x) = \sqrt[3]{x}$ when $f(x)$ is replaced by $af(x)$, $f(bx)$, $f(x - c)$, and $f(x) + d$ for specific positive and negative real values of a, b, c, and d. Also A2.2.A, A2.7.I

Explore Graphing and Analyzing the Parent Cube Root Function

The cube root parent function is $f(x) = \sqrt[3]{x}$. To graph $f(x)$, choose values of x and find corresponding values of y. Choose both negative and positive values of x.

Graph the function $f(x) = \sqrt[3]{x}$. Identify the domain and range of the function.

(A) Make the table of values.

x	y	x, y
−8		
−1		
0		
1		
8		

(B) Use the table to graph the function.

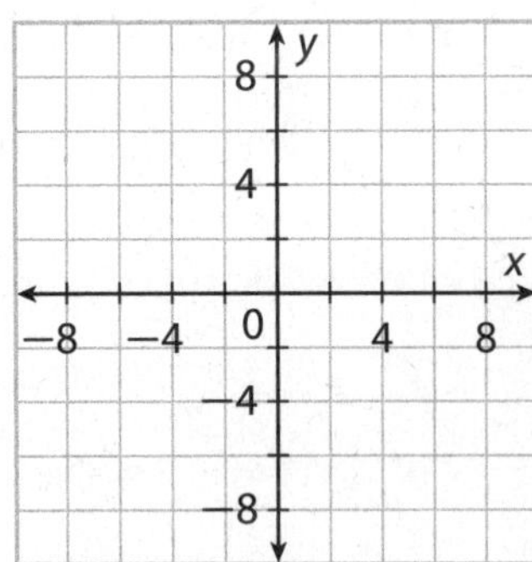

(C) Identify the domain and range of the function.

The domain is the ______________________.

The range is ______________________.

(D) Does the graph of $f(x) = \sqrt[3]{x}$ have any symmetry?

The graph has ______________.

Reflect

1. Can the radicand in a cube root function be negative?

____

Explore 2 Predicting the Effects of Parameters on the Graphs of Cube Root Functions

Given the parent function $f(x) = \sqrt[3]{x}$, predict the effect of parameters on the graphs of cube root functions.

(A) Predict the effect of the parameter, h, on the graph of $g(x) = \sqrt[3]{x - h}$ for each function.

a. $g(x) = \sqrt[3]{x - 3}$: The graph is a ______________ of the graph of $f(x)$ right/left/up/down 3 units.

b. $g(x) = \sqrt[3]{x + 3}$: The graph is a ______________ of the graph of $f(x)$ right/left/up/down 3 units.

Check your answers using a graphing calculator.

(B) Predict the effect of the parameter k on the graph of $g(x) = \sqrt[3]{x} + k$ for each function.

a. $g(x) = \sqrt[3]{x} + 3$: The graph is a ______________ of the graph of $f(x)$ right/up/left/down 3 units.

b. $g(x) = \sqrt[3]{x} - 3$: The graph is a ______________ of the graph of $f(x)$ right/up/left/down 3 units.

Check your answers using a graphing calculator.

(C) Predict the effect of the parameter a on the graph of $g(x) = a\sqrt[3]{x}$ for each function.

a. $g(x) = 3\sqrt[3]{x}$: The graph is a ______________ stretch of the graph of $f(x)$ by a factor of ______.

b. $g(x) = \frac{1}{3}\sqrt[3]{x}$: The graph is a ______________ compression of the graph of $f(x)$ by a factor of ______.

c. $g(x) = -\frac{1}{3}\sqrt[3]{x}$: The graph is a ______________ compression of the graph of $f(x)$ by a factor of ______ as well as a ______________ across the ______________.

d. $g(x) = -3\sqrt[3]{x}$: The graph is a ______________ stretch of the graph of $f(x)$ by a factor of ______ as well as a ______________ across the ______________.

Check your answers using a graphing calculator.

(D) Predict the effect of the parameter, b, on the graph of $g(x) = \sqrt[3]{\frac{1}{b}x}$ for each function.

a. $g(x) = \sqrt[3]{\frac{1}{3}x}$: The graph is a ______________ stretch of the graph of $f(x)$ by a factor of ______.

b. $g(x) = \sqrt[3]{3x}$: The graph is a ______________ compression of the graph of $f(x)$ by a factor of ______.

c. $g(x) = \sqrt[3]{-\frac{1}{3}x}$: The graph is a ______________ stretch of the graph of $f(x)$ by a factor of ______ as well as a ______________ across the ______________.

d. $g(x) = \sqrt[3]{-3x}$: The graph is a ______________ compression of the graph of $f(x)$ by a factor of ______ as well as a ______________ across the ______________.

Check your answers using a graphing calculator.

Reflect

2. In $g(x) = \sqrt[3]{x - h} + k$ how do h and k effect the graphs of cube root functions?

Explain 1 Graphing Cube Root Functions

Transformations of the Cube Root Parent Function $f(x) = \sqrt[3]{x}$

Transformation	$f(x)$ Notation	Examples
Vertical translation	$f(x) + k$	$y = \sqrt[3]{x} + 3$ 3 units up $y = \sqrt[3]{x} - 4$ 4 units down
Horizontal translation	$f(x - h)$	$y = \sqrt[3]{x - 2}$ 2 units right $y = \sqrt[3]{x + 1}$ 1 units left
Vertical stretch/compression	$af(x)$	$y = 6\sqrt[3]{x}$ vertical stretch by a factor of 6 $y = \frac{1}{2}\sqrt[3]{x}$ vertical compression by a factor of $\frac{1}{2}$
Horizontal stretch/ compression	$f\left(\frac{1}{b}x\right)$	$y = \sqrt[3]{\frac{1}{5}x}$ horizontal stretch by a factor of 5 $y = \sqrt[3]{3x}$ horizontal compression by a factor of $\frac{1}{3}$
Reflection	$-f(x)$ $f(-x)$	$y = -\sqrt[3]{x}$ across x-axis $y = \sqrt[3]{-x}$ across y-axis

For the function $f(x) = a\sqrt[3]{x - h} + k$, (h, k) is the graph's point of symmetry. Use the values of a, h, and k to draw each graph. Note that the point $(1, 1)$ on the graph of the parent function becomes the point $(1 + h, a + k)$ on the graph of the given function.

For the function $f(x) = \sqrt[3]{\frac{1}{b}(x - h)} + k$, (h, k) remains the graph's point of symmetry. Note that the point $(1, 1)$ on the graph of the parent function becomes the point $(b + h, 1 + k)$ on the graph of the given function.

Example 1 **Graph the cube root functions.**

(A) Graph $g(x) = 2\sqrt[3]{x-3} + 5$.

The transformations of the graph of $f(x) = \sqrt[3]{x}$ that produce the graph of $g(x)$ are:

- a vertical stretch by a factor of 2
- a translation of 3 units to the right and 5 units up

Choose points on $f(x) = \sqrt[3]{x}$ and find the transformed corresponding points on $g(x) = 2\sqrt[3]{x-3} + 5$.

Graph $g(x) = 2\sqrt[3]{x-3} + 5$ using the transformed points.

$f(x) = \sqrt[3]{x}$	$g(x) = 2\sqrt[3]{x-3} + 5$
(−8, −2)	(−5, 1)
(−1, −1)	(2, 3)
(0, 0)	(3, 5)
(1, 1)	(4, 7)
(8, 2)	(11, 9)

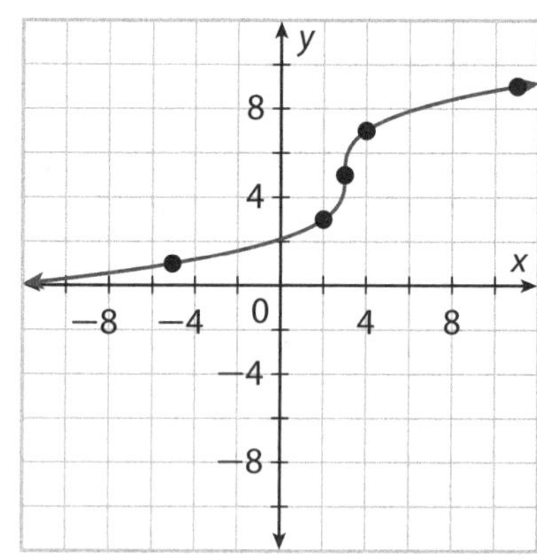

(B) Graph $g(x) = \sqrt[3]{\frac{1}{2}(x-10)} + 4$.

The transformations of the graph of $f(x) = \sqrt[3]{x}$ that produce the graph of $g(x)$ are:

- a horizontal stretch by a factor of 2
- a translation of 10 units to the right and 4 units up

Choose points on $f(x) = \sqrt[3]{x}$ and find the transformed corresponding points on $g(x) = \sqrt[3]{\frac{1}{2}(x-10)} + 4$.

Graph $g(x) = \sqrt[3]{\frac{1}{2}(x-10)} + 4$ using the transformed points.

$f(x) = \sqrt[3]{x}$	$g(x) = \sqrt[3]{\frac{1}{2}(x-10)} + 4$
(−8, −2)	
(−1, −1)	
(0, 0)	
(1, 1)	
(8, 2)	

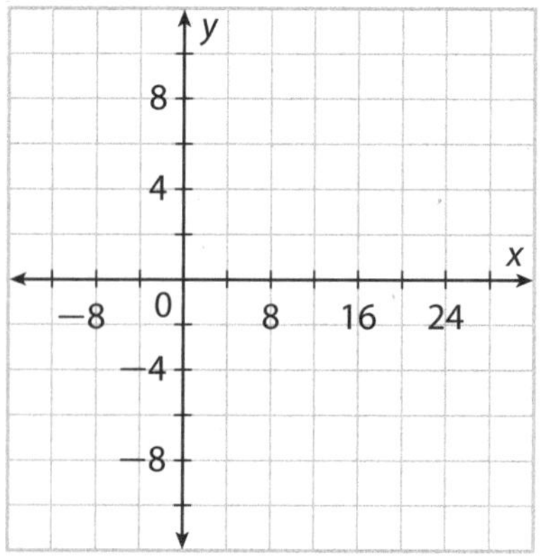

Your Turn

Graph the cube root functions.

3. Graph $g(x) = \sqrt[3]{x-3} + 6$.

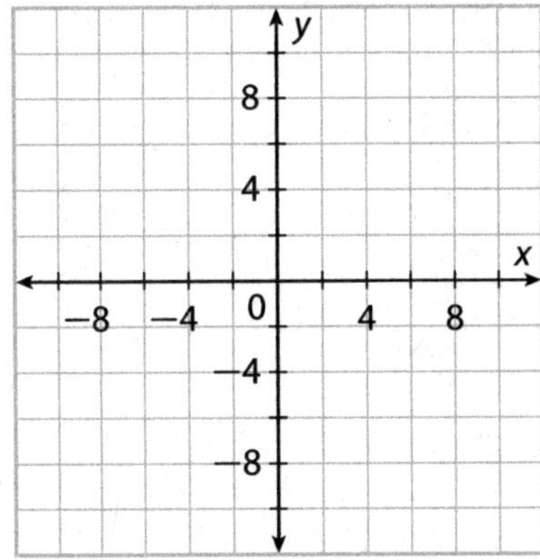

$f(x) = \sqrt[3]{x}$	$g(x) = \sqrt[3]{x-3} + 6$
$(-8, -2)$	
$(-1, -1)$	
$(0, 0)$	
$(1, 1)$	
$(8, 2)$	

4. Graph $g(x) = \sqrt[3]{x+3} - 7$.

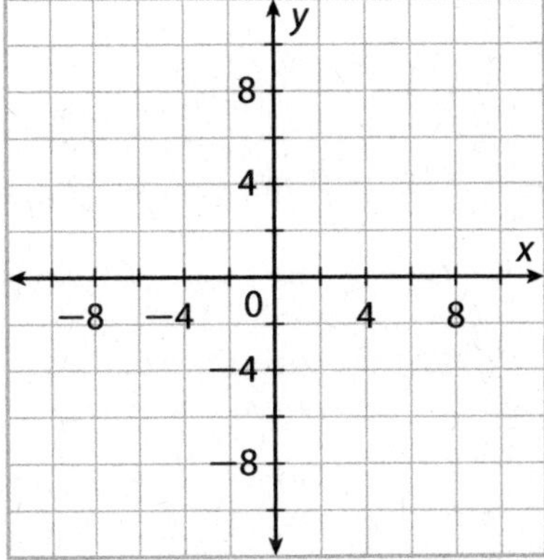

$f(x) = \sqrt[3]{x}$	$g(x) = \sqrt[3]{x+3} - 7$
$(-8, -2)$	
$(-1, -1)$	
$(0, 0)$	
$(1, 1)$	
$(8, 2)$	

Explain 2 Writing Cube Root Functions

Given the graph of the transformed function $g(x) = a\sqrt[3]{\frac{1}{b}(x-h)} + k$, you can determine the values of the parameters by using the reference points $(-1, 1)$, $(0, 0)$, and $(1, 1)$ that you used to graph $g(x)$ in the previous example.

Example 2 **For the given graphs, write a cube root function.**

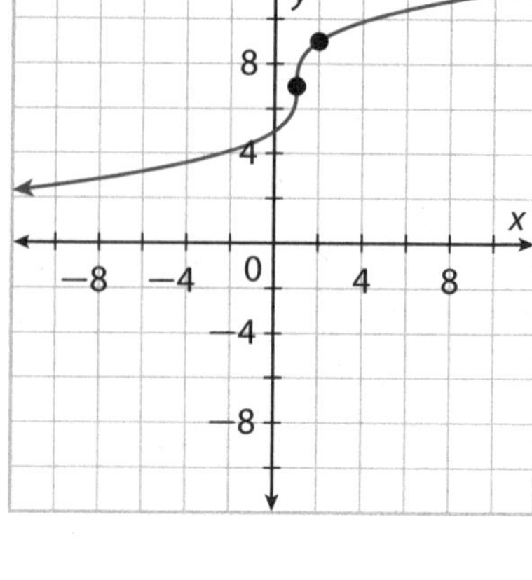

Ⓐ Write the function in the form $g(x) = a\sqrt[3]{x-h} + k$.

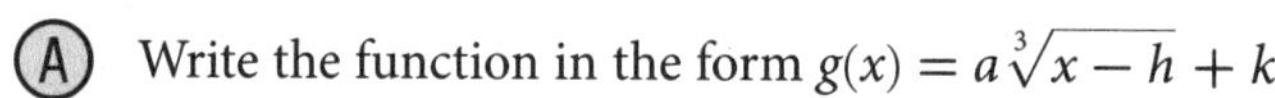

Identify the values of a, h, and k.

Identify the values of h and k from the point of symmetry.

$(h, k) = (1, 7)$, so $h = 1$ and $k = 7$.

Identify the value of a from either of the other two reference points $(-1, 1)$ or $(1, 1)$.

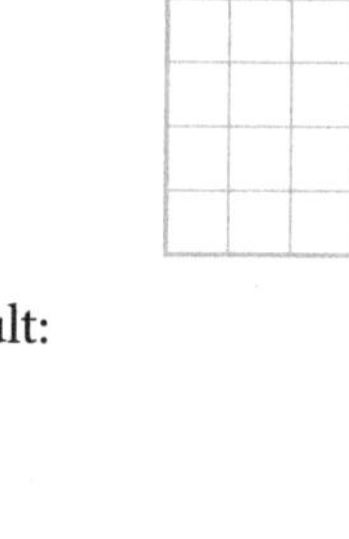

The image of the reference point $(1, 1)$ has general coordinates $(h + 1, a + k)$. Substituting 1 for h and 7 for k and setting the general coordinates equal to the actual coordinates gives this result:

$(h + 1, a + k) = (2,\ a + 7) = (2, 9)$, so $a = 2$.

$a = 2$ $\qquad\qquad$ $h = 1$ $\qquad\qquad$ $k = 7$

The function is $g(x) = 2\sqrt[3]{x-1} + 7$.

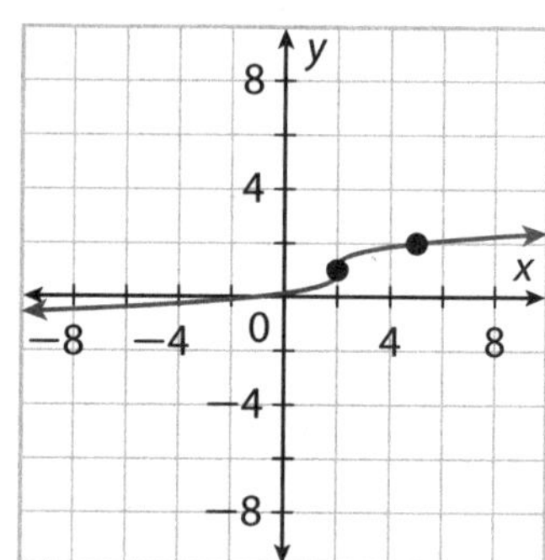

Ⓑ Write the function in the form $g(x) = \sqrt[3]{\frac{1}{b}(x-h)} + k$.

Identify the values of b, h, and k.

Identify the values of h and k from the point of symmetry.

$(h, k) = \left(2, \square\right)$ so $h = 2$ and $k = \square$.

Identify the value of b from either of the other two reference points.

The rightmost reference point has general coordinates $(b + h, 1 + k)$. Substituting 2 for h and ____ for k and setting the general coordinates equal to the actual coordinates gives this result:

$\left(b + h, 1 + \square\right) = \left(b + 2, \square\right) = (5, 2)$, so $b = \square$.

$b = \square$ $\qquad\qquad$ $h = \square$ $\qquad\qquad$ $k = \square$

The function is $g(x) =$ ________.

Your Turn

For the given graphs, write a cube root function.

5. Write the function in the form $g(x) = a\sqrt[3]{x - h} + k$.

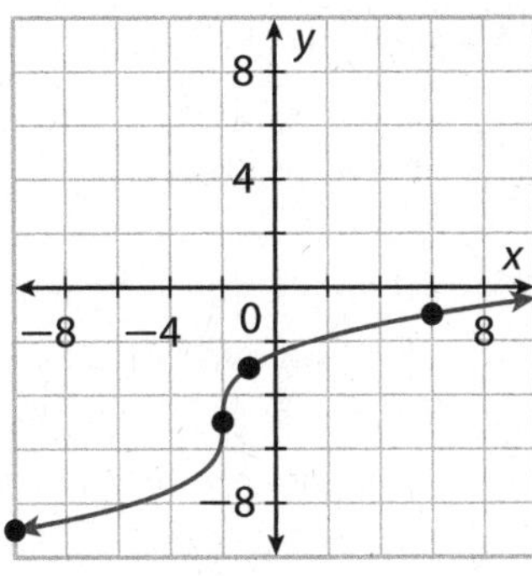

6. Write the function in the form $g(x) = \sqrt[3]{\frac{1}{b}(x - h)} + k$.

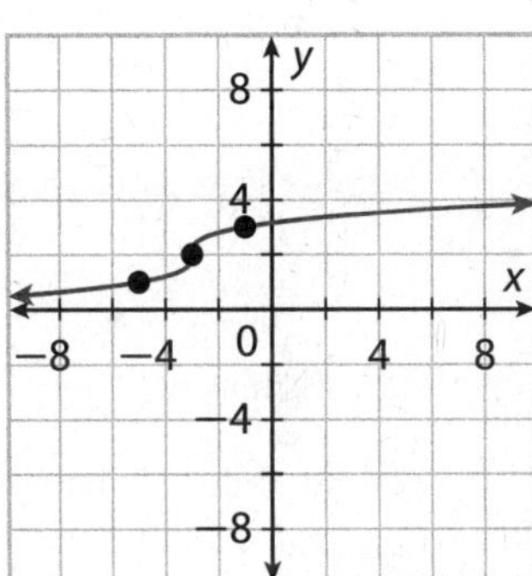

Explain 3 Modeling with Cube Root Functions

You can use cube root functions to model real-world situations.

Example 3

(A) The shoulder height h (in centimeters) of a particular elephant is modeled by the function $h(t) = 62.1\sqrt[3]{t} + 76$, where t is the age (in years) of the elephant. Graph the function and examine its average rate of change over the equal t-intervals $(0, 20)$, $(20, 40)$, and $(40, 60)$. What is happening to the average rate of change as the t-values of the intervals increase? Use the graph to find the height when $t = 35$.

Graph $h(t) = 62.1\sqrt[3]{t} + 76$.

The graph is the graph of $f(x) = \sqrt[3]{x}$ translated up 76 and stretched vertically by a factor of 62.1. Graph the transformed points $(0, 76)$, $(8, 200.2)$, $(27, 262.3)$, and $(64, 324.4)$. Connect the points with a smooth curve.

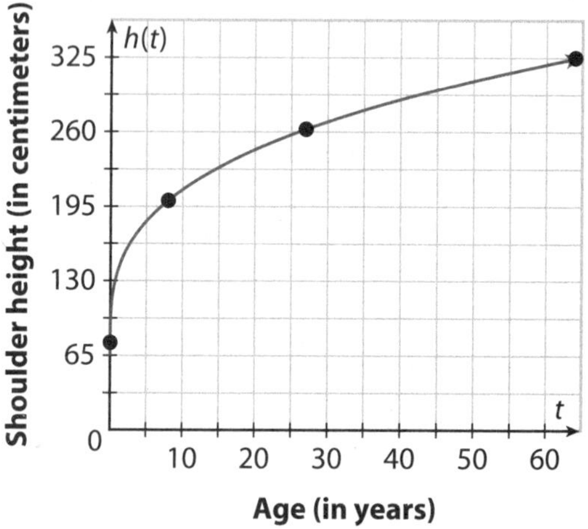

First interval:

$$\text{Average Rate of change} \approx \frac{244.6 - 76}{20 - 0}$$
$$= 8.43$$

Second interval:

$$\text{Average Rate of change} \approx \frac{288.4 - 244.6}{40 - 20}$$
$$= 2.19$$

Third interval:

$$\text{Average Rate of change} \approx \frac{319.1 - 288.4}{60 - 40}$$
$$= 1.54$$

The average rate of change is becoming less.

Drawing a vertical line up from 35 gives a value of about 280 cm.

(B) The velocity of a 1400-kilogram car at the end of a 400-meter run is modeled by the function $v = 15.2\sqrt[3]{p}$, where v is the velocity in kilometers per hour and p is the power of its engine in horsepower. Graph the function and examine its average rate of change over the equal p-intervals $(0,60)$, $(60,120)$, and $(120,180)$. What is happening to the average rate of change as the p-values of the intervals increase? Use the function to find the velocity when p is 100 horsepower.

Graph $v = 15.2\sqrt[3]{p}$.

The graph is the graph of $f(x) = \sqrt[3]{x}$ stretched ________ by a factor of 15.2. Graph the transformed points $(0, 0)$, $(8, \_\_\_\_)$, $(27, \_\_\_\_)$, $(64, \_\_\_\_)$, $(125, \_\_\_\_)$, and $(216, \_\_\_\_)$.

Connect the points with a smooth curve.

v
90
80
70
60
50
40
30
20
10
0
Velocity (km/h)
20 60 100 140 180
p
Power (hp)

The average rate of change over the interval $(0, 60)$ is

$\frac{\square - \square}{60 - 0}$ which is about ________.

The average rate of change over the interval $(60, 120)$ is $\frac{\square - \square}{120 - 60}$ which is about ________.

The average rate of change over the interval $(120, 180)$ is $\frac{\square - \square}{180 - 120}$ which is about ________.

The average rate of change is becoming ________.

Substitute $p = 100$ in the function.

$v = 15.2\sqrt[3]{p}$

$v = 15.2\sqrt[3]{\square}$

$v \approx 15.2(\square)$

$v \approx \square$

The velocity is about ________ km/h.

Your Turn

7. The fetch is the length of water over which the wind is blowing in a certain direction. The function $s(f) = 7.1\sqrt[3]{f}$, relates the speed of the wind s in kilometers per hour to the fetch f in kilometers. Graph the function and examine its average rate of change over the intervals $(20, 80)$, $(80, 140)$, and $(140, 200)$. What is happening to the average rate of change as the f-values of the intervals increase? Use the function to find the speed of the wind when $f = 64$.

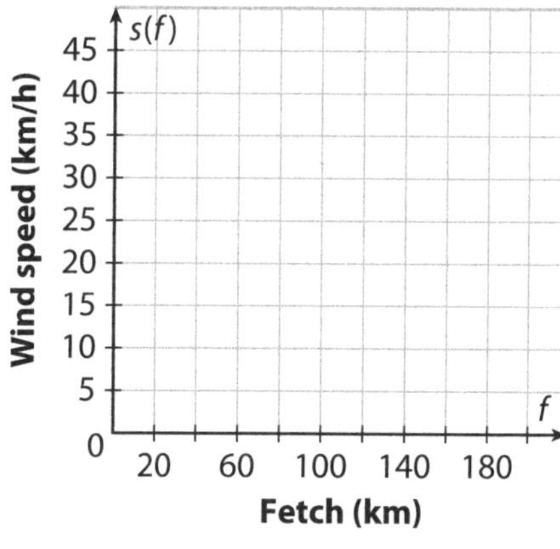

Elaborate

8. Discussion Why is the domain of $f(x) = \sqrt[3]{x}$ all real numbers?

9. Identify which transformations (stretches or compressions, reflections, and translations) of $f(x) = \sqrt[3]{x}$ change the following attributes of the function.

a. Location of the point of symmetry

b. Symmetry about a point

10. Essential Question Check-In How do parameters a, b, h, and k effect the graphs of $f(x) = a\sqrt[3]{(x - h)} + k$ and $g(x) = \sqrt[3]{\frac{1}{b}(x - h)} + k$?

Evaluate: Homework and Practice

- Online Homework
- Hints and Help
- Extra Practice

1. Graph the function $g(x) = \sqrt[3]{x} + 3$. Identify the domain and range of the function.

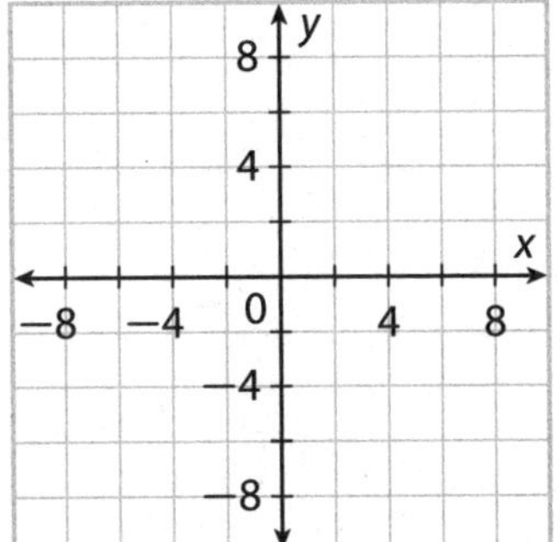

2. Graph the function $g(x) = \sqrt[3]{x} - 5$. Identify the domain and range of the function.

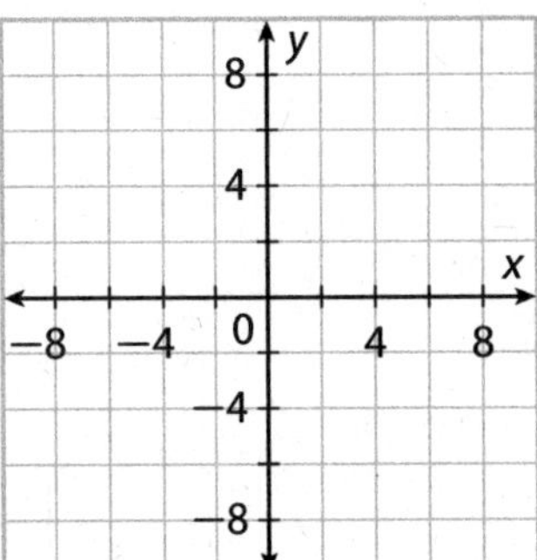

Describe how the graph of the function compares to the graph of $f(x) = \sqrt[3]{x}$.

3. $g(x) = \sqrt[3]{x} + 6$

4. $g(x) = \sqrt[3]{x - 5}$

5. $g(x) = \frac{1}{3}\sqrt[3]{-x}$

6. $g(x) = \sqrt[3]{5x}$

7. $g(x) = -2\sqrt[3]{x} + 3$

8. $g(x) = \sqrt[3]{x + 4} - 3$

Graph the cube root functions.

9. $g(x) = 3\sqrt[3]{x+4}$

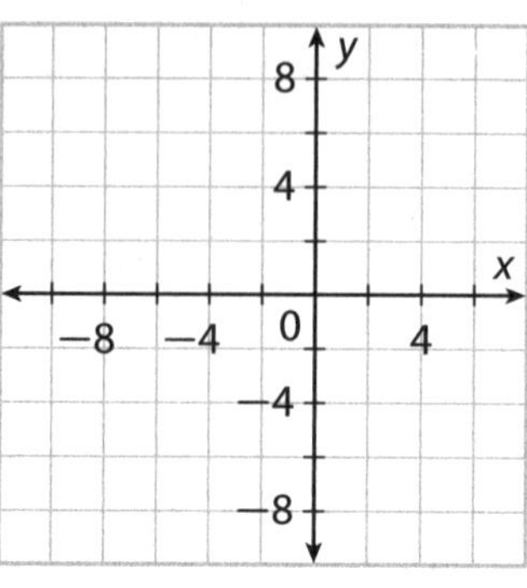

10. $g(x) = 2\sqrt[3]{x} + 3$

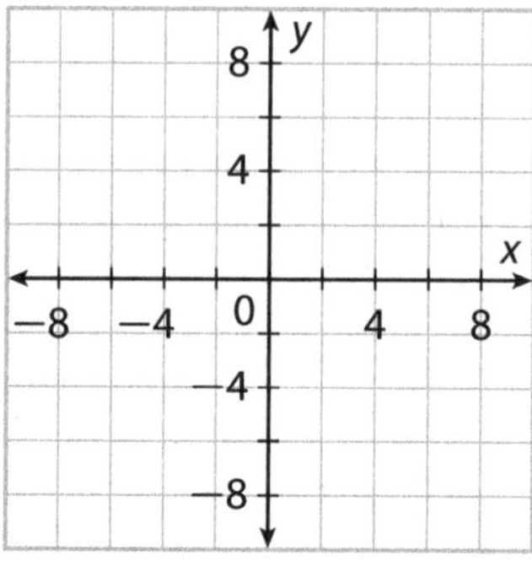

11. $g(x) = \sqrt[3]{x-3} + 2$

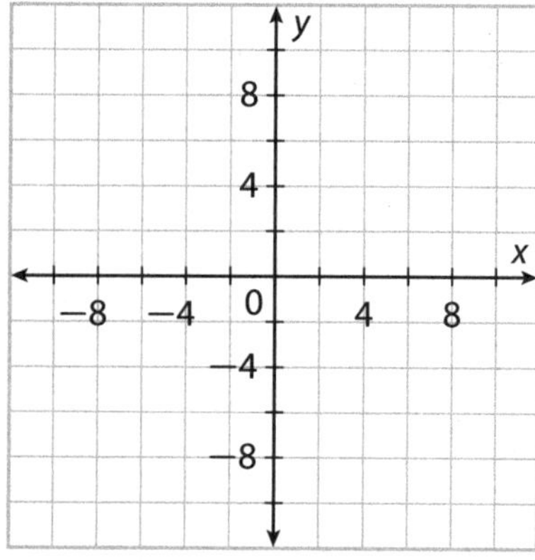

For the given graphs, write a cube root function.

12. Write the function in the form $g(x) = a\sqrt[3]{x - h} + k$.

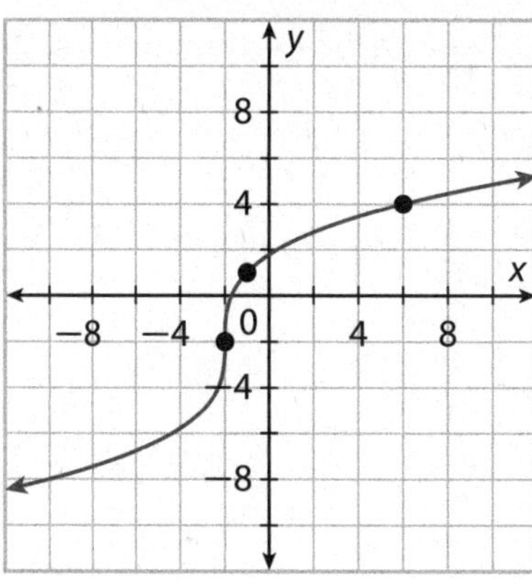

13. Write the function in the form $g(x) = a\sqrt[3]{x - h} + k$.

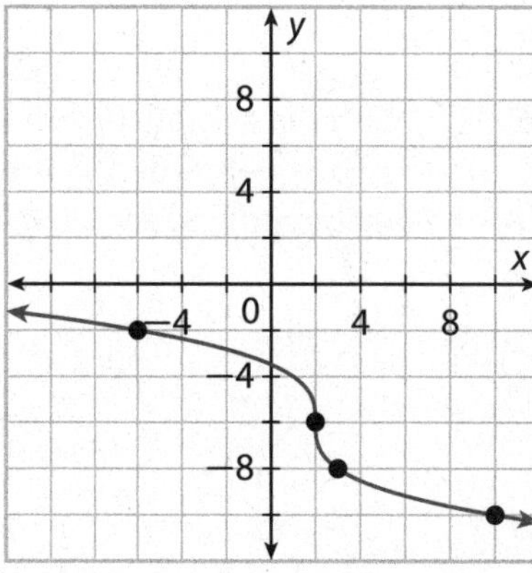

14. Write the function in the form $g(x) = \sqrt[3]{\frac{1}{b}(x - h)} + k$.

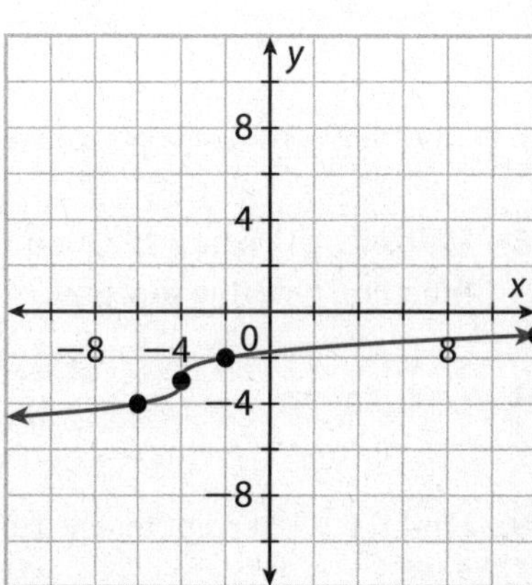

15. The length of the side of a cube is modeled by $s = \sqrt[3]{V}$. Graph the function. Use the graph to find s when $V = 48$.

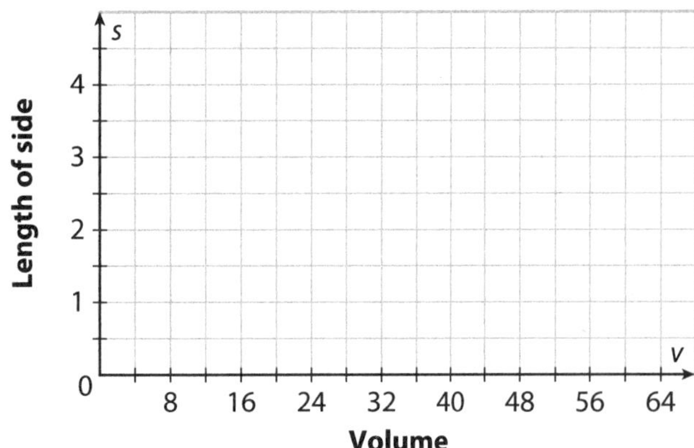

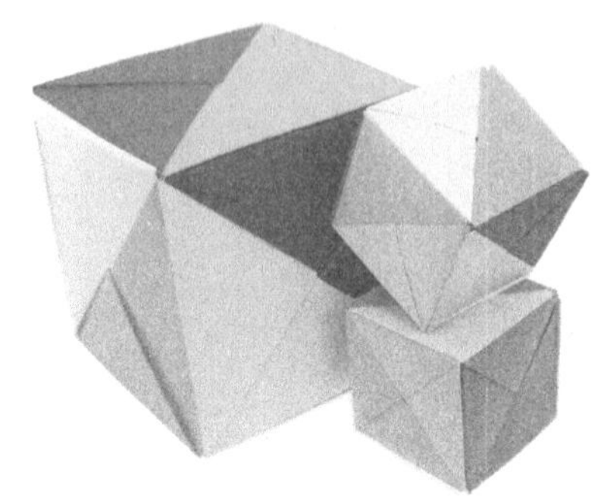

16. The radius of a stainless steel ball can be modeled by $r(m) = 0.31\sqrt[3]{m}$, where m is the mass of the ball. Use the function to find r when $m = 125$.

17. Describe the steps for graphing $g(x) = \sqrt[3]{x + 8} - 11$.

18. **Modeling** Write a situation that can be modeled by a cube root function. Give the function.

19. Find the y-intercept for the function $y = a\sqrt[3]{x - h} + k$.

20. Find the x-intercept for the function $y = a\sqrt[3]{x - h} + k$.

21. Describe the translation(s) used to get $g(x) = \sqrt[3]{x - 9} + 12$ from $f(x) = \sqrt[3]{x}$. Select all that apply.

A. translated 9 units right

B. translated 9 units left

C. translated 9 units up

D. translated 9 units down

E. translated 12 units right

F. translated 12 units left

G. translated 12 units up

H. translated 12 units down

H.O.T. Focus on Higher Order Thinking

22. Explain the Error Tim says that to graph $g(x) = \sqrt[3]{x - 6} + 3$, you need to translate the graph of $f(x) = \sqrt[3]{x}$ 6 units to the left and then 3 units up. What mistake did he make?

23. Communicate Mathematical Ideas Why does the square root function have a restricted domain but the cube root function does not?

24. Justify Reasoning Does a horizontal translation and a vertical translation of the function $f(x) = \sqrt[3]{x}$ affect the function's domain or range? Explain.

Lesson Performance Task

The side length of a 243-gram copper cube is 3 centimeters. Use this information to write a model for the radius of a copper sphere as a function of its mass. Then, find the radius of a copper sphere with a mass of 50 grams. How would changing the material affect the function?

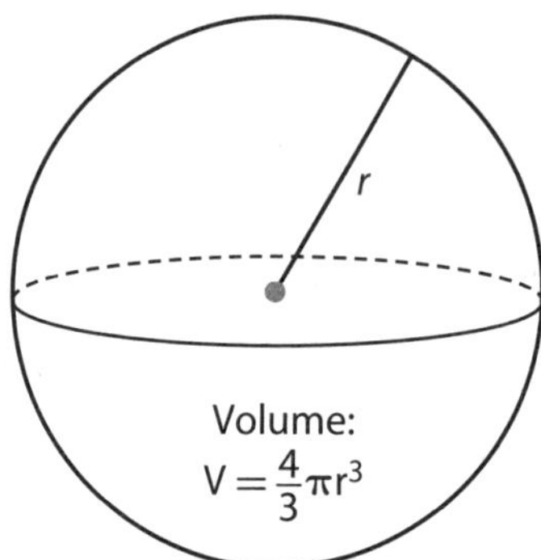

Radical Functions

Essential Question: How can you use radical functions to solve real-world problems?

Key Vocabulary
cube-root function *(función de raíz cúbica)*
index *(índice)*
inverse function *(función inversa)*
square-root function *(función de raíz cuadrada)*

KEY EXAMPLE (Lesson 11.2)

Graph $y = -\sqrt{x-3} + 2$. Describe the domain and range.

Sketch the graph of $y = -\sqrt{x}$.

It begins at the origin and passes through $(1, -1)$.

For $y = -\sqrt{x-3} + 2$, $h = 3$ and $k = 2$.
Shift the graph of $y = -\sqrt{x}$ right 3 units and up 2 units. The graph begins at $(3, 2)$ and passes through $(4, 1)$.

Domain: $\{x|x \geq 3\}$ Range: $\{y|y \leq 2\}$

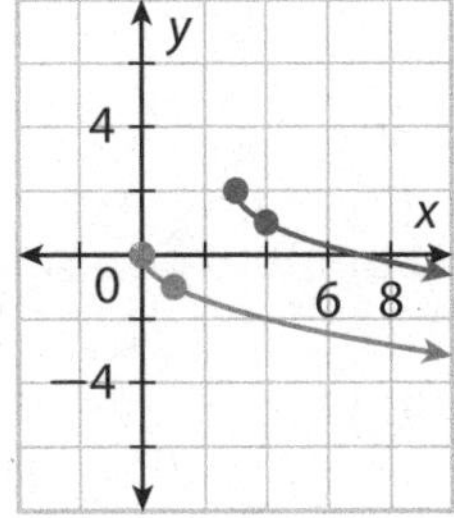

KEY EXAMPLE (Lesson 11.4)

Graph $y = \sqrt[3]{x+2} - 4$.

Sketch the graph of $y = \sqrt[3]{x}$.

It passes through $(-1, -1)$, $(0, 0)$, and $(1, 1)$.

For $y = \sqrt[3]{x+2} - 4$, $h = -2$ and $k = -4$.
Shift the graph of $y = \sqrt[3]{x}$ left 2 units and down 4 units. The graph passes through $(-3, -5)$, $(-2, -4)$, and $(-1, -3)$.

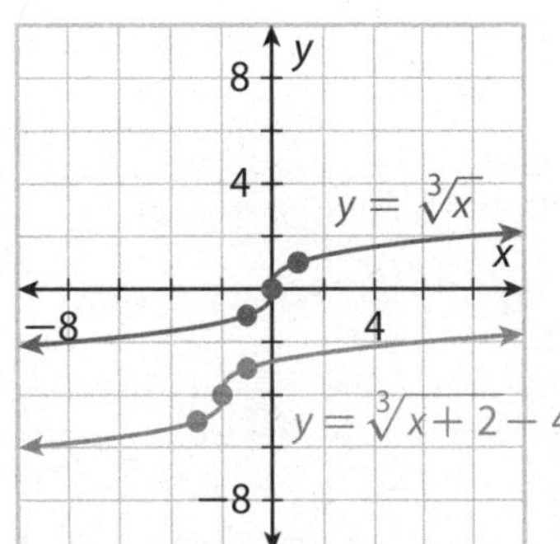

EXERCISES

Find the inverse of each function. Restrict the domain where necessary. *(Lesson 11.1)*

1. $f(x) = 16x^2$

2. $f(x) = x^3 - 20$

Identify the transformations of the graph $f(x) = \sqrt{x}$ that produce the graph of the function. *(Lesson 11.2)*

3. $g(x) = -\sqrt{4x}$

4. $h(x) = \frac{1}{2}\sqrt{x} + 1$

Use the vertex and a second point to fit a square root function to the data. *(Lesson 11.3)*

5. Points: $(3, 5.2)$, $(4, 10)$, $(5, 12.9)$, $(6, 13.1)$, $(7, 15.3)$, $(8, 15.6)$

Identify the transformations of the graph $f(x) = \sqrt[3]{x}$ that produce the graph of the function. *(Lesson 11.4)*

6. $g(x) = 4\sqrt[3]{x}$

7. $h(x) = \sqrt[3]{x - 5} + 3$

MODULE PERFORMANCE TASK

We Have Liftoff!

A rocket scientist is designing a rocket to visit the planets in the solar system. The velocity that is needed to escape a planet's gravitational pull is called the escape velocity. The escape velocity depends on the planet's radius and its mass, according to the equation $V_{escape} = \sqrt{2gR}$, where R is the radius and g is the gravitational constant for the particular planet. The rocket's maximum velocity is exactly double Earth's escape velocity. For which planets will the rocket have enough velocity to escape the planet's gravity?

Planet	Radius (m)	Mass (kg)	g (m/s^2)
Mercury	2.43×10^6	3.20×10^{23}	3.61
Venus	6.07×10^6	4.88×10^{24}	8.83
Mars	3.38×10^6	6.42×10^{23}	3.75
Jupiter	6.98×10^7	1.90×10^{27}	26.0
Saturn	5.82×10^7	5.68×10^{26}	11.2
Uranus	2.35×10^7	8.68×10^{25}	10.5
Neptune	2.27×10^7	1.03×10^{26}	13.3

Begin by listing in the space below any additional information you will need to solve the problem. Then use your own paper to complete the task. Be sure to write down all your data and assumptions. Then use graphs, numbers, words, or algebra to explain how you reached your conclusions.

Ready to Go On?

11.1–11.4 Radical Functions

- Online Homework
- Hints and Help
- Extra Practice

Find the inverse of each function. State any restrictions on the domain. *(Lesson 11.1)*

1. $f(x) = x^2 + 9$

2. $f(x) = -7x^3$

Identify the transformations of the graph $f(x) = \sqrt{x}$ or $h(x) = \sqrt[3]{x}$ that produce the graph of the function. *(Lessons 11.2, 11.4)*

3. $g(x) = \frac{1}{3}\sqrt{x - 5} - 4$

4. $g(x) = \sqrt[3]{4x} + 3$

Solve. *(Lesson 11.3)*

5. Use the vertex and a second point to fit a square root function to the data.

x	y
16.7	8
16.8	9
17.1	10
17.5	11
17.9	12

ESSENTIAL QUESTION

6. How do you use a parent square root or cube root function to graph a transformation of the function? *(Lessons 11.2, 11.4)*

MODULE 11
MIXED REVIEW

Assessment Readiness

1. Which function is graphed below?

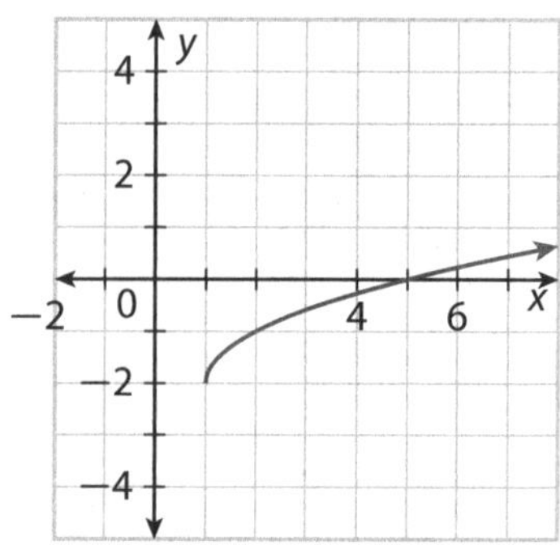

A. $y = \sqrt{x-1} - 2$

B. $y = \sqrt{x-1} + 2$

C. $y = \sqrt{x+1} - 2$

D. $y = \sqrt{x+1} + 2$

2. What is the inverse of $f(x) = x^3 - 16$?

A. $f^{-1}(x) = \sqrt[3]{x-16}$

B. $f^{-1}(x) = \sqrt[3]{x+16}$

C. $f^{-1}(x) = \sqrt[3]{x} + 16$

D. $f^{-1}(x) = \sqrt[3]{x} - 16$

3. A plane's average speed when flying from one city to another is 550 mi/h and is 430 mi/h on the return flight. To the nearest mile per hour, what is the plane's average speed for the entire trip?

A. 360 mi/h

B. 453 mi/h

C. 483 mi/h

D. 490 mi/h

4. The kinetic energy E (in joules) of a 1250 kilogram compact car is given by the equation $E = 625s^2$ where s is the speed of the car (in meters per second). Write an inverse model that gives the speed of the car as a function of its kinetic energy. If the kinetic energy doubles, will the speed double? Explain why or why not.

Radical Expressions and Equations

MODULE 12

Essential Question: How can you use radical expressions and equations to solve real-world problems?

TEKS

REAL WORLD VIDEO
A field biologist studying howler monkeys can use radical functions to calculate sound intensity, which decreases faster than linearly with distance.

MODULE PERFORMANCE TASK PREVIEW

Don't Disturb the Neighbors!

The loudness of a sound is subjective and depends on the listener's sensitivity to the frequencies of the sound waves. An objective measure, sound intensity, can be used to measure sounds. Sound intensity decreases the farther you get from the source of a sound. How far away do your neighbors have to be so that a loud band does not bother them? Let's find out!

Are YOU Ready?

Complete these exercises to review skills you will need for this module.

Personal Math Trainer

- Online Homework
- Hints and Help
- Extra Practice

Exponents

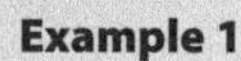

Example 1 Simplify $(x^3)^2 + x \cdot x^3 + 3x^4$.

$(x^3)^2 + x \cdot x^3 + 3x^4 = (x^3)(x^3) + x \cdot x^3 + 3x^4$ Start with the raised power.

$= x^{3+3} + x^{1+3} + 3x^4$ Add exponents.

$= x^6 + x^4 + 3x^4$ Simplify.

$= x^6 + 4x^4$ Add like terms.

Simplify each expression.

1. $(-x^5)^2$ ________ **2.** $(3x^2)^3 - x^4 \cdot x^2$ ________ **3.** $3x(2x)^2$ ________

Inverse Linear Functions

Example 2 Write the inverse function of $y = 10x - 4$.

$y - 4 = 10x$ Isolate the x-term.

$\frac{y-4}{10} = \frac{10x}{10}$ Divide.

$\frac{y-4}{10} = x$

$\frac{x-4}{10} = y$ Switch x and y.

The inverse function of $y = 10x - 4$ is $y = \frac{x-4}{10}$.

Write the inverse function.

4. $y = 3x + 1$ **5.** $y = 2(x - 9)$ **6.** $y = \frac{1}{4}(3x + 4)$

________ ________ ________

Rational and Radical Exponents

Example 3 Write $\sqrt[9]{a^3}$ using a rational exponent.

$\sqrt[9]{a^3} = a^{\frac{3}{9}} = a^{\frac{1}{3}}$

Write each radical expression using a rational exponent.

7. $\sqrt[2]{x^5}$ ________ **8.** $\sqrt[4]{a^2b}$ ________ **9.** $\sqrt[4]{p^8q^2}$ ________

Name ______________________ Class ______________ Date __________

12.1 Radical Expressions and Rational Exponents

Resource Locker

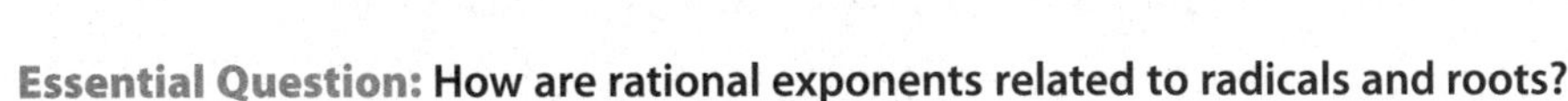

Essential Question: How are rational exponents related to radicals and roots?

TEKS A2.7.G Rewrite radical expressions that contain variables to equivalent forms.

Explore Defining Rational Exponents in Terms of Roots

Remember that a number a is an nth root of a number b if $a^n = b$. As you know, a square root is indicated by $\sqrt{}$ and a cube root by $\sqrt[3]{}$. In general, the nth root of a real number a is indicated by $\sqrt[n]{a}$, where n is the **index** of the radical and a is the radicand. (Note that when a number has more than one real root, the radical sign indicates only the principal, or positive, root.)

A *rational exponent* is an exponent that can be expressed as $\frac{m}{n}$, where m is an integer and n is a natural number. You can use the definition of a root and properties of equality and exponents to explore how to express roots using rational exponents.

(A) How can you express a square root using an exponent? That is, if $\sqrt{a} = a^m$, what is m?

Given	$\sqrt{a} = a^m$
Square both sides.	$(\sqrt{a})^2 = (a^m)^2$
Definition of square root	$\square = (a^m)^2$
Power of a power property	$a = a^{\square}$
Definition of first power	$a^{\square} = a^{2m}$
The bases are the same, so equate exponents.	$\square = \square$
Solve.	$m = \square$
So,	$\sqrt{a} = a^{\square}$.

(B) How can you express a cube root using an exponent? That is, if $\sqrt[3]{a} = a^m$, what is m?

Given	$\sqrt[3]{a} = a^m$
Cube both sides.	$(\sqrt{a})^3 = (a^m)^3$
Definition of cube root	$\square = \square$
Power of a power property	$\square = \square$

Definition of first power $\square = \square$

The bases are the same, so equate exponents. $\square = \square$

Solve. $m = \square$

So, $\sqrt[3]{a} = a^{\square}$.

Reflect

1. **Discussion** Examine the reasoning in Steps A and B. Can you apply the same reasoning for any nth root, $\sqrt[n]{a}$, where n is a natural number? Explain. What can you conclude?

2. For a positive number a, under what condition on n will there be only one real nth root? two real nth roots? Explain.

3. For a negative number a, under what condition on n will there be no real nth roots? one real nth root? Explain.

Explain 1 Translating Between Radical Expressions and Rational Exponents

In the Explore, you found that a rational exponent $\frac{m}{n}$ with $m = 1$ represents an nth root, or that $a^{\frac{1}{n}} = \sqrt[n]{a}$ for positive values of a. This is also true for negative values of a when the index is odd. When $m \neq 1$, you can think of the numerator m as the power and the denominator n as the root. The following ways of expressing the exponent $\frac{m}{n}$ are equivalent.

Rational Exponents

For any natural number n, integer m, and real number a when the nth root of a is real:

Words	Numbers	Algebra
The exponent $\frac{m}{n}$ indicates the mth power of the nth root of a quantity.	$27^{\frac{2}{3}} = (\sqrt[3]{27})^2 = 3^2 = 9$	$a^{\frac{m}{n}} = (\sqrt[n]{a})^m$
The exponent $\frac{m}{n}$ indicates the nth root of the mth power of a quantity.	$4^{\frac{3}{2}} = \sqrt{4^3} = \sqrt{64} = 8$	$a^{\frac{m}{n}} = \sqrt[n]{a^m}$

Notice that you can evaluate each example in the "Numbers" column using the equivalent definition.

$$27^{\frac{2}{3}} = \sqrt[3]{27^2} = \sqrt[3]{729} = 9 \qquad 4^{\frac{3}{2}} = \left(\sqrt{4}\right)^3 = 2^3 = 8$$

Example 1 **Translate radical expressions into expressions with rational exponents, and vice versa. Simplify numerical expressions when possible. Assume all variables are positive.**

(A) **a.** $(-125)^{\frac{4}{3}}$ **b.** $x^{\frac{11}{8}}$ **c.** $\sqrt[5]{6^4}$ **d.** $\sqrt[4]{x^3}$

a. $(-125)^{\frac{4}{3}} = \left(\sqrt[3]{-125}\right)^4 = (-5)^4 = 625$

b. $x^{11/8} = \sqrt[8]{x^{11}}$ or $\left(\sqrt[8]{x}\right)^{11}$

c. $\sqrt[5]{6^4} = 6^{\frac{4}{5}}$

d. $\sqrt[4]{x^3} = x^{\frac{3}{4}}$

(B) **a.** $\left(\frac{81}{16}\right)^{\frac{3}{4}}$ **b.** $(xy)^{\frac{5}{3}}$ **c.** $\sqrt[3]{11^6}$ **d.** $\sqrt[3]{\left(\frac{2x}{y}\right)^5}$

a. $\left(\frac{81}{16}\right)^{\frac{3}{4}} = \left(\sqrt[\square]{\frac{81}{16}}\right)^{\square} = \left(\square\right)^3 = \square$

b. $(xy)^{\frac{5}{3}} = \sqrt[\square]{(xy)^{\square}}$ or $\left(\sqrt[\square]{xy}\right)^{\square}$

c. $\sqrt[3]{11^6} = 11^{\square} = 11^{\square} = \square$

d. $\sqrt[3]{\left(\frac{2x}{y}\right)^5} = \left(\frac{2x}{y}\right)^{\square}$

Reflect

4. How can you use a calculator to show that evaluating $0.001728^{\frac{4}{3}}$ as a power of a root and as a root of a power are equivalent methods?

Your Turn

5. Translate radical expressions into expressions with rational exponents, and vice versa. Simplify numerical expressions when possible. Assume all variables are positive.

a. $\left(-\frac{32}{243}\right)^{\frac{2}{5}}$

b. $(3y)^{\frac{b}{c}}$

c. $\sqrt[3]{0.5^9}$

d. $\left(\sqrt[u]{st}\right)^v$

Explain 2 Modeling with Power Functions

The following functions all involve a given power of a variable.

$A = \pi r^2$ (area of a circle)

$V = \frac{4}{3}\pi r^3$ (volume of a sphere)

$T = 1.11 \cdot L^{\frac{1}{2}}$ (the time T in seconds for a pendulum of length L feet to complete one back-and-forth swing)

These are all examples of *power functions.* A power function has the form $y = ax^b$ where a is a real number and b is a rational number.

Example 2 **Solve each problem by modeling with power functions.**

Ⓐ **Biology** The function $R = 73.3\sqrt[4]{M^3}$, known as Kleiber's law, relates the basal metabolic rate R in Calories per day burned and the body mass M of a mammal in kilograms. The table shows typical body masses for some members of the cat family.

Typical Body Mass	
Animal	**Mass (kg)**
House cat	4.5
Cheetah	55
Lion	170

a. Rewrite the formula with a rational exponent.

b. What is the value of R for a cheetah to the nearest 50 Calories?

c. From the table, the mass of the lion is about 38 times that of the house cat. Is the lion's metabolic rate more or less than 38 times the cat's rate? Explain.

```
73.3*55^(3/4)
          1480.389485
```

a. Because $\sqrt[n]{a^m} = a^{\frac{m}{n}}$, $\sqrt[4]{M^3} = M^{\frac{3}{4}}$, so the formula is $R = 73.3M^{\frac{3}{4}}$.

b. Substitute 55 for M in the formula and use a calculator.

The cheetah's metabolic rate is about 1500 Calories.

c. Less; find the ratio of R for the lion to R for the house cat.

$$\frac{73.3(170)^{\frac{3}{4}}}{73.3(4.5)^{\frac{3}{4}}} = \frac{170^{\frac{3}{4}}}{4.5^{\frac{3}{4}}} \approx \frac{47.1}{3.1} \approx 15$$

The metabolic rate for the lion is only about 15 times that of the house cat.

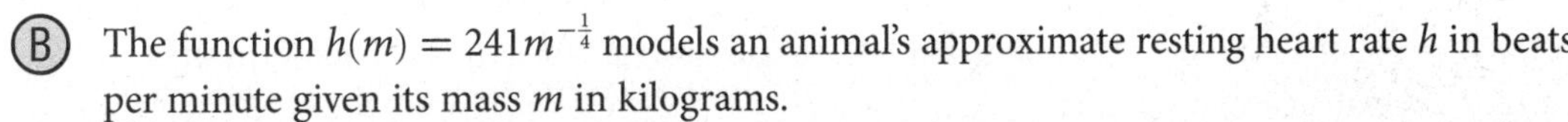

Ⓑ The function $h(m) = 241m^{-\frac{1}{4}}$ models an animal's approximate resting heart rate h in beats per minute given its mass m in kilograms.

a. A common shrew has a mass of only about 0.01 kg. To the nearest 10, what is the model's estimate for this shrew's resting heart rate?

b. What is the model's estimate for the resting heart rate of an American elk with a mass of 300 kg?

c. Two animal species differ in mass by a multiple of 10. According to the model, about what percent of the smaller animal's resting heart rate would you expect the larger animal's resting heart rate to be?

a. Substitute ______ for m in the formula and use a calculator.

$$h(m) = 241\left(\square\right)^{-\frac{1}{4}} \approx \square$$

The model estimates the shrew's resting heart rate to be is about ______ beats per minute.

b. Substitute ______ for m in the formula and use a calculator.

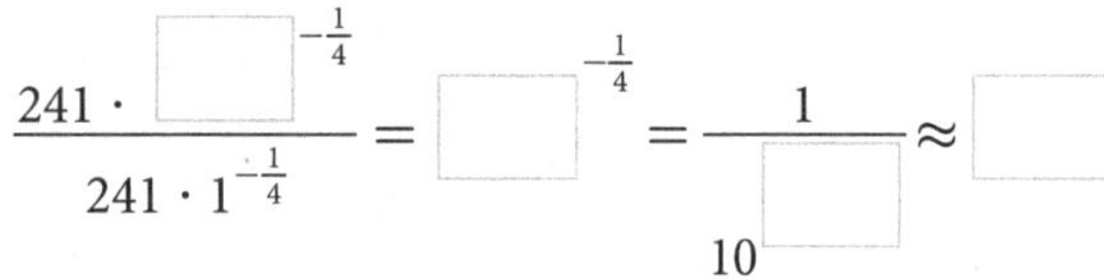

$$h(m) = 241\left(\square\right)^{-\frac{1}{4}} \approx \square$$

The model estimates the elk's resting heart rate to be about ____ beats per minute.

c. Find the ratio of $h(m)$ for the ________ animal to the ________ animal. Let 1 represent the mass of the smaller animal.

$$\frac{241 \cdot \square^{-\frac{1}{4}}}{241 \cdot 1^{-\frac{1}{4}}} = \square^{-\frac{1}{4}} = \frac{1}{10^{\square}} \approx \square$$

You would expect the larger animal's resting heart rate to be about ______ of the smaller animal's resting heart rate.

Reflect

6. What is the difference between a power function and an exponential function?

7. In Part B, the exponent is negative. Are the results consistent with the meaning of a negative exponent that you learned for integers? Explain.

Your Turn

8. Use Kleiber's law from Part A.

a. Find the basal metabolic rate for a 170 kilogram lion to the nearest 50 Calories. Then find the formula's prediction for a 70 kilogram human.

b. Use your metabolic rate result for the lion to find what the basal metabolic rate for a 70 kilogram human *would* be *if* metabolic rate and mass were directly proportional. Compare the result to the result from Part a.

Elaborate

9. Explain how can you use a radical to write and evaluate the power $4^{2.5}$.

10. When $y = kx$ for some constant k, y varies directly as x. When $y = kx^2$, y varies directly as the square of x; and when $y = k\sqrt{x}$, y varies directly as the square root of x. How could you express the relationship $y = kx^{\frac{3}{5}}$ for a constant k?

11. **Essential Question Check-In** Which of the following are true? Explain.

- To evaluate an expression of the form $a^{\frac{m}{n}}$, first find the nth root of a. Then raise the result to the mth power.
- To evaluate an expression of the form $a^{\frac{m}{n}}$, first find the mth power of a. Then find the nth root of the result.

Evaluate: Homework and Practice

- Online Homework
- Hints and Help
- Extra Practice

Translate expressions with rational exponents into radical expressions. Simplify numerical expressions when possible. Assume all variables are positive.

1. $64^{\frac{5}{3}}$

2. $x^{\frac{p}{q}}$

3. $(-512)^{\frac{2}{3}}$

4. $3^{\frac{2}{7}}$

5. $-\left(\frac{729}{64}\right)^{\frac{5}{6}}$

6. $0.125^{\frac{4}{3}}$

7. $vw^{\frac{2}{3}}$

8. $(-32)^{0.6}$

Translate radical expressions into expressions with rational exponents. Simplify numerical expressions when possible. Assume all variables are positive.

9. $\sqrt[7]{y^5}$

10. $\sqrt[7]{-6^6}$

11. $\sqrt[3]{3^{15}}$

12. $\sqrt[4]{(\pi z)^3}$

13. $\sqrt[6]{(bcd)^4}$

14. $\sqrt{6^6}$

15. $\sqrt[5]{32^2}$

16. $\sqrt[3]{\left(\frac{4}{x}\right)^9}$

17. Music Frets are small metal bars positioned across the neck of a guitar so that the guitar can produce the notes of a specific scale. To find the distance a fret should be placed from the bridge, multiply the length of the string by $2^{-\frac{n}{12}}$, where n is the number of notes higher than the string's root note. Where should a fret be placed to produce a F note on a B string (6 notes higher) given that the length of the string is 64 cm?

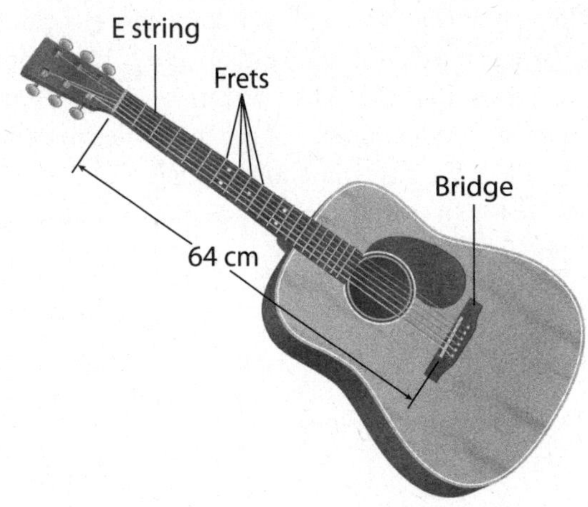

18. Meteorology The function $W = 35.74 + 0.6215T - 35.75V^{\frac{4}{25}} + 0.4275TV^{\frac{4}{25}}$ relates the windchill temperature W to the air temperature T in degrees Fahrenheit and the wind speed V in miles per hour. Use a calculator to find the wind chill temperature to the nearest degree when the air temperature is 28 °F and the wind speed is 35 miles per hour.

19. Astronomy New stars can form inside a cloud of interstellar gas when a cloud fragment, or *clump*, has a mass M greater than the *Jean's mass* M_J. The Jean's mass is $M_J = 100n^{-\frac{1}{2}}(T + 273)^{\frac{3}{2}}$ where n is the number of gas molecules per cubic centimeter and T is the gas temperature in degrees Celsius. A gas clump has $M = 137$, $n = 1000$, and $T = -263$. Will the clump form a star? Justify your answer.

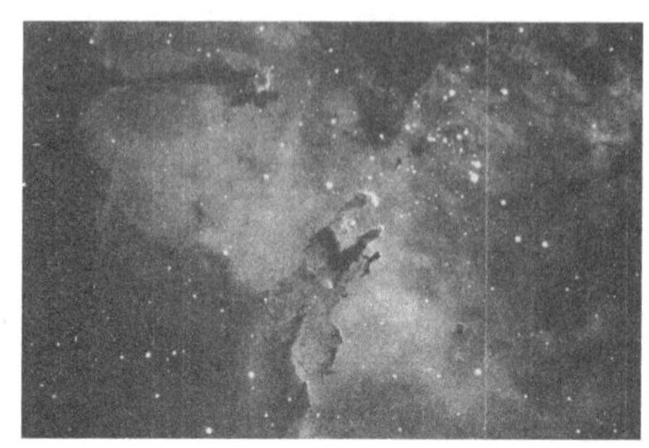

20. Urban geography The total wages W in a metropolitan area compared to its total population p can be approximated by a power function of the form $W = a \cdot p^{\frac{9}{8}}$ where a is a constant. About how many times greater does the model predict the total earnings for a metropolitan area with 3,000,000 people will be compared to a metropolitan area with 750,000?

21. Which statement is true?

A. In the expression $8x^{\frac{3}{4}}$, $8x$ is the radicand.

B. In the expression $(-16)x^{\frac{4}{5}}$, 4 is the index.

C. The expression $1024^{\frac{m}{n}}$ represents the nth root of the mth power of 1024.

D. $50^{-\frac{2}{5}} = -50^{\frac{2}{5}}$

E. $\sqrt{(xy)^3} = xy^{\frac{3}{2}}$

H.O.T. Focus on Higher Order Thinking

22. Critical Thinking For a negative real number a, under what condition(s) on m and n $(n \neq 0)$ is $a^{\frac{m}{n}}$ a real number? Explain. (Assume $\frac{m}{n}$ is written in simplest form.)

23. Explain the Error A teacher asked students to evaluate $10^{-\frac{3}{5}}$ using their graphing calculators. The calculator entries of several students are shown below. Which entry will give the incorrect result? Explain.

$10^{(-3/5)}$ $\sqrt[5]{10^{-3}}$ $10^{-.6}$ $1/10^{5/3}$ $(1/10^{1/5})^3$

24. Critical Thinking The graphs of three functions of the form $y = ax^{\frac{m}{n}}$ are shown for a specific value of a, where m and n are natural numbers. What can you conclude about the relationship of m and n for each graph? Explain.

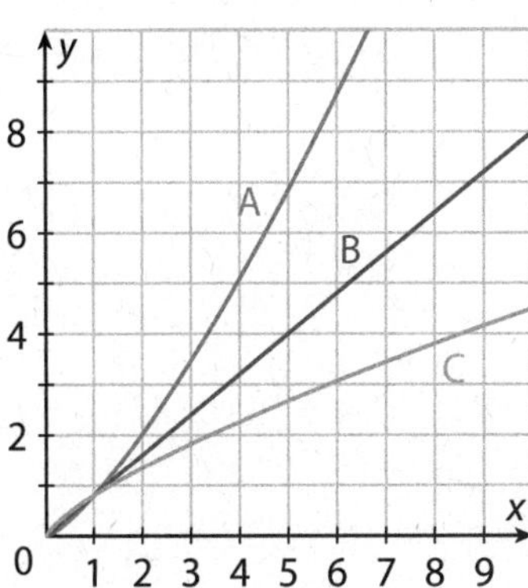

Lesson Performance Task

The formula $W = 35.74 + 0.6215T - 35.75V^{\frac{4}{25}} + 0.4275TV^{\frac{4}{25}}$ relates the wind chill temperature W to the air temperature T in degrees Fahrenheit and the wind speed V in miles per hour. Find the wind chill to the nearest degree when the air temperature is 40 °F and the wind speed is 35 miles per hour. If the wind chill is about 23 °F to the nearest degree when the air temperature is 40 °F, what is the wind speed to the nearest mile per hour?

Name__________________ Class__________ Date______

12.2 Simplifying Radical Expressions

Essential Question: How can you simplify expressions containing rational exponents or radicals involving *n*th roots?

Resource Locker

TEKS A2.7.G Rewrite radical expressions that contain variables to equivalent forms.

Explore Establishing the Properties of Rational Exponents

In previous courses, you have used properties of integer exponents to simplify and evaluate expressions, as shown here for a few simple examples:

$4^2 \cdot 4^3 = 4^{2+3} = 4^5 = 1024$

$(4 \cdot x)^2 = 4^2 \cdot x^2 = 16x^2$

$(4^2)^3 = 4^{2 \cdot 3} = 4^6 = 4096$

$\frac{4^2}{4^3} = 4^{2-3} = 4^{-1} = \frac{1}{4}$

$\left(\frac{4}{x}\right)^3 = \frac{4^3}{x^3} = \frac{64}{x^3}$

Now that you have been introduced to expressions involving rational exponents, you can explore the properties that apply to simplifying them.

Ⓐ Let $a = 64$, $b = 4$, $m = \frac{1}{3}$, and $n = \frac{3}{2}$. Evaluate each expression by substituting and simplifying inside parentheses and applying exponents individually, as shown.

Expression	Substitute	Simplify	Result
$a^m \cdot a^n$	$64^{\frac{1}{3}} \cdot 64^{\frac{3}{2}}$	$4 \cdot 512$	2048
$(a \cdot b)^n$	$(64 \cdot 4)^{\frac{3}{2}}$	$256^{\frac{3}{2}}$	
$(a^m)^n$			
$\frac{a^n}{a^m}$			
$\left(\frac{a}{b}\right)^n$			

(B) Complete the table again. This time, however, apply the rule of exponents that you would use for integer exponents.

Expression	Apply Rule and Substitute	Simplify	Result
$a^m \cdot a^n$	$64^{\frac{1}{3}+\frac{3}{2}}$	$64^{\frac{11}{6}}$	
$(a \cdot b)^n$			
$(a^m)^n$			
$\frac{a^n}{a^m}$			
$\left(\frac{a}{b}\right)^n$			

Reflect

1. Compare your results in Steps A and B. What can you conclude?

2. In Steps A and B, you evaluated $\frac{a^n}{a^m}$ two ways. Now evaluate $\frac{a^m}{a^n}$ two ways, using the definition of negative exponents. Are your results consistent with your previous conclusions about integer and rational exponents?

Explain 1 Simplifying Rational-Exponent Expressions

Rational exponents have the same properties as integer exponents.

Properties of Rational Exponents

For all nonzero real numbers *a* and *b* and rational numbers *m* and *n*

Words	Numbers	Algebra
Product of Powers Property To multiply powers with the same base, add the exponents.	$12^{\frac{1}{2}} \cdot 12^{\frac{3}{2}} = 12^{\frac{1}{2}+\frac{3}{2}} = 12^2 = 144$	$a^m \cdot a^n = a^m + n$
Quotient of Powers Property To divide powers with the same base, subtract the exponents.	$\frac{125^{\frac{2}{3}}}{125^{\frac{1}{3}}} = 125^{\frac{2}{3}-\frac{1}{3}} = 125^{\frac{1}{3}} = 5$	$\frac{a^m}{a^n} = a^{m-n}$
Power of a Power Property To raise one power to another, multiply the exponents.	$\left(8^{\frac{2}{3}}\right)^3 = 8^{\frac{2}{3}\cdot 3} = 8^2 = 64$	$(a^m)^n = a^{m\cdot n}$
Power of a Product Property To find a power of a product, distribute the exponent.	$(16 \cdot 25)^{\frac{1}{2}} = 16^{\frac{1}{2}} \cdot 25^{\frac{1}{2}} = 4 \cdot 5 = 20$	$(ab)^m = a^m b^m$
Power of a Quotient Property To find the power of a qoutient, distribute the exponent.	$\left(\frac{16}{81}\right)^{\frac{1}{4}} = \frac{16^{\frac{1}{4}}}{81^{\frac{1}{4}}} = \frac{2}{3}$	$\left(\frac{a}{b}\right)^m = \frac{a^m}{b^m}$

Example 1 **Simplify the expression. Assume that all variables are positive. Exponents in simplified form should all be positive.**

(A) **a.** $25^{\frac{3}{5}} \cdot 25^{\frac{7}{5}}$

$= 25^{\frac{3}{5}+\frac{7}{5}}$ Product of Powers Prop.

$= 25^2$ Simplify.

$= 625$

b. $\frac{8^{\frac{1}{3}}}{8^{\frac{2}{3}}}$

$= 8^{\frac{1}{3}-\frac{2}{3}}$ Quotient of Powers Prop.

$= 8^{-\frac{1}{3}}$ Simplify.

$= \frac{1}{8^{\frac{1}{3}}}$ Definition of neg. power

$= \frac{1}{2}$ Simplify.

(B) **a.** $\left(\dfrac{y^{\frac{4}{3}}}{16y^{\frac{2}{3}}}\right)^{\frac{3}{2}}$

$= \left(\dfrac{y^{\frac{4}{3}-\frac{2}{3}}}{16}\right)^{\frac{3}{2}}$ $\square$ Prop.

$= \left(\dfrac{\square}{16}\right)^{\frac{3}{2}}$ Simplify.

$= \dfrac{\left(y^{\frac{2}{3}}\right)^{\frac{3}{2}}}{16^{\frac{3}{2}}}$ $\square$ Prop.

$= \dfrac{y^{\frac{2}{3}\cdot\frac{3}{2}}}{16^{\frac{3}{2}}}$ $\square$ Prop.

=  Simplify.

b. $\left(27x^{\frac{3}{4}}\right)^{\frac{2}{3}}$

$= \square^{\frac{2}{3}}\left(\square\right)^{\frac{2}{3}}$ Power of a Product Prop.

$= 27^{\frac{2}{3}}\left(x^{\square}\right)$ Power of a Power Prop.

= 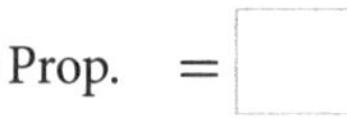 Simplify.

Your Turn

Simplify the expression. Assume that all variables are positive. Exponents in simplified form should all be positive.

3. $\left(12^{\frac{2}{3}} \cdot 12^{\frac{4}{3}}\right)^{\frac{3}{2}}$

4. $\dfrac{\left(6x^{\frac{1}{3}}\right)^2}{x^{\frac{5}{3}}y}$

Explain 2 Simplifying Radical Expressions Using the Properties of Exponents

When you are working with radical expressions involving *n*th roots, you can rewrite the expressions using rational exponents and then simplify them using the properties of exponents.

Example 2 **Simplify the expression by writing it using rational exponents and then using the properties of rational exponents. Assume that all variables are positive. Exponents in simplified form should all be positive.**

(A) $x\left(\sqrt[3]{2y}\right)\left(\sqrt[3]{4x^2y^2}\right)$

$= x(2y)^{\frac{1}{3}}\left(4x^2y^2\right)^{\frac{1}{3}}$ Write using rational exponents.

$= x\left(2y \cdot 4x^2y^2\right)^{\frac{1}{3}}$ Power of a Product Property

$= x\left(8x^2y^3\right)^{\frac{1}{3}}$ Product of Powers Property

$= x\left(2x^{\frac{2}{3}}y\right)$ Power of a Product Property

$= 2x^{\frac{5}{3}}y$ Product of Powers Property

(B) $\dfrac{\sqrt{64y}}{\sqrt[3]{64y}}$

$= \dfrac{(64y)^{\frac{1}{2}}}{(64y)^{\frac{1}{3}}}$ Write using rational exponents.

$= (64y)^{\square}$ Quotient of Powers Property

$= (64y)^{\square}$ Simplify.

$= \square$ Power of a Product Property

$= \square$ Simplify.

Your Turn

Simplify the expression by writing it using rational exponents and then using the properties of rational exponents.

5. $\dfrac{\sqrt{x^3}}{\sqrt[3]{x^2}}$

6. $\sqrt[6]{16^3} \cdot \sqrt[4]{4^6} \cdot \sqrt[3]{8^2}$

Explain 3 Simplifying Radical Expressions Using the Properties of *n*th Roots

From working with square roots, you know, for example, that $\sqrt{8} \cdot \sqrt{2} = \sqrt{8 \cdot 2} = \sqrt{16} = 4$ and $\dfrac{\sqrt{8}}{\sqrt{2}} \cdot = \sqrt{\dfrac{8}{2}} = \sqrt{4} = 2$. The corresponding properties also apply to *n*th roots.

Properties of *n*th Roots

For $a > 0$ and $b > 0$

Words	Numbers	Algebra
Product Property of Roots The *n*th root of a product is equal to the product of the *n*th roots.	$\sqrt[3]{16} = \sqrt[3]{8} \cdot \sqrt[3]{2} = 2\sqrt[3]{2}$	$\sqrt[n]{ab} = \sqrt[n]{a} \cdot \sqrt[n]{b}$
Quotient Property of Roots The *n*th root of a quotient is equal to the quotient of the *n*th roots.	$\sqrt{\dfrac{25}{16}} = \dfrac{\sqrt{25}}{\sqrt{16}} = \dfrac{5}{4}$	$\sqrt[n]{\dfrac{a}{b}} = \dfrac{\sqrt[n]{a}}{\sqrt[n]{b}}$

Example 3 **Simplify the expression using the properties of nth roots. Assume that all variables are positive. Rationalize any irrational denominators.**

(A) $\sqrt[3]{256x^3y^7}$

$$\sqrt[3]{256x^3y^7}$$

Write 256 as a power. $= \sqrt[3]{2^8 \cdot x^3y^7}$

Product Property of Roots $= \sqrt[3]{2^6 \cdot x^3y^6} \cdot \sqrt[3]{2^2 \cdot y}$

Factor out perfect cubes. $= \sqrt[3]{2^6} \cdot \sqrt[3]{x^3} \cdot \sqrt[3]{y^6} \cdot \sqrt[3]{4y}$

Simplify. $= 4xy^2\sqrt[3]{4y}$

(B) $\sqrt[4]{\frac{81}{x}}$

$$\sqrt[4]{\frac{81}{x}}$$

[ ] $= \frac{\sqrt[4]{81}}{\sqrt[4]{x}}$

Simplify. $= \frac{\square}{\sqrt[4]{x}}$

Rationalize the denominator. $= \frac{3}{\sqrt[4]{x}} \cdot \frac{\square}{\sqrt[4]{x^3}}$

[ ] $= \frac{3\sqrt[4]{x^3}}{\sqrt[4]{x^4}}$

Simplify. $= \square$

Reflect

7. In Part B, why was $\sqrt[4]{x^3}$ used when rationalizing the denominator? What factor would you use to rationalize a denominator of $\sqrt[5]{4y^3}$?

Your Turn

Simplify the expression using the properties of *n*th roots. Assume that all variables are positive.

8. $\sqrt[3]{216x^{12}y^{15}}$

9. $\sqrt[4]{\frac{16}{x^{14}}}$

Explain 4 Rewriting a Radical-Function Model

When you find or apply a function model involving rational powers or radicals, you can use the properties in this lesson to help you find a simpler expression for the model.

Ⓐ **Manufacturing** A can that is twice as tall as its radius has the minimum surface area for the volume it contains. The formula $S = 6\pi\left(\frac{V}{2\pi}\right)^{\frac{2}{3}}$ expresses the surface area of a can with this shape in terms of its volume.

a. Use the properties of rational exponents to simplify the expression for the surface area. Then write the approximate model with the coefficient rounded to the nearest hundredth.

b. Graph the model using a graphing calculator. What is the surface area in square centimeters for a can with a volume of 440 cm^3?

a. $S = 6\pi\left(\frac{V}{2\pi}\right)^{\frac{2}{3}}$

Power of a Quotient Property $= 6\pi \cdot \frac{V^{\frac{2}{3}}}{(2\pi)^{\frac{2}{3}}}$

Group Powers of 2π. $= \frac{3(2\pi)}{(2\pi)^{\frac{2}{3}}} \cdot V^{\frac{2}{3}}$

Quotient of Powers Property $= 3(2\pi)^{1-\frac{2}{3}} \cdot V^{\frac{2}{3}}$

Simplify. $= 3(2\pi)^{\frac{1}{3}} \cdot V^{\frac{2}{3}}$

Use a calculator. $\approx 5.54V^{\frac{2}{3}}$

A simplified model is $S = 3(2\pi)^{\frac{1}{3}} \cdot V^{\frac{2}{3}}$, which gives $S \approx 5.54V^{\frac{2}{3}}$.

b.

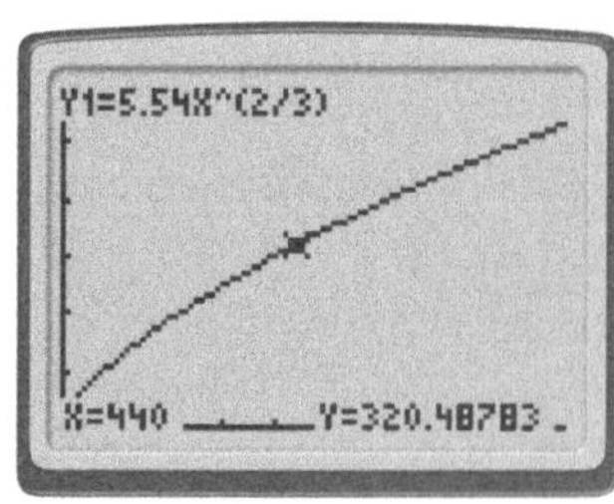

The surface area is about 320 cm^2.

Ⓑ **Commercial fishing** The buoyancy of a fishing float in water depends on the volume of air it contains. The radius of a spherical float as a function of its volume is given by $r = \sqrt[3]{\frac{3V}{4\pi}}$.

a. Use the properties of roots to rewrite the expression for the radius as the product of a coefficient term and a variable term. Then write the approximate formula with the coefficient rounded to the nearest hundredth.

b. What should the radius be for a float that needs to contain 4.4 ft^3 of air to have the proper buoyancy?

a. $r = \sqrt[3]{\frac{3V}{4\pi}}$

Rewrite radicand. $= \sqrt[3]{\frac{3}{4\pi} \cdot \square}$

Product Property of Roots $= \sqrt[3]{\frac{3}{4\pi}} \cdot \square$

Use a calculator $\approx \square$

The rewritten formula is $r = \square$, which gives $r \approx \square$.

b. Substitute 4.4 for V.

$r = 0.62\sqrt[3]{4.4} \approx \square$

The radius is about ____ feet.

Reflect

10. Discussion What are some reasons you might want to rewrite an expression involving radicals into an expression involving rational exponents?

Your Turn

11. The surface area as a function of volume for a box with a square base and a height that is twice the side length of the base is $S = 10\left(\frac{V}{2}\right)^{\frac{2}{3}}$. Use the properties of rational exponents to simplify the expression for the surface area so that no fractions are involved. Then write the approximate model with the coefficient rounded to the nearest hundredth.

Elaborate

12. In problems with a radical in the denominator, you rationalized the denominator to remove the radical. What can you do to remove a rational exponent from the denominator? Explain by giving an example.

13. Show why $\sqrt[n]{a^n}$ is equal to a for all natural numbers a and n using the definition of nth roots and using rational exponents.

14. Show that the Product Property of Roots is true using rational exponents.

15. **Essential Question Check-In** Describe the difference between applying the Power of a Power Property and applying the Power of a Product Property for rational exponents using an example that involves both properties.

Evaluate: Homework and Practice

- Online Homework
- Hints and Help
- Extra Practice

Simplify the expression. Assume that all variables are positive. Exponents in simplified form should all be positive.

1. $\left(\left(\frac{1}{16}\right)^{-\frac{2}{3}}\right)^{\frac{3}{4}}$

2. $\dfrac{x^{\frac{1}{3}} \cdot x^{\frac{5}{6}}}{x^{\frac{1}{6}}}$

3. $\dfrac{9^{\frac{3}{2}} \cdot 9^{\frac{1}{2}}}{9^{-2}}$

4. $\left(\dfrac{16^{\frac{5}{3}}}{16^{\frac{5}{6}}}\right)^{\frac{9}{5}}$

5. $\dfrac{2xy}{\left(x^{\frac{1}{3}}y^{\frac{2}{3}}\right)^{\frac{3}{2}}}$

6. $\dfrac{3y^{\frac{3}{4}}}{2xy^{\frac{3}{2}}}$

Simplify the expression by writing it using rational exponents and then using the properties of rational exponents. Assume that all variables are positive. Exponents in simplified form should all be positive.

7. $\sqrt[4]{25} \cdot \sqrt[3]{5}$

8. $\dfrac{\sqrt[4]{2^{-2}}}{\sqrt[6]{2^{-9}}}$

9. $\dfrac{\sqrt[4]{3^3} \cdot \sqrt[3]{x^2}}{\sqrt{3x}}$

10. $\dfrac{\sqrt[4]{x^4y^6} \cdot \sqrt{x^6}}{y}$

11. $\dfrac{\sqrt[6]{s^4t^9}}{\sqrt[3]{st}}$

12. $\sqrt[4]{27} \cdot \sqrt{3} \cdot \sqrt[6]{81^3}$

Simplify the expression using the properties of *n*th roots. Assume that all variables are positive. Rationalize any irrational denominators.

13. $\dfrac{\sqrt[4]{36} \cdot \sqrt[4]{216}}{\sqrt[4]{6}}$

14. $\sqrt[4]{4096x^6y^8}$

15. $\dfrac{\sqrt[3]{x^8y^4}}{\sqrt[3]{x^2y}}$

16. $\sqrt[5]{\dfrac{125}{w^6}} \cdot \sqrt[5]{25v}$

17. **Weather** The volume of a sphere as a function of its surface area is given by $V = \frac{4\pi}{3}\left(\frac{S}{4\pi}\right)^{\frac{3}{2}}$.

a. Use the properties of roots to rewrite the expression for the volume as the product of a simplified coefficient term (with positive exponents) and a variable term. Then write the approximate formula with the coefficient rounded to the nearest thousandth.

b. A spherical weather balloon has a surface area of 500 ft^2. What is the approximate volume of the balloon?

18. Amusement parks An amusement park has a ride with a free fall of 128 feet. The formula $t = \sqrt{\frac{2d}{g}}$ gives the time t in seconds it takes the ride to fall a distance of d feet. The formula $v = \sqrt{2gd}$ gives the velocity v in feet per second after the ride has fallen d feet. The letter g represents the gravitational constant.

a. Rewrite each formula, factoring out $\sqrt{d}$. Then simplify each formula using the fact that $g \approx 32$ ft/s^2.

b. Find the time it takes the ride to fall halfway and its velocity at that time. Then find the time and velocity for the full drop.

c. What is the ratio of the time it takes for the whole drop to the time it takes for the first half? What is the ratio of the velocity after the second half of the drop to the velocity after the first half? What do you notice?

19. Which choice(s) is/are equivalent to $\sqrt{2}$?

A. $\left(\sqrt[8]{2}\right)^4$

B. $\dfrac{2^3}{2^{-\frac{5}{2}}}$

C. $\left(4^{\frac{2}{3}} \cdot 2^{\frac{2}{3}}\right)^{\frac{1}{4}}$

D. $\dfrac{\sqrt[3]{2^2}}{\sqrt[6]{2}}$

E. $\dfrac{\sqrt{2^{-\frac{3}{4}}}}{\sqrt{2^{-\frac{7}{4}}}}$

20. Home Heating A propane storage tank for a home is shaped like a cylinder with hemispherical ends, and a cylindrical portion length that is 4 times the radius. The formula $S = 12\pi\left(\frac{3V}{16\pi}\right)^{\frac{2}{3}}$ expresses the surface area of a tank with this shape in terms of its volume.

a. Use the properties of rational exponents to rewrite the expression for the surface area by factoring out the power of V. Then write the approximate model with the coefficient rounded to the nearest hundredth.

b. Graph the model using a graphing calculator. What is the surface area in square feet for a tank with a volume of 150 ft^3 ?

H.O.T. Focus on Higher Order Thinking

21. Critique Reasoning Aaron's work in simplifying an expression is shown. What mistake(s) did Aaron make? Show the correct simplification.

$$625^{-\frac{1}{3}} \div 625^{-\frac{4}{3}}$$
$$= 625^{-\frac{1}{3}-\left(-\frac{4}{3}\right)}$$
$$= 625^{-\frac{1}{3}\left(-\frac{3}{4}\right)}$$
$$= 625^{\frac{1}{4}}$$
$$= 5$$

22. Critical Thinking Use the definition of nth root to show that the Product Property of Roots is true, that is, that $\sqrt[n]{ab} = \sqrt[n]{a} \cdot \sqrt[n]{b}$. (Hint: Begin by letting x be the nth root of a and letting y be the nth root of b.)

23. Critical Thinking For what real values of a is $\sqrt[4]{a}$ greater than a? For what real values of a is $\sqrt[5]{a}$ greater than a?

Lesson Performance Task

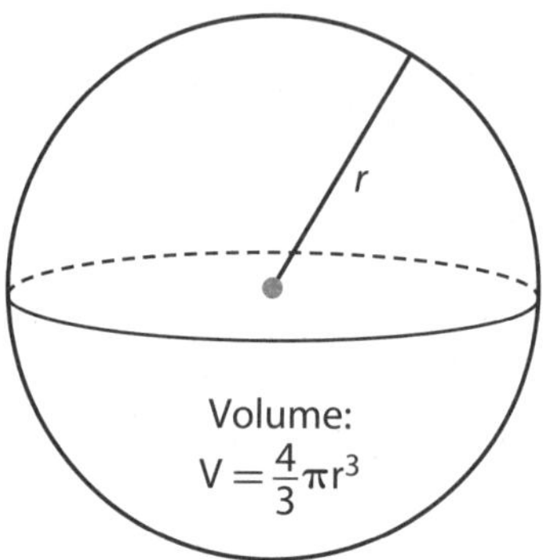

You've been asked to help decorate for a school dance, and the theme chosen is "The Solar System." The plan is to have a bunch of papier-mâché spheres serve as models of the planets, and your job is to paint them. All you're told are the volumes of the individual spheres, but you need to know their surface areas so you can get be sure to get enough paint. How can you write a simplified equation using rational exponents for the surface area of a sphere in terms of its volume?

(The formula for the volume of a sphere is $V = \frac{4}{3}\pi r^3$ and the formula for the surface area of a sphere is $A = 4\pi r^2$.)

Name________________________ Class________________ Date__________

12.3 Solving Radical Equations

Resource Locker

Essential Question: How can you solve equations involving square roots and cube roots?

TEKS A2.4.F Solve...square root equations. Also A2.4.G, A2.6.B, A2.7.H

Explore Investigating Solutions of Square Root Equations

When solving quadratic equations, you have learned that the number of real solutions depends upon the values in the equation, with different equations having 0, 1, or 2 real solutions. How many real solutions does a square root equation have? In the Explore, you will investigate graphically the numbers of real solutions for different square root equations.

(A) Remember that you can graph the two sides of an equation as separate functions to find solutions of the equation: a solution is any x-value where the two graphs intersect.

The graph of $y = \sqrt{x - 3}$ is shown on a calculator window of $-4 \leq x \leq 16$ and $-2 \leq y \leq 8$. Reproduce the graph on your calculator. Then add the graph of $y = 2$.

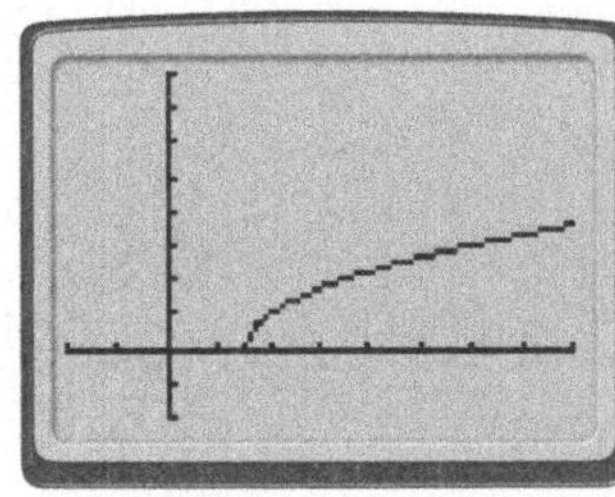

How many solutions does the equation $\sqrt{x - 3} = 2$ have? ______ How do you know?

____

On your calculator, replace the graph of $y = 2$ with the graph of $y = -1$.

How many solutions does the equation $\sqrt{x - 3} = -1$ have? ______ How do you know?

____

(B) Graph $y = \sqrt{x - 3} + 2$ on your calculator (you can use the same viewing window as in Step A).

Add the graph of $y = 3$ to the graph of $y = \sqrt{x - 3} + 2$.

How many solutions does $\sqrt{x - 3} + 2 = 3$ have? ______

Replace the graph of $y = 3$ with the graph of $y = 1$.

How many solutions does $\sqrt{x - 3} + 2 = 1$ have? ______

(C) Graph both sides of $\sqrt{4x - 4} = x + 1$ as separate functions on your calculator.

How many solutions does $\sqrt{4x - 4} = x + 1$ have? ______

Replace the graph of $y = x + 1$ with the graph of $y = \frac{1}{2}x$.

How many solutions does $\sqrt{4x - 4} = \frac{1}{2}x$ have? ______

Replace the graph of $y = \frac{1}{2}x$ with the graph of $y = 2x - 5$.

How many solutions does $\sqrt{4x - 4} = 2x - 5$ have? ______

(D) Graph both sides of $\sqrt{2x - 3} = \sqrt{x}$ as separate functions on your calculator.

How many solutions does $\sqrt{2x - 3} = \sqrt{x}$ have? ______

Replace the graph of $y = \sqrt{x}$ with the graph of $y = \sqrt{2x + 3}$.

How many solutions does $\sqrt{2x - 3} = \sqrt{2x + 3}$ have? ______

Reflect

1. For a square root equation of the form $\sqrt{bx - h} = c$, what can you conclude about the number of solutions based on the sign of c?

2. For a square root equation of the form $\sqrt{bx - h} + k = c$, what can you conclude about the number of solutions based on the values of k and c?

3. For a cube root equation of the form $\sqrt[3]{bx - h} = c$, will the number of solutions depend on the sign of c? Explain.

4. The graphs in the second part of Step D appear to be get closer and closer as x increases. How can you be sure that they never meet, that is, that $\sqrt{2x - 3} = \sqrt{2x + 3}$ really has no solutions?

Explain 1 Solving Square Root and $\frac{1}{2}$-Power Equations

A *radical equation* contains a variable within a radical or a variable raised to a (non-integer) rational power. To solve a square root equation, or, equivalently, an equation involving the power $\frac{1}{2}$, you can square both sides of the equation and solve the resulting equation.

Because opposite numbers have the same square, squaring both sides of an equation may introduce an apparent solution that is not an actual solution (an extraneous solution). For example, while the only solution of $x = 2$ is 2, the equation that is the square of each side, $x^2 = 4$, has two solutions, -2 and 2. But -2 is not a solution of the original equation.

Example 1 **Solve the equation. Check for extraneous solutions.**

Ⓐ $2 + \sqrt{x + 10} = x$

	$2 + \sqrt{x + 10} = x$
Isolate the radical.	$\sqrt{x + 10} = x - 2$
Square both sides.	$\left(\sqrt{x + 10}\right)^2 = (x - 2)^2$
Simplify.	$x + 10 = x^2 - 4x + 4$
Simplify.	$0 = x^2 - 5x - 6$
Factor.	$0 = (x - 6)(x + 1)$
Zero Product Property	$x = 6$ or $x = -1$

Check:

$2 + \sqrt{x + 10} = x$	$2 + \sqrt{x + 10} = x$
$2 + \sqrt{6 + 10} \stackrel{?}{=} 6$	$2 + \sqrt{-1 + 10} \stackrel{?}{=} -1$
$2 + \sqrt{16} \stackrel{?}{=} 6$	$2 + \sqrt{9} \stackrel{?}{=} -1$
$6 = 6$ ✓	$5 \neq -1$
$x = 6$ is a solution.	$x = -1$ is not a solution.

The solution is $x = 6$.

 $(x + 6)^{\frac{1}{2}} - (2x - 4)^{\frac{1}{2}} = 0$

Rewrite with radicals.	$\sqrt{x + 6} - \sqrt{2x - 4} = 0$
Isolate radicals on each side.	$\sqrt{x + 6} = \square$
Square both sides.	$\left(\sqrt{x + 6}\right)^2 = \left(\square\right)^2$
Simplify.	$\square = \square$
Solve.	$\square = x$

Check:

$$\sqrt{x + 6} - \sqrt{2x - 4} = 0$$

$$\sqrt{10 + 6} - \sqrt{2\left(\square\right) - 4} \stackrel{?}{=} 0$$

$$\square - \square \stackrel{?}{=} 0$$

$$\square = 0$$

The solution is ________.

Reflect

5. The graphs of each side of the equation from Part A are shown on the graphing calculator screen below. How can you tell from the graph that one of the two solutions you found algebraically is extraneous?

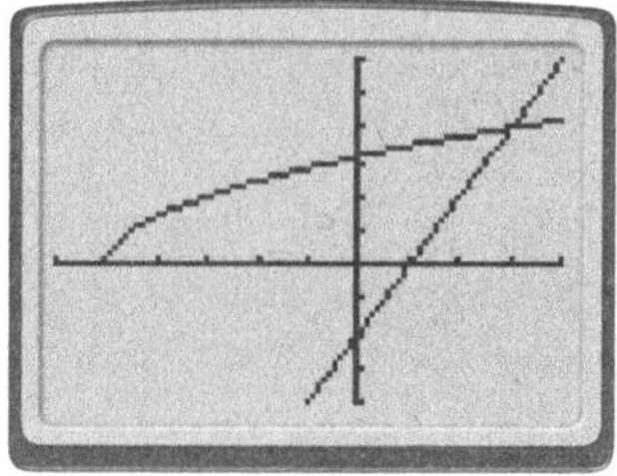

Your Turn

6. Solve $(x + 5)^{\frac{1}{2}} - 2 = 1$.

Explain 2 Solving Cube Root and $\frac{1}{3}$-Power Equations

You can solve radical equations that involve roots other than square roots by raising both sides to the index of the radical. So, to solve a cube root equation, or, equivalently, an equation involving the power $\frac{1}{3}$, you can cube both sides of the equation and solve the resulting equation.

Example 2 **Solve the equation.**

(A) $\sqrt[3]{x + 2} + 7 = 5$

$$\sqrt[3]{x + 2} + 7 = 5$$

Isolate the radical. $\sqrt[3]{x + 2} = -2$

Cube both sides. $\left(\sqrt[3]{x + 2}\right)^3 = (-2)^3$

Simplify. $x + 2 = -8$

Solve for x $x = -10$

The solution is $x = -10$.

(B) $\sqrt[3]{x-5} = x+1$

$$\sqrt[3]{x-5} = x+1$$

Cube both sides. $\left(\sqrt[3]{x-5}\right)^3 = (x+1)^3$

Simplify $\boxed{} = \boxed{}$

Simplify. $0 = \boxed{}$

Begin to factor by grouping. $0 = x^2\left(\boxed{}\right) + 2\left(\boxed{}\right)$

Complete factoring $0 = (x^2+2)\left(\boxed{}\right)$

By the Zero Product Property, $\boxed{} = 0$ or $\boxed{} = 0$.

Because there are no real values of x for which $x^2 = \boxed{}$, the only solution is $x = \boxed{}$.

Reflect

7. **Discussion** Example 1 shows checking for extraneous solutions, while Example 2 does not. While it is always wise to check your answers, can a cubic equation have an extraneous solution? Explain your answer.

Your Turn

8. Solve $2(x-50)^{\frac{1}{3}} = -10$.

Explain 3 Solving a Real-World Problem

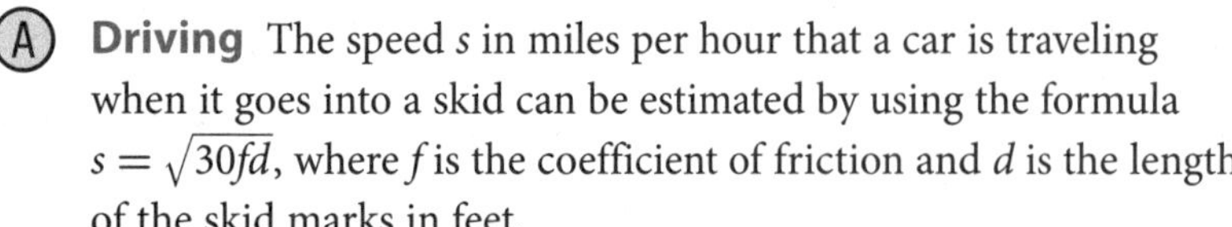

Ⓐ **Driving** The speed s in miles per hour that a car is traveling when it goes into a skid can be estimated by using the formula $s = \sqrt{30fd}$, where f is the coefficient of friction and d is the length of the skid marks in feet.

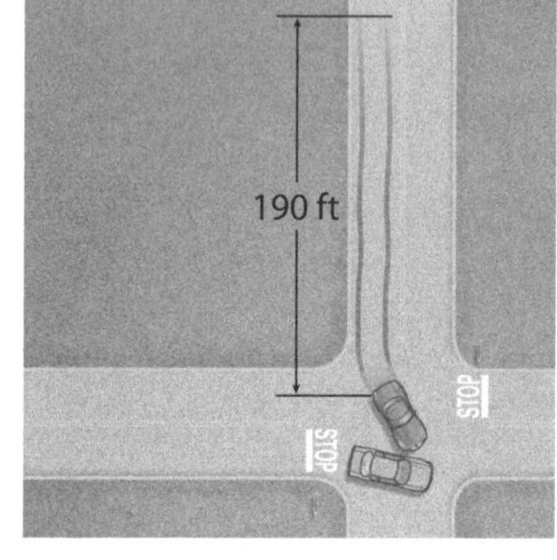

After an accident, a driver claims to have been traveling the speed limit of 55 mi/h. The coefficient of friction under the conditions at the time of the accident was 0.6, and the length of the skid marks is 190 feet. Is the driver telling the truth about the car's speed? Explain.

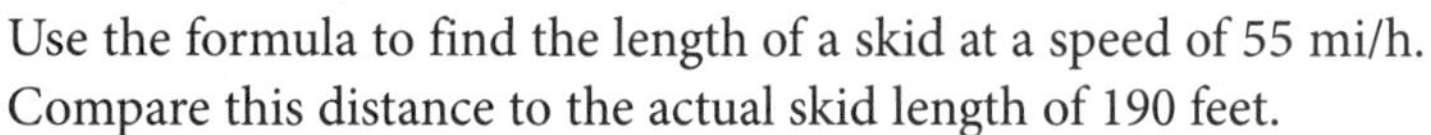

Use the formula to find the length of a skid at a speed of 55 mi/h. Compare this distance to the actual skid length of 190 feet.

$$s = \sqrt{30fd}$$

Substitute 55 for s and 0.6 for f $\quad 55 = \sqrt{30(0.6)d}$

Simplify. $\quad 55 = \sqrt{18d}$

Square both sides. $\quad 55^2 = \left(\sqrt{18d}\right)^2$

Simplify. $\quad 3025 = 18d$

Solve for d. $\quad 168 \approx d$

If the driver had been traveling at 55 mi/h, the skid marks would measure about 168 feet. Because the skid marks actually measure 190 feet, the driver must have been driving faster than 55 mi/h.

Ⓑ **Construction** The diameter d in inches of a rope needed to lift a weight of w tons is given by the formula $d = \frac{\sqrt{15w}}{\pi}$. How much weight can be lifted with a rope with a diameter of 1.0 inch?

Use the formula for the diameter as a function of weight, and solve for the weight given the diameter.

$$d = \frac{\sqrt{15w}}{\pi}$$

Substitute. $\quad \square = \frac{\sqrt{15w}}{\pi}$

Square both sides. $\quad \square = \square$

Isolate the radical. $\quad \left(\square\right)^2 = \left(\sqrt{15w}\right)^2$

Simplify. $\quad \square = 15w$

Solve for w. $\quad \square \approx w$

A rope with a diameter of 1.0 can hold about ______ ton, or about ______ pounds.

Your Turn

9. **Biology** The trunk length (in inches) of a male elephant can be modeled by $l = 23\sqrt[3]{t} + 17$, where t is the age of the elephant in years. If a male elephant has a trunk length of 100 inches, about what is his age?

Elaborate

10. A student asked to solve the equation $\sqrt{4x + 8} + 9 = 1$ isolated the radical, squared both sides, and solved for x to obtain $x = 14$, only to find out that the apparent solution was extraneous. Why could the student have stopped trying to solve the equation after isolating the radical?

11. When you see a cube root equation with the radical expression isolated on one side and a constant on the other, what should you expect for the number of solutions? Explain. What are some reasons you should check your answer anyway?

12. **Essential Question Check-In** Solving a quadratic equation of the form $x^2 = a$ involves taking the square root of both sides. Solving a square root equation of the form $\sqrt{x} = b$ involves squaring both sides of the equation. Which of these operations can create an equation that is not equivalent to the original equation? Explain how this affects the solution process.

Evaluate: Homework and Practice

- Online Homework
- Hints and Help
- Extra Practice

Solve the equation.

1. $\sqrt{x-9}=5$

2. $\sqrt{3x}=6$

3. $\sqrt{x+3}=x+1$

4. $\sqrt{(15x+10)}=2x+3$

5. $(x+4)^{\frac{1}{2}}=6$

6. $(45-9x)^{\frac{1}{2}}=x-5$

7. $(x-6)^{\frac{1}{2}}=x-2$

8. $4(x-2)^{\frac{1}{2}}=(x+13)^{\frac{1}{2}}$

9. $5-\sqrt[3]{x-4}=2$

10. $2\sqrt[3]{3x+2}=\sqrt[3]{4x-9}$

11. $\sqrt[3]{69x + 35} = x + 5$

12. $\sqrt[3]{x + 5} = x - 1$

13. $(x + 7)^{\frac{1}{3}} = (4x)^{\frac{1}{3}}$

14. $(5x + 1)^{\frac{1}{4}} = 4$

15. $(-9x - 54)^{\frac{1}{3}} = -2x + 3$

16. $2(x - 1)^{\frac{1}{5}} = (2x - 17)^{\frac{1}{5}}$

17. Driving The formula for the speed versus skid length in Example 3A assumes that all 4 wheel brakes are working at top efficiency. If this is not true, another parameter is included in the equation so that the equation becomes $s = \sqrt{30fdn}$ where n is the percent braking efficiency as a decimal. Accident investigators know that the rear brakes failed on a car, reducing its braking efficiency to 60%. On a dry road with a coefficient of friction of 0.7, the car skidded 250 feet. Was the car going above the speed limit of 60 mi/h when the skid began?

18. Anatomy The surface area S of a human body in square meters can be approximated by $S = \sqrt{\frac{hm}{36}}$ where h is height in meters and m is mass in kilograms. A basketball player with a height of 2.1 meters has a surface area of about 2.7 m^2. What is the player's mass?

19. **Biology** The approximate antler length L (in inches) of a deer buck can be modeled by $L = 9\sqrt[3]{t} + 15$ where t is the age in years of the buck. If a buck has an antler length of 36 inches, what is its age?

20. **Amusement Parks** For a spinning amusement park ride, the velocity v in meters per second of a car moving around a curve with radius r meters is given by $v = \sqrt{ar}$ where a is the car's acceleration in m/s^2. If the ride has a maximum acceleration of 30 m/s^2 and the cars on the ride have a maximum velocity of 12 m/s, what is the smallest radius that any curve on the ride may have?

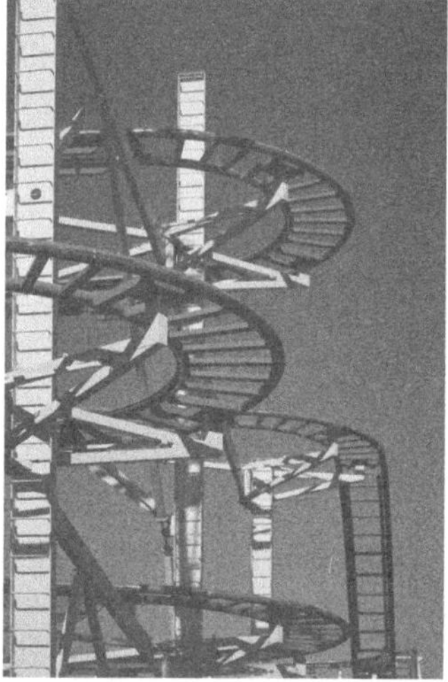

21. For each radical equation, state the number of solutions that exist.

A. $\sqrt{x-4} = -5$

B. $\sqrt{x-4} + 6 = 11$

C. $4 = -2\sqrt[3]{x+2}$

D. $\sqrt{x+40} = 0$

E. $\sqrt[3]{2x+5} = -18$

H.O.T. Focus on Higher Order Thinking

22. Critical Thinking For an equation of the form $\sqrt{x+a} = b$ where b is a constant, does the sign of a affect whether or not there is a solution for a given value of b? If so, how? If not, why not?

23. Explain the Error Below is a student's work in solving the equation $2\sqrt{3x+3} = 12$. What mistake did the student make? What is the correct solution?

$$2\sqrt{3x+3} = 12$$

$$2\left(\sqrt{3x+3}\right)^2 = 12^2$$

$$2(3x+3) = 144$$

$$6x+6 = 144$$

$$x = 23$$

24. Communicate Mathematical Ideas Describe the key difference between solving radical equations for which you solve by raising both sides to an even power and those you solve by raising both sides to an odd power.

25. Critical Thinking How could you solve an equation for which one side is a rational power expression and the other side is a constant? Give an example. Under what condition would you have to be especially careful to check your solutions?

Lesson Performance Task

For many years scientists have used a scale known as the Fujita Scale to categorize different types of tornados in relation to the velocity of the winds produced. The formula used to generate the scale is given by $V = k(F + 2)^{\frac{3}{2}}$. The scale employs a constant, k, and the tornado's category number to determine wind speed. If you wanted to determine the different category numbers, how could you solve the radical equation for the variable F? (The value for k is about 14.1.) Solve the equation for F then verify the different categories using the minimum wind velocity. Do the values seem reasonable given the value for k?

Fujita Tornado Scale			
Damage Level	**Category**	**Minimum Wind Velocity (mi/h)**	**Calculations**
Moderate	F1	73	
Significant	F2	113	
Severe	F3	158	
Devastating	F4	207	
Incredible	F5	261	

STUDY GUIDE REVIEW

Radical Expressions and Equations

MODULE 12

Essential Question: How can you use radical expressions and equations to solve real-world problems?

Key Vocabulary

extraneous solution
(solución extraña)

radical expression
(expresión radical)

rational exponent
(exponente racional)

KEY EXAMPLE *(Lesson 12.1)*

Evaluate the expression.

$$\left(\sqrt[4]{16}\right)^5 = 2^5 = 32$$

$$27^{\frac{4}{3}} = \left(\sqrt[3]{27}\right)^4 = 3^4 = 81$$

KEY EXAMPLE *(Lesson 12.2)*

Write the expression in simplest form. Assume all variables are positive.

$$\sqrt[3]{48} = \sqrt[3]{8 \cdot 6} = \sqrt[3]{8} \cdot \sqrt[3]{6} = 2\sqrt[3]{6}$$

$$\left(\frac{x^4}{y^8}\right)^{\frac{1}{2}} = \frac{\left(x^4\right)^{\frac{1}{2}}}{\left(y^8\right)^{\frac{1}{2}}} = \frac{x^{4 \cdot \frac{1}{2}}}{y^{8 \cdot \frac{1}{2}}} = \frac{x^2}{y^4}$$

KEY EXAMPLE *(Lesson 12.3)*

Solve the equation. $\sqrt{x + 15} = x - 5$

$\left(\sqrt{x + 15}\right)^2 = (x - 5)^2$ Square both sides.

$x + 15 = x^2 - 10x + 25$

$x^2 - 11x + 10 = 0$ Write in standard form.

$(x - 10)(x - 1) = 0$ Factor.

$x = 10$ or $x = 1$ Solve for *x*.

$\sqrt{10 + 15} \stackrel{?}{=} 10 - 5$ $\quad$ $\sqrt{1 + 15} \stackrel{?}{=} 1 - 5$ Check.

$5 = 5$ $\quad$ $4 \neq -4$

The solution $x = 1$ is extraneous. The only solution is $x = 10$.

EXERCISES

Evaluate the expression. *(Lesson 12.1)*

1. $\sqrt[3]{-64}$

2. $81^{\frac{1}{4}}$

3. $256^{\frac{3}{4}}$

Write the expression in simplest form. Assume that all variables are positive. *(Lesson 12.2)*

4. $\sqrt[3]{80}$

5. $(3^4 \cdot 5^4)^{-\frac{1}{4}}$

6. $(25a^{10}b^{16})^{\frac{1}{2}}$

7. $\sqrt[5]{\frac{c}{d^8}}$

Solve each equation. *(Lesson 12.3)*

8. $\sqrt[3]{5x-4} = 2$

9. $\sqrt{x+6} - 7 = -2$

MODULE PERFORMANCE TASK

Don't Disturb the Neighbors!

The faintest sound an average person can detect has an intensity of 1×10^{-12} watts per square meter, where watts are a unit of power. The intensity of a sound is given by $I = \frac{P}{4\pi d^2}$, where P is the power of the sound and d is the distance from the source. Yolanda wants to throw a party at her house and plans to invite a band to perform. The power of the sound from the band's speakers is typically 3.0 watts. Yolanda's neighborhood has a rule that between 7 p.m. and 11 p.m., a sound intensity up to $I = 5.0 \times 10^{-5}$ W/m^2 is acceptable; after 11 p.m., the acceptable intensity is $I = 5.0 \times 10^{-7}$ W/m^2. How far away would Yolanda's closest neighbors need to be for the band to play till 11 p.m.? How far would they need to be for the band to play all night?

Start by listing in the space below the information you will need to solve the problem. Then use your own paper to complete the task. Be sure to write down all your data and assumptions. Then use graphs, numbers, words, or algebra to explain how you reached your conclusion.

Ready to Go On?

12.1–12.3 Radical Expressions and Equations

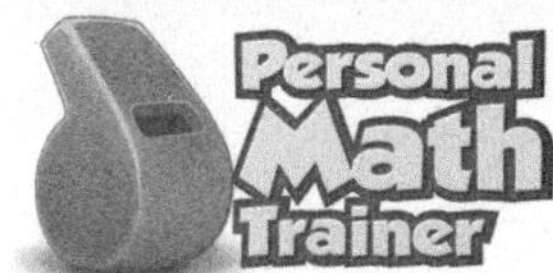

- Online Homework
- Hints and Help
- Extra Practice

Simplify each expression. Assume that all variables are positive. *(Lessons 12.1, 12.2)*

1. $32^{\frac{1}{5}}$

2. $\left(\sqrt[3]{64}\right)^4$

3. $\sqrt[3]{27x^6}$

4. $\sqrt[4]{2x^6y^8}$

Solve each equation. *(Lesson 12.3)*

5. $\sqrt{10x} = 3\sqrt{x+1}$

6. $\sqrt[3]{2x-2} = 6$

7. $(4x+7)^{\frac{1}{2}} = 3$

8. $(x+3)^{\frac{1}{3}} = -6$

ESSENTIAL QUESTION

9. How do you solve a radical equation and identify any extraneous roots?

MODULE 12
MIXED REVIEW

Assessment Readiness

1. Which expression can be simplified to a rational number?

A. $\sqrt{2} + \sqrt{2}$

B. $\sqrt{4} \cdot \sqrt{20}$

C. $\left(\sqrt{12}\right)^2$

D. $\sqrt{\frac{16}{2}}$

2. What is the solution of the equation $(3x + 1)^{\frac{1}{3}} = -2$?

A. -9

B. -3

C. 3

D. 9

3. Which is equivalent to $(4 - 3i)(2 - i)$?

A. $5 - 10i$

B. $5 + 10i$

C. $8 - 10i$

D. $8 + 10i$

4. At what point does the graph of $f(x) = \frac{2x^2 - 5x - 3}{x - 3}$ have a hole?

A. $(-0.5, 0)$

B. $(-0.5, -3.5)$

C. $(3, 0)$

D. $(3, 7)$

5. The formula $s = \sqrt{\frac{A}{4.828}}$ can be used to approximate the side length s of a regular octagon with area A. A stop sign is shaped like a regular octagon with a side length of 12.4 in. To the nearest square inch, what is the area of the stop sign? Explain how you got your answer.

UNIT 5

MIXED REVIEW

Assessment Readiness

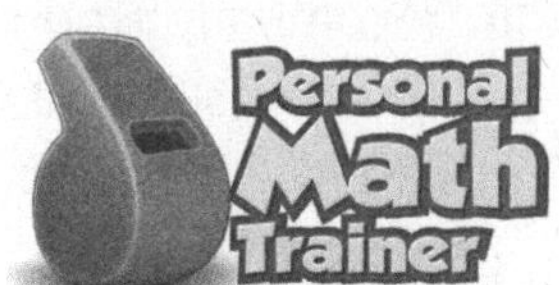

• Online Homework
• Hints and Help
• Extra Practice

1. How is the graph of $g(x) = \frac{1}{3}\sqrt{x-1}$ related to the graph of $f(x) = \sqrt{x}$?

A. A vertical stretch by a factor of 3 and a translation 1 unit right

B. A vertical compression by a factor of $\frac{1}{3}$ and a translation 1 unit up

C. A vertical compression by a factor of $\frac{1}{3}$ and a translation 1 unit right

D. A vertical compression by a factor of $\frac{1}{3}$ and a translation 1 unit left

2. Which function represents the transformation of the graph of $f(x) = \sqrt[3]{x}$ by a vertical stretch by a factor of 5 and a translation 1 unit left and 2 units down?

A. $g(x) = 2\sqrt[3]{x+1} - 5$

B. $g(x) = \frac{1}{5}\sqrt[3]{x+1} - 2$

C. $g(x) = 5\sqrt[3]{x-1} + 2$

D. $g(x) = 5\sqrt[3]{x+1} - 2$

3. What is the inverse of $f(x) = \frac{1}{2}x^3 + 5$?

A. $f^{-1}(x) = \sqrt[3]{2(x-5)}$

B. $f^{-1}(x) = \sqrt[3]{\frac{1}{2}(x-5)}$

C. $f^{-1}(x) = \sqrt[3]{2(x+5)}$

D. $f^{-1}(x) = \sqrt[3]{2x} - 10$

4. Which expression can be simplified to 25?

A. $\left(\sqrt{125}\right)^3$

B. $\left(\sqrt[3]{125}\right)^2$

C. $\left(\sqrt{125}\right)^{\frac{1}{3}}$

D. $\sqrt[3]{125}$

5. Which expression is equivalent to $\frac{2x^2}{\sqrt[3]{y}}$? Assume that all variables are positive.

A. $\sqrt[3]{\frac{8x^6}{y}}$

B. $\sqrt{\frac{8x^6}{y^3}}$

C. $\sqrt[3]{\frac{6x^6}{y}}$

D. $\sqrt[3]{\frac{8x^9}{3y}}$

6. A company produces canned tomatoes. They want the height h of each can to be twice the diameter d. Write an equation for the surface area A of the can in terms of the radius r. If the company wants to use no more than 90 square inches of metal for each can, what is the maximum radius of the can they will produce? *(Lesson 11.1)*

7. The period T of a pendulum in seconds is given by $T = 2\pi\sqrt{\frac{L}{9.81}}$, where L is the length of the pendulum in meters. If you want to use a pendulum as a clock, with a period of 2 seconds, how long should the pendulum be? *(Lesson 12.3)*

Performance Tasks

★ **8.** The formula $P = 73.3\sqrt[4]{m^3}$, known as Kleiber's law, relates the metabolism rate P of an organism in Calories per day and the body mass m of the organism in kilograms. The table shows the typical body mass of several members of the cat family.

Typical Body Mass	
Animal	**Mass(kg)**
House cat	4.5
Cheetah	55.0
Lion	170.0

A. What is the metabolism rate of a cheetah to the nearest Calorie per day?

B. Approximately how many more Calories of food does a lion need to consume each day than a house cat does?

★★ **9.** On a clear day, the approximate distance d in miles that a person can see is given by $d = 1.2116\sqrt{h}$, where h is the person's height in feet above the ocean.

A. To the nearest tenth of a mile, how far can the captain on a clipper ship 15 feet above the ocean see?

B. How much farther, to the nearest tenth of a mile, will a sailor at the top of a mast 120 feet above the ocean be able to see than will the captain?

C. A pirate ship is approaching the clipper ship at a relative speed of 10 miles per hour. Approximately how many minutes sooner will the sailor be able to see the pirate ship than will the captain?

★★★10. The time it takes a pendulum to make one complete swing back and forth depends on its string length, as shown in the table.

String Length (m)	2	4	6	8	10
Time (s)	2.8	4.0	4.9	5.7	6.3

A. Graph the relationship between string length and time, and identify the parent function which best describes the data.

B. The function $T(x) = 2\pi\sqrt{\frac{x}{9.8}}$ gives the period in seconds of a pendulum of length x. The period of a pendulum is the time it takes the pendulum to complete one back-and-forth swing. Describe the graph of T as a transformation of $f(x) = \sqrt{x}$.

C. By what factor must the length of a pendulum be increased to double its period?

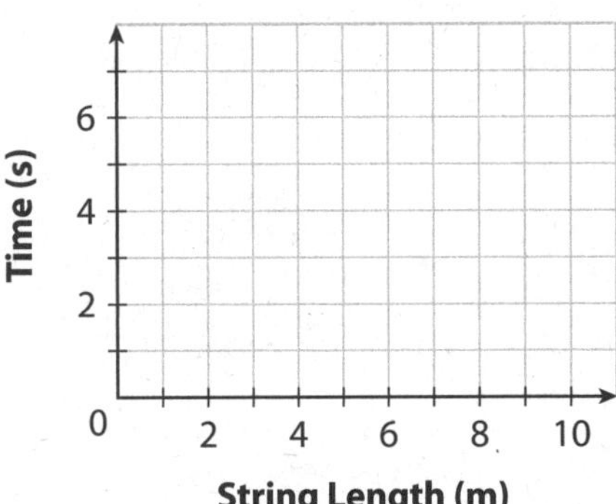

MATH IN CAREERS

Nutritionist Body mass index (BMI) is a measure used to determine healthy body mass based on a person's height. BMI is calculated by dividing a person's mass in kilograms by the square of his or her height in meters. The median BMI measures for a group of boys, ages 2 to 10 years are given in the chart below.

Age of Boys	2	3	4	5	6	7	8	9	10
Median BMI	16.6	16.0	15.6	15.4	15.4	15.5	15.8	16.2	16.6

a. Create a scatter plot for the data in the table, treating age as the independent variable x and median BMI as the dependent variable y.

b. Use a calculator to find a quadratic regression model for the data. What is your model?

c. Give the domain of $f(x)$ based on the data set. Because $f(x)$ is quadratic, it is not one-to-one and its inverse is not a function. Restrict the domain of $f(x)$ to values of x for which $f(x)$ is increasing so that its inverse will be a function. What is the restricted domain of $f(x)$?

d. Find and graph the inverse of $f(x)$. What does $f^{-1}(x)$ model?

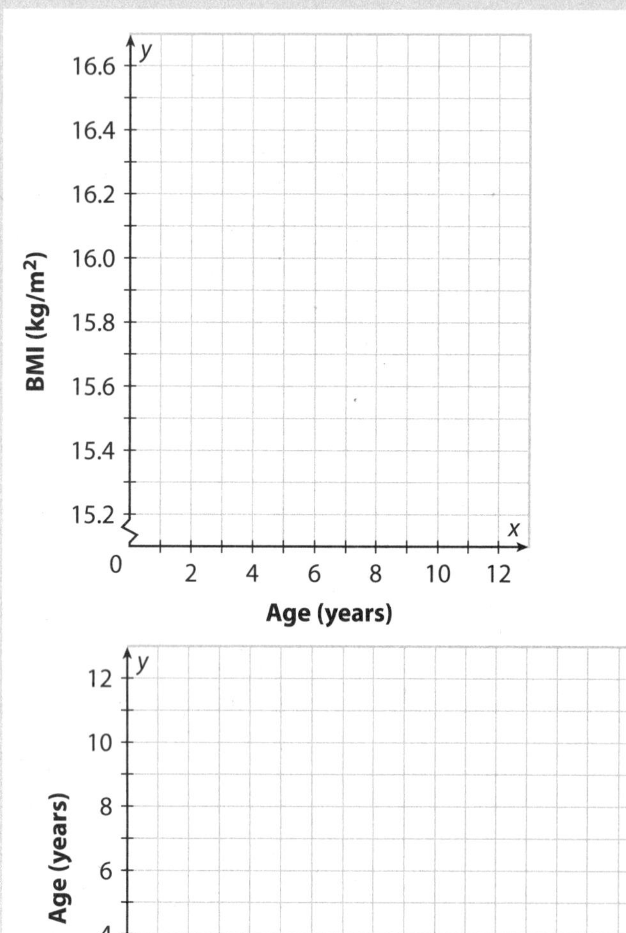

UNIT 6

Exponential and Logarithmic Functions and Equations

MODULE 13

Exponential Functions

TEKS A2.2.A, A2.5.A, A2.5.B

MODULE 14

Modeling with Exponential and Other Functions

TEKS A2.8.A, A2.8.B, A2.8.C

MODULE 15

Logarithmic Functions

TEKS A2.2.A, A2.2B, A2.2.C, A2.5.A, A2.5.B, A2.5.C, A2.7.I

MODULE 16

Logarithmic Properties and Exponential and Logarithmic Equations

TEKS A2.5.B, A2.5.C, A2.5.D

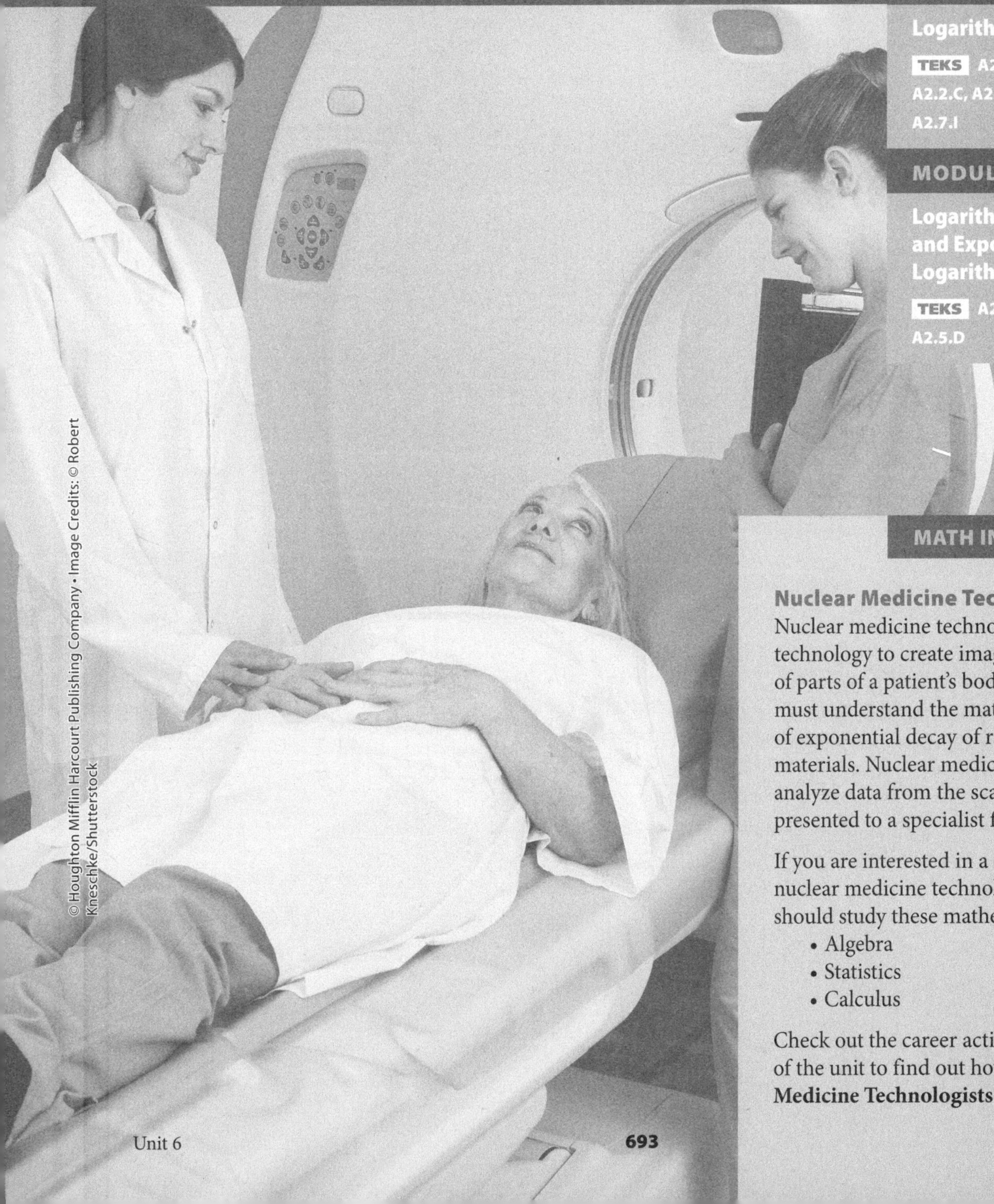

MATH IN CAREERS

Nuclear Medicine Technologist
Nuclear medicine technologists use technology to create images, or *scans*, of parts of a patient's body. They must understand the mathematics of exponential decay of radioactive materials. Nuclear medicine technologists analyze data from the scan, which are presented to a specialist for diagnosis.

If you are interested in a career as a nuclear medicine technologist, you should study these mathematical subjects:

- Algebra
- Statistics
- Calculus

Check out the career activity at the end of the unit to find out how **Nuclear Medicine Technologists** use math.

Reading Start-Up

Vocabulary

Review Words

- asymptote (asíntota)
- inverse function (función inversa)
- ✔ translation (traslación)
- ✔ vertical compression (compresión vertical)
- ✔ vertical stretch (estiramiento vertical)

Preview Words

- explicit formula (fórmula explícita)
- geometric sequence (sucesión geométrica)
- logarithmic function (función logarítmica)
- recursive formula (fórmula recurrente)

Visualize Vocabulary

Use the ✓ words to complete the graphic.

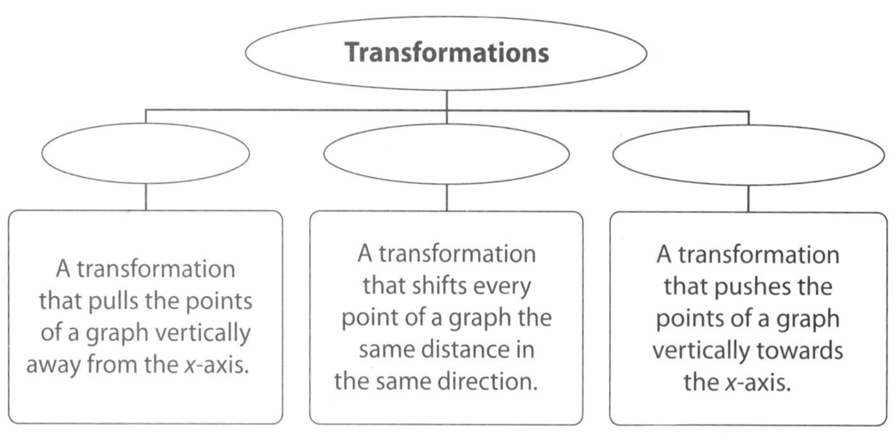

Understand Vocabulary

To become familiar with some of the vocabulary terms in the module, consider the following. You may refer to the module, the glossary, or a dictionary.

1. A sequence in which the ratio of successive terms is a constant r, where $r \neq 0$ or 1, is a ____________________.

2. A ____________________ is a rule for a sequence in which one or more previous terms are used to generate the next term.

3. A ____________________ is the inverse of an exponential function.

Active Reading

Booklet Before beginning the unit, create a booklet to help you organize what you learn. Each page of the booklet should correspond to a lesson in the unit and summarize the most important information from the lesson. Include definitions, examples, and graphs to help you recall the main elements of the lesson. Highlight the main idea and use colored pencils to help you illustrate any important concepts or differences.

Exponential Functions

MODULE 13

Essential Question: How can you use exponential functions to solve real-world problems?

TEKS

LESSON 13.1
Geometric Sequences
TEKS A2.5.B

LESSON 13.2
Exponential Growth Functions
TEKS A2.2.A, A2.5.A, A2.5.B

LESSON 13.3
Exponential Decay Functions
TEKS A2.2.A, A2.5.B

LESSON 13.4
The Base *e*
TEKS A2.2.A, A2.5.A, A2.5.B

REAL WORLD VIDEO
Check out how exponential functions can be used to model the path of a bouncing ball and other real-world patterns.

MODULE PERFORMANCE TASK PREVIEW

That's the Way the Ball Bounces

The height that a ball reaches after bouncing off a hard surface depends on several factors, including the height from which the ball was dropped and the material the ball is made of. The height of each successive bounce will be less than the previous bounce. How can mathematical modeling be used to represent a bouncing ball? Let's find out!

Are YOU Ready?

Complete these exercises to review skills you will need for this module.

- Online Homework
- Hints and Help
- Extra Practice

Real Numbers

Example 1 Write the multiplicative inverse of 0.3125.

$0.3125 = \frac{3125}{10,000}$ Write the decimal as a fraction.

$= \frac{5}{16} \rightarrow \frac{16}{5}$ Simplify, then take the reciprocal.

The multiplicative inverse of 0.3125 is $\frac{16}{5}$.

Write the multiplicative inverse.

1. 1.125 ______ **2.** −0.6875 ______ **3.** 0.444 ______

Exponential Functions

Example 2 Evaluate $12(1.5)^{x-5}$ for $x = 8$.

$12(1.5)^{8-5}$ Substitute.

$12(1.5)^3$ Simplify.

$12(3.375) = 40.5$ Evaluate the exponent. Multiply.

Evaluate each expression.

4. $30(1.1)^{5x}$; $x = 0.4$ ______ **5.** $8(2.5)^{\frac{x}{3}}$; $x = 12$ ______ **6.** $2(x)^{2x+9}$; $x = -3$ ______

Geometric Sequences

Example 3 List the first four terms of the sequence if $t_1 = 2$ and $t_n = -3t_{n-1}$.

$t_2 = -3t_{2-1} = -3t_1 = -3 \cdot 2 = -6$ Find the second term.

$t_3 = -3t_{3-1} = -3t_2 = -3 \cdot (-6) = 18$ Find the third term.

$t_4 = -3t_{4-1} = -3t_3 = -3 \cdot 18 = -54$ Find the fourth term.

The first four terms are 2, –6, 18, and –54.

List the first four terms of the sequence.

7. $t_1 = 0.02$

$t_n = 5t_{n-1}$ ______

8. $t_1 = 3$

$t_n = 0.5t_{n-1}$ ______

Name________________________ Class______________ Date________

13.1 Geometric Sequences

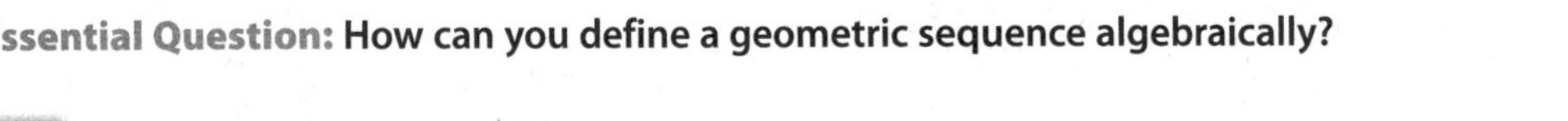

Essential Question: How can you define a geometric sequence algebraically?

A2.5.B Formulate exponential … equations that model real-world situations, including exponential relationships written in recursive notation. Also A2.5.A, A2.7.I

Resource Locker

Explore Investigating Geometric Sequences

As a tree grows, limbs branch off of the trunk, then smaller limbs branch off these limbs and each branch splits off into smaller and smaller copies of itself the same way throughout the entire tree. A mathematical object called a *fractal tree* resembles this growth.

Start by drawing a vertical line at the bottom of a piece of paper. This is Stage 0 of the fractal tree and is considered to be one 'branch'. The length of this branch defines 1 unit.

For Stage 1, draw 2 branches off of the top of the first branch. For this fractal tree, each smaller branch is $\frac{1}{2}$ the length of the previous branch and is at a 45-degree angle from the direction of the parent branch. The first four iterations, Stages 0-3, are shown.

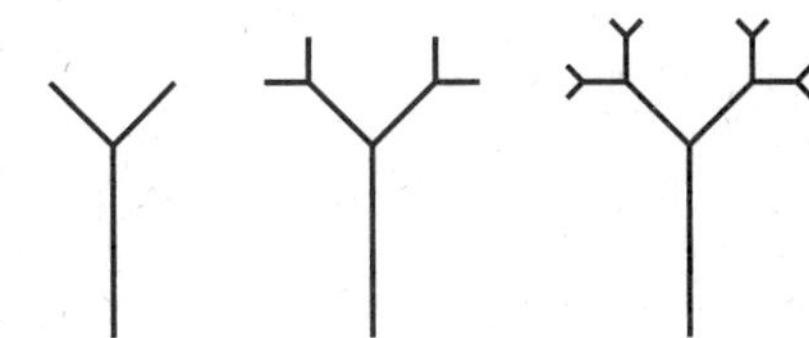

Ⓐ In Stage 2, there are 2 branches drawn on the end of each of the 2 branches drawn in Stage 1. There are ☐ new branches in Stage 2. Each one of these branches will be $\frac{1}{2}$ the length of its predecessors or ☐ unit in length.

Ⓑ For Stage 3, there are 8 new branches in total. To draw Stage 4, a total of ☐ branches must be drawn and to draw Stage 5, a total of ☐ branches must be drawn. Thus, each stage adds ☐ times as many branches as the previous stage did.

Ⓒ Complete the table.

Stage	New Branches	Pattern	New Branches as a Power
Stage 0	1	1	2^0
Stage 1	2	$2 \cdot 1$	2^1
Stage 2	4	$2 \cdot 2$	2^2
Stage 3	8	$2 \cdot$	
Stage 4	16	$2 \cdot$	
Stage 5	32	$2 \cdot$	
Stage 6	64	$2 \cdot$	

(D) The procedure for each stage after Stage 0 is to draw ______ branches on ______ branch added in the previous step.

(E) Using the description above, write an equation for the number of new branches in a stage given the previous stage. Represent stage s as N_s; Stage 3 will be N_3.

$N_4 = \square \cdot \square$ $N_6 = \square \cdot \square$

$N_5 = \square \cdot \square$ $N_s = \square \cdot \square$

(F) Rewrite the rule for Stage s as a function $N(s)$ that has a stage number as an input and the number of new branches in the stage as an output.

$N(s) = \square$

(G) Recall that the domain of a function is the set of all numbers for which the function is defined. $N(s)$ is a function of s and s is the stage number. Since the stage number refers to the ______ the tree has branched, it has to be ______.

Write the domain of $N(s)$ in set notation.

$\{s \mid s \text{ is a } \_\_\_\_\_\_ \text{ number}\}$

(H) Similarly, the range of a function is the set of all possible values that the function can output over the domain. Let $N(s) = b$, the ______.

The range of $N(s)$ is $\{1,2,4 \square, \square, \square, \ldots\}$.

The range of $N(s)$ is $\{N \mid N = 2^n$, where n is ______$\}$.

(I) Graph the first five values of $N(s)$ on the axes provided.

The first value has been graphed for you.

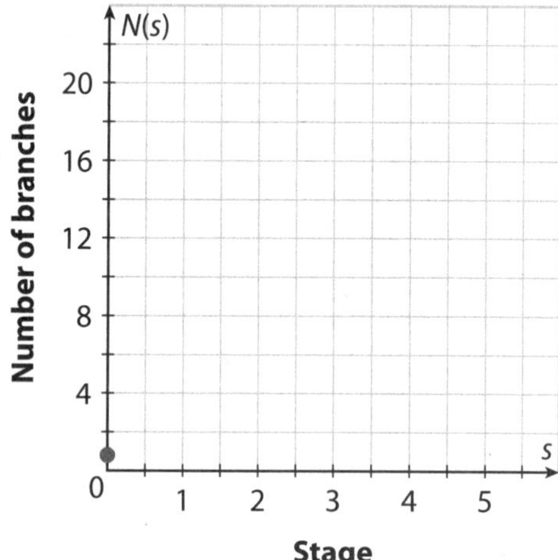

(J) As s increases, $N(s)$ ______.

$N(s)$ is ______ function.

(K) Complete the table for branch length.

Stage Number	Number of Branches	Branch Length
0	1	1
1	2	$\frac{1}{2}$
2	4	$\frac{1}{4}$
3	8	$\frac{1}{\square}$
4	16	$\frac{1}{\square}$
5	32	$\frac{1}{\square}$
⋮	⋮	⋮
n	2^n	$\frac{1}{\square}$

(L) Write $L(s)$ expressing the branch length as a function of the Stage.

$$L(s) = \left(\frac{1}{\square}\right)^{\square}$$

(M) Write the domain and range of $L(s)$ in set notation.

The domain of $L(s)$ is $\{s \mid s$ is a ________ number$\}$.

The range of $L(s)$ is $\left\{1, \frac{1}{2}, \frac{1}{4}, \square, \square, \square, \ldots\right\}$.

The range of $L(s)$ is $\left\{L \mid L = \left(\frac{1}{2}\right)^n\right.$, where n is ________ $\}$.

(N) Graph the first five values of $L(s)$ on the axes provided. The fifth point has been graphed for you.

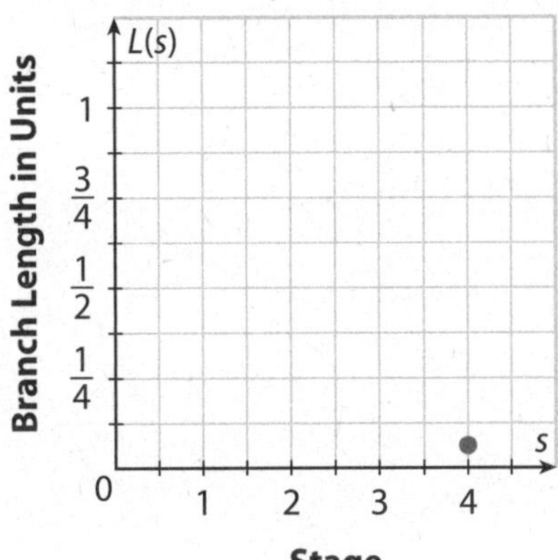

(O) As s increases, $L(s)$ ________. $L(s)$ is ________ function.

Reflect

1. What is the total length added at each stage?

____

____

____

2. Is the total length of all the branches a sequence? If so, identify the sequence.

____

Writing Explicit and Recursive Rules for Geometric Sequences

A sequence is a set of numbers related by a common rule. All sequences start with an initial term. In a **geometric sequence**, the ratio of any term to the previous term is constant. This constant ratio is called the **common ratio** and is denoted by r $(r \neq 1)$. In the **explicit form** of the sequence, each term is found by evaluating the function $f(n) = ar^n$ or $f(n) = ar^{n-1}$ where a is the initial value and r is the common ratio, for some whole number n. Note that there are two forms of the explicit rule because it is permissible to call the initial value the first term or to call ar the first term.

A geometric sequence can also be defined recursively by $f(n) = r \cdot f(n-1)$ where either $f(0) = a$ or $f(1) = a$, again depending on the way the terms of the sequence are numbered. $f(n) = r \cdot f(n-1)$ is called the **recursive rule** for the sequence.

Example 1 **Write the explicit and recursive rules for a geometric sequence given a table of values.**

(A)

n	0	1	2	3	4	...	$j-1$	j	...
$f(n)$	3	6	12	24	48	...	$ar^{(j-1)}$	ar^j	...

Determine a and r, then write the explicit and recursive rules.

Find the common ratio: $\dfrac{f(n)}{f(n-1)} = r$. $\quad \dfrac{f(1)}{f(0)} = \dfrac{6}{3} = 2 = r$

Find the initial value, $a = f(0)$, from the table. $\quad f(0) = 3 = a$

Find the explicit rule: $f(n) = ar^n$. $\quad f(n) = 3 \cdot (2)^n$

Write the recursive rule. $\quad f(n) = 2 \cdot f(n-1), n \geq 1$ and $f(0) = 3$

The explicit rule is $f(n) = 3 \cdot (2)^n$ and the recursive rule is $f(n) = 2 \cdot f(n-1), n \geq 1$ and $f(0) = 3$.

n	1	2	3	4	5	...	$j-1$	j	...
$f(n)$	$\frac{1}{25}$	$\frac{1}{5}$	1	5	25	...	$ar^{(j-1)}$	ar^j	...

Determine a and r, then write the explicit and recursive rules.

Find the common ratio: $\dfrac{f(n)}{f(n-1)} = r$. $\quad \dfrac{f(\square)}{f(\square)} = \dfrac{\square}{\square} = \square = r$

Find the initial value, $a = f(1)$, from the table. $\quad f(1) = \square = a$

Find the explicit rule: $f(n) = ar^{n-1}$. $\quad f(n) = \square \cdot (\square)^{n-1}$

Write the recursive rule. $\quad f(n) = \square \cdot f(n-1), n \geq \square$ and $f(1) = \square$

The explicit rule is $f(n) = \square$ and the recursive rule is $f(n) = \square \cdot f(n-1), n \geq \square$ where $f(1) = \square$.

Reflect

3. **Discussion** If you were told that a geometric sequence had an initial value of $f(5) = 5$, could you write an explicit and a recursive rule for the function? What would the explicit rule be?

Your Turn

Write the explicit and recursive rules for a geometric sequence given a table of values.

4.

n	0	1	2	3	4	5	6	...
$f(n)$	$\frac{1}{27}$	$\frac{1}{9}$	$\frac{1}{3}$	1	3	9	27	...

5.

n	1	2	3	4	5	6	7	...
$f(n)$	0.001	0.01	0.1	1	10	100	1000	...

Explain 2 Graphing Geometric Sequences

To graph a geometric sequence given an explicit or a recursive rule you can use the rule to generate a table of values and then graph those points on a coordinate plane. Since the domain of a geometric sequence consists only of whole numbers, its graph consists of individual points, not a smooth curve.

Example 2 **Given either an explicit or recursive rule for a geometric sequence, use a table to generate values and draw the graph of the sequence.**

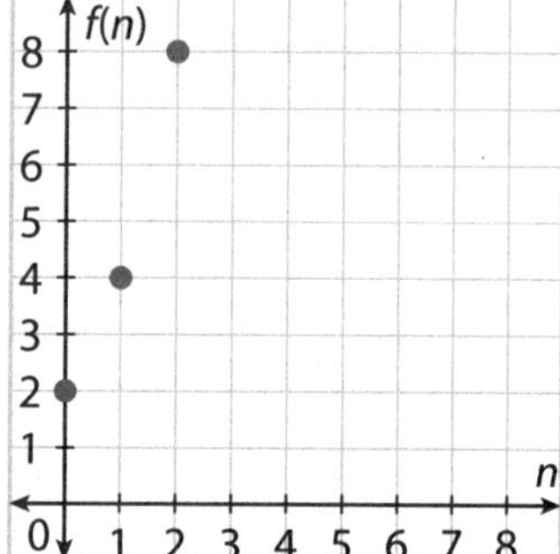

(A) Explicit rule: $f(n) = 2 \cdot 2^n, n \geq 0$

Use a table to generate points.

n	0	1	2	3	4	5	...
f(n)	2	4	8	16	32	64	...

Plot the first three points on the graph.

(B) Recursive rule: $f(n) = 0.5 \cdot f(n - 1), n \geq 1$ and $f(0) = 16$

Use a table to generate points.

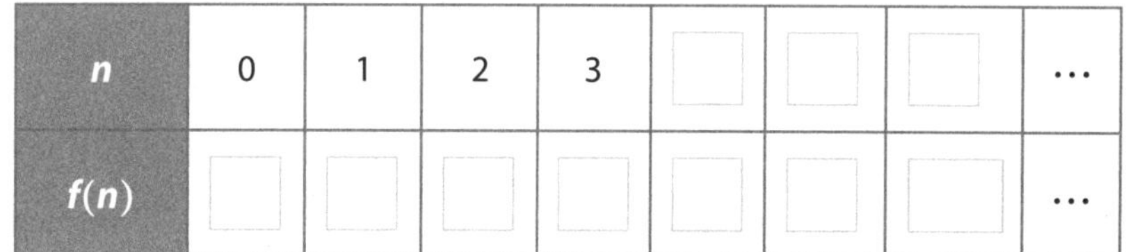

n	0	1	2	3				...
f(n)								...

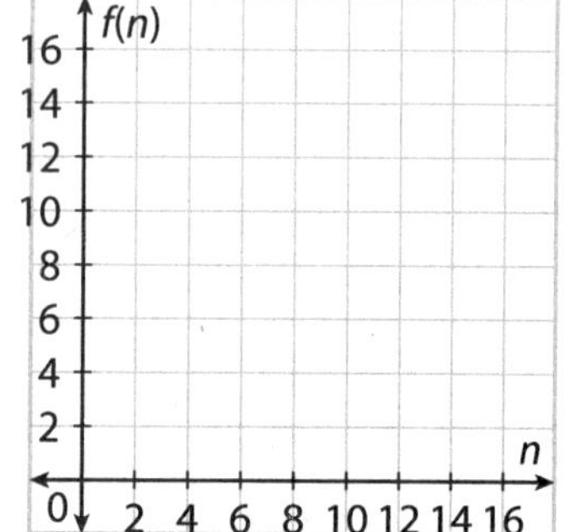

Your Turn

Given either an explicit or recursive rule for a geometric sequence, use a table to generate values and draw the graph of the sequence.

6. $f(n) = 3 \cdot 2^{n-1}, n \geq 1$

n	1	2	3	4	5	...
f(n)						...

7. $f(n) = 3 \cdot f(n - 1), n \geq 2$ and $f(1) = 2$

n	1	2	3	4	5	...
f(n)						...

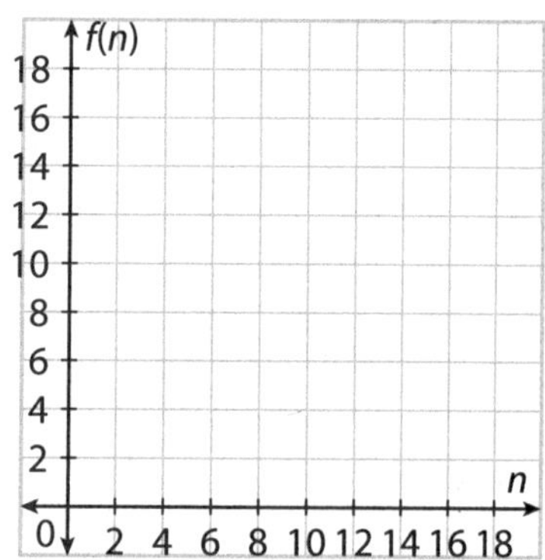

Explain 3 Modeling With a Geometric Sequence

Given a real-world situation that can be modeled by a geometric sequence, you can use an explicit or a recursive rule to answer a question about the situation.

Example 3 **Write both an explicit and recursive rule for the geometric sequence that models the situation. Use the sequence to answer the question asked about the situation.**

(A) The Wimbledon Ladies' Singles Championship begins with 128 players. Each match, two players play and only one moves to the next round. The players compete until there is one winner. How many rounds must the winner play?

Analyze Information

Identify the important information:

- The first round requires ______ matches, so $a =$ ☐.
- The next round requires half as many matches, so $r =$ ☐.

Formulate a Plan

Using the fact that the domain starts at ______ and the first round has 128 players, create the explicit rule and the recursive rule for the tournament. The final round will have ______ match(es), so substitute this value into the explicit rule and solve for n.

Solve

The explicit rule is $f(n)$ ☐, $n = \geq 1$.

The recursive rule is $f(n) =$ ☐ $\cdot f(n-1)$, $n \geq 2$ and $f(1) =$ ☐.

The final round will have 1 match, so substitute 1 for $f(n)$ into the explicit rule and solve for n.

$$f(n) = 64 \cdot \left(\frac{1}{2}\right)^{n-1}$$

$$\square = 64 \cdot \left(\frac{1}{2}\right)^{n-1}$$

$$\square = \left(\frac{1}{2}\right)^{n-1}$$

$$\left(\frac{1}{2}\right)^{\square} = \left(\frac{1}{2}\right)^{n-1}$$

Two powers with the same positive base other than 1 are equal if and only if the exponents are equal.

$$\left(\frac{1}{2}\right)^{\square} = \left(\frac{1}{2}\right)^{n-1}$$

$$\square = n - 1$$

$$\square = n$$

The winner must play in ______ rounds.

Justify and Evaluate

The answer of 7 rounds makes sense because using the explicit rule gives $f(7) =$ ______ and the final round will have 1 match(es). This result can be checked using the recursive rule, which again results in $f(7) =$ ______.

Your Turn

Write both an explicit and recursive rule for the geometric sequence that models the situation. Use the sequence to answer the question asked about the situation.

8. A particular type of bacteria divides into two new bacteria every 20 minutes. A scientist growing the bacteria in a laboratory begins with 200 bacteria. How many bacteria are present 4 hours later?

Elaborate

9. Describe the difference between an explicit rule for a geometric sequence and a recursive rule.

10. How would you decide to use $n = 0$ or $n = 1$ as the starting value of n for a geometric sequence modeling a real-world situation?

11. **Essential Question Check-In** How can you define a geometric sequence in an algebraic way? What information do you need to write these rules?

Evaluate: Homework and Practice

- Online Homework
- Hints and Help
- Extra Practice

You are creating self-similar fractal trees. You start with a trunk of length 1 unit (at Stage 0). Then the trunk splits into two branches each one-third the length of the trunk. Then each one of these branches splits into two new branches, with each branch one-third the length of the previous one.

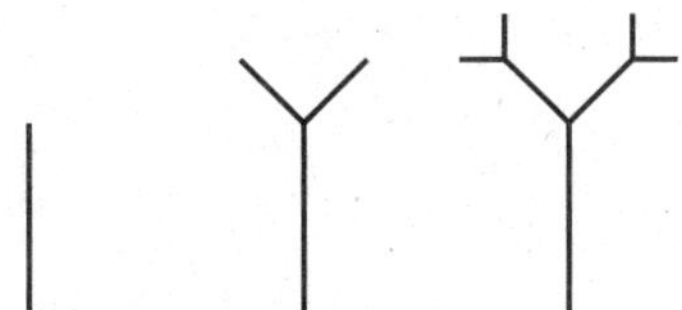

1. Can the length of the new branches at each stage be described with a geometric sequence? Explain. If so, find the explicit form of the length of each branch.

2. Can the number of new branches at each stage be described with a geometric sequence? Explain. If so, find the recursive rule for the number of new branches.

3. Can the total length of the new branches at each stage be modeled with a geometric sequence? Explain. (The total length of the new branches is the sum of the lengths of all the new branches.)

Write the explicit and recursive rules for a geometric sequence given a table of values.

4.

n	0	1	2	3	4	...
$f(n)$	0.1	0.3	0.9	2.7	8.1	...

5.

n	0	1	2	3	4	...
$f(n)$	100	10	1	0.1	0.01	...

6.

n	1	2	3	4	5	...
$f(n)$	1000	100	10	1	0.1	...

7.

n	1	2	3	4	5	...
$f(n)$	10^{50}	10^{47}	10^{44}	10^{41}	10^{38}	...

Given either an explicit or recursive rule for a geometric sequence, use a table to generate values and draw the graph of the sequence.

8. $f(n) = \left(\frac{1}{2}\right) \cdot 4^n, n \geq 0$

n	0	1	2	3	4	...
f(n)						...

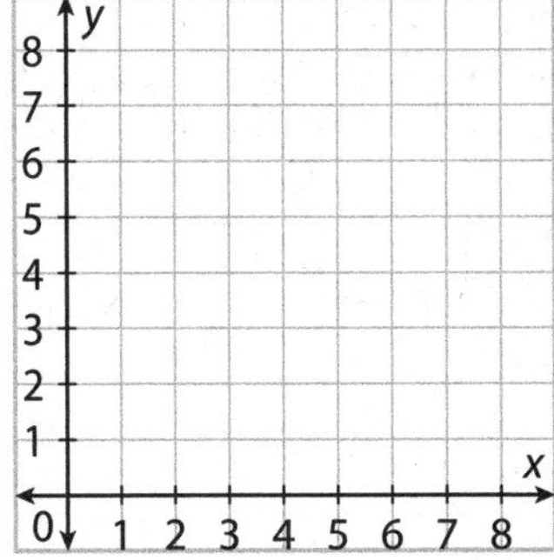

9. $f(n) = 2f(n-1), n \geq 1$ and $f(0) = 0.5$

n	0	1	2	3	4	...
f(n)						...

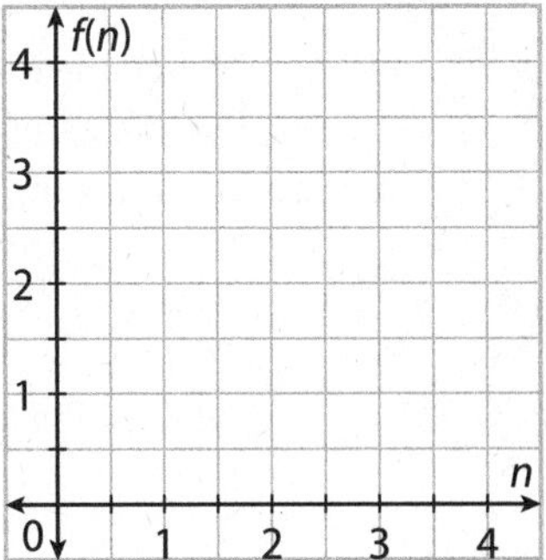

10. $f(n) = 0.5 \cdot f(n-1), n \geq 2$ and $f(1) = 8$

n	1	2	3	4	5	...
f(n)						...

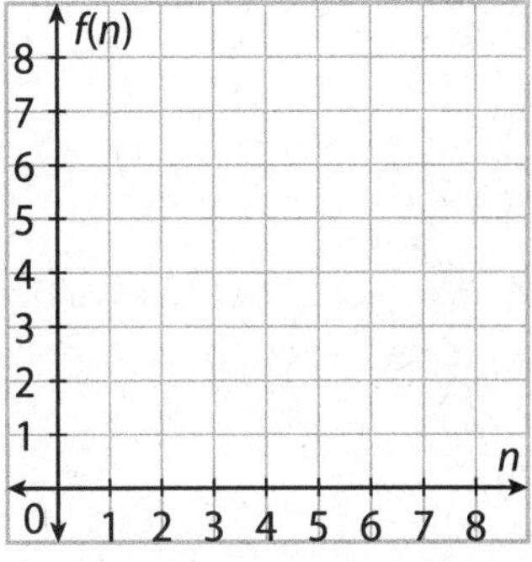

11. $f(n) = \frac{2}{3} \cdot f(n-1), n \geq 2$ and $f(1) = 1$

n	1	2	3	4	5	...
f(n)						...

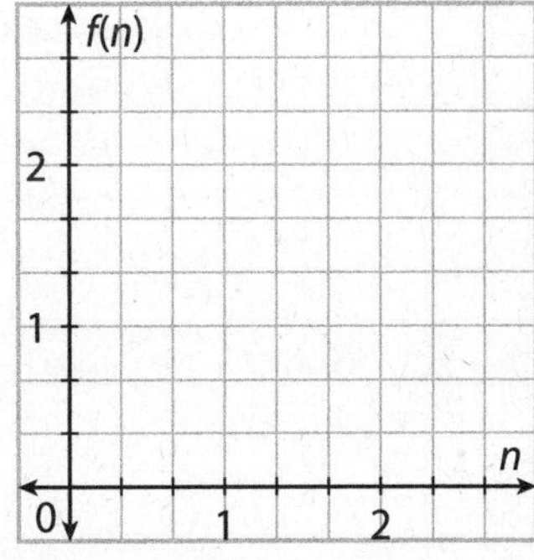

Write both an explicit and recursive rule for the geometric sequence that models the situation. Use the sequence to answer the question asked about the situation.

12. The Alphaville Youth Basketball committee is planning a single-elimination tournament. The committee wants the winner to play 4 games. How many teams should the committee invite?

13. An online video game tournament begins with 1024 players. Four players play in each game. In each game there is only one winner, and only the winner advances to the next round. How many games will the winner play?

14. **Genealogy** You have 2 biological parents, 4 biological grandparents, and 8 biological great-grandparents.

a. How many direct ancestors do you have if you trace your ancestry back 6 generations? How many direct ancestors do you have if you go back 12 generations?

b. **What if...?** How does the explicit rule change if you are considered the first generation?

15. Fractals Waclaw Sierpinski designed various fractals. He would take a geometric figure, shade it in, and then start removing the shading to create a fractal pattern.

a. The Sierpinski triangle is a fractal based on a triangle. In each iteration, the center of each shaded triangle is removed.

Given that the area of the original triangle is 1 square unit, write a sequence for the area of the nth iteration of the Sierpinski triangle. (The first iteration is the original triangle.)

b. The Sierpinski carpet is a fractal based on a square. In each iteration, the center of each shaded square is removed.

Given that the area of the original square is 1 square unit, write a sequence for the area of the nth iteration of the Sierpinski carpet. (The first iteration is the original square.)

c. Find the shaded area of the fourth iteration of the Sierpinski carpet.

16. A piece of paper is 0.1 millimeter thick. When folded, the paper is twice as thick.

a. Find both the explicit and recursive rule for this geometric sequence.

b. Studies have shown that you can fold a piece of paper a maximum of 7 times. How thick will the paper be if it is folded on top of itself 7 times?

c. Assume that you could fold the paper as many times as you want. How many folds would be required for the paper to be taller than Mount Everest at 8850 meters? (Hint: Use a calculator to generate two large powers of 2 and check if the required number of millimeters is between those two powers. Continue to refine your guesses.)

H.O.T. Focus on Higher Order Thinking

17. Justify Reasoning Suppose you have the following table of points of a geometric sequence. The table is incomplete so you do not know the initial value. Determine whether each of the following can or cannot be the rule for the function in the table. If a function cannot be the rule for the sequence, explain why.

n	...	4	5	6	7	...
$f(n)$	...	6	12	24	48	...

A. $f(n) = 2^n$

B. $f(n) = \frac{3}{8} \cdot (2)^n$

C. $f(n) = 2 \cdot f(n-1),\ n \geq 1$ and $f(0) = 6$

D. $f(n) = \frac{3}{4} \cdot (2)^{n-1}$

E. $f(n) = 2 \cdot f(n-1),\ n \geq 1$ and $f(0) = \frac{3}{8}$

F. $f(n) = 2 \cdot f(n-1),\ n \geq 1$ and $f(1) = \frac{3}{4}$

G. $f(n) = (1.5) \cdot (2)^{n-2}$

H. $f(n) = 3 \cdot (2)^{n-3}$

18. Communicate Mathematical Ideas Show that the rules $f(n) = ar^n$ for $n \geq 0$ and $f(n) = ar^{n-1}$ for $n \geq 1$ for a geometric sequence are equivalent.

Lesson Performance Task

Have you ever heard of musical octaves? An octave is the interval between a musical note and the same musical note in the next higher or lower pitch. The frequencies of the sound waves of successive octaves of a note form a geometric sequence. For example, the table shows the frequencies in hertz (Hz), or cycles per second, produced by playing the note D in ascending octaves, D_0 being the lowest D note audible to the human ear.

Scale of D's	
Note	**Frequency (Hz)**
D_0	18.35
D_1	36.71
D_2	73.42
D_3	146.83

a. Explain how to write an explicit rule and a recursive rule for the frequency of D notes in hertz, where $n = 1$ represents D_1.

b. The note commonly called "middle D" is D_4. Use the explicit rule or the recursive rule from part **a** to predict the frequency for middle D.

c. Humans generally cannot hear sounds with frequencies greater than 20,000 Hz. What is the first D note that humans cannot hear? Explain.

Name _______________ Class _______________ Date _______________

13.2 Exponential Growth Functions

Essential Question: How is the graph of $g(x) = ab^{x-h} + k$ where $b > 1$ related to the graph of $f(x) = b^x$?

A2.5.A Determine the effects on the key attributes on the graphs of $f(x) = b^x \ldots$ where b is 2, 10, … when $f(x)$ is replaced by $af(x)$, $f(x) + d$, and $f(x - c)$ for specific positive and negative real values of a, c, and d. Also A2.2.A, A2.5.B, A2.5.D, A2.7.I

Explore 1 Graphing and Analyzing $f(x) = 2^x$ and $f(x) = 10^x$

A *parent exponential function* is a function of the form $f(x) = b^x$, where b is a positive constant other than 1 and the exponent x is a variable. Notice that there is no single parent exponential function because each choice of the base b determines a different parent function.

(A) Complete the input-output table for each of the parent exponential functions below.

x	$f(x) = 2^x$
−3	
−2	
−1	
0	
1	
2	
3	

x	$p(x) = 10^x$
−3	
−2	
−1	
0	
1	
2	
3	

(B) Graph the parent functions $f(x) = 2^x$ and $p(x) = 10^x$ by plotting points.

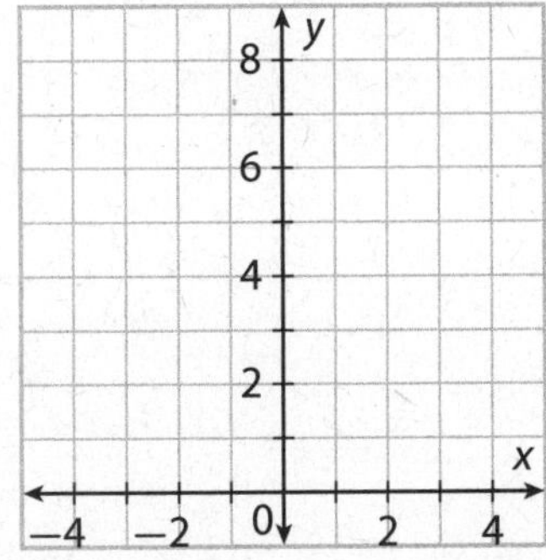

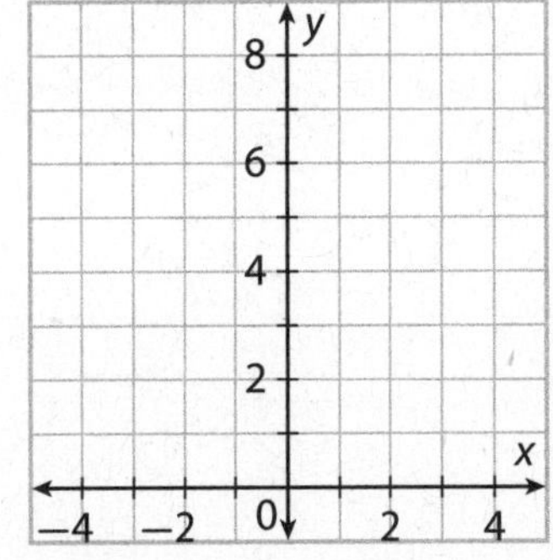

(C) What is the domain of each function?

Domain of $f(x) = 2^x$: $\{x|$ ______ $\}$

Domain of $p(x) = 10^x$: $\{x|$ ______ $\}$

(D) What is the range of each function?

Range of $f(x) = 2^x$: $\{y|$ ______ $\}$

Range of $p(x) = 10^x$: $\{y|$ ______ $\}$

(E) What is the y-intercept of each function?

y-intercept of $f(x) = 2^x$: $(0, \square)$

y-intercept of $p(x) = 10^x$: $(0, \square)$

(F) What is the trend of each function?

In both $f(x) = 2^x$ and $p(x) = 10^x$, as the value of x increases, the value of y increases/decreases.

Reflect

1. Will the domain be the same for every parent exponential function? Why or why not?

______

2. Will the range be the same for every parent exponential function. Why or why not?

______

______

3. Will the value of the y-intercept be the same for every parent exponential function? Why or why not?

______

Explore 2 Predicting Transformations of the Graphs of $f(x) = 2^x$ and $f(x) = 10^x$

Based on your experience with transforming the parent function $f(x)$ in previous lessons, make predictions about the effect of varying the parameters in $g(x) = af(x - c) + d$. Confirm your predictions using a graphing calculator.

The graph of $f(x) = 2^x$ is shown, as is a separate graph of $p(x) = 10^x$.

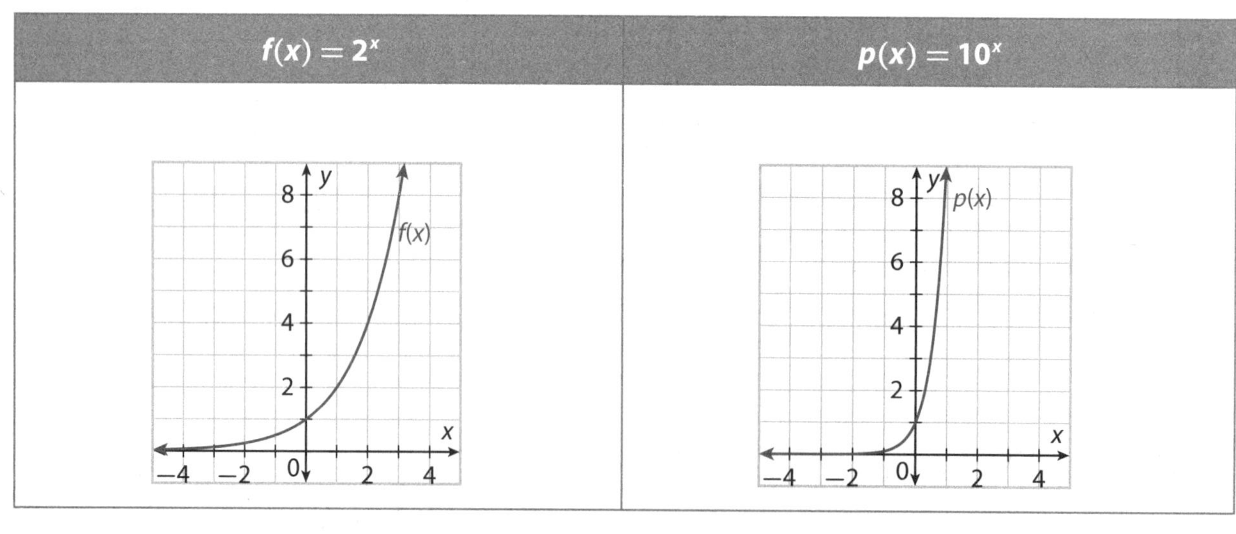

Predict what the graph of each function will look like, and then sketch the graph based on the graph shown on your calculator.

(A) The graph of $g_1(x) = \frac{1}{4}(2^x)$ will be the graph of $f(x) = 2^x$ vertically ________ by a factor of ____.

The graph of $g_2(x) = 3(2^x)$ will be the graph of $f(x) = 2^x$ vertically ________ by a factor of ____.

The graph of $q_1(x) = 2(10^x)$ will be the graph of $p(x) = 10^x$ vertically ________ by a factor of ____.

The graph of $q_2(x) = \frac{1}{5}(10^x)$ will be the graph of $p(x) = 10^x$ vertically ________ by a factor of ____.

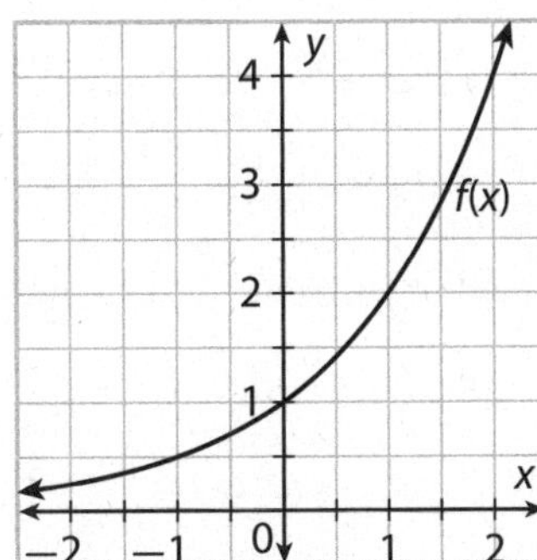

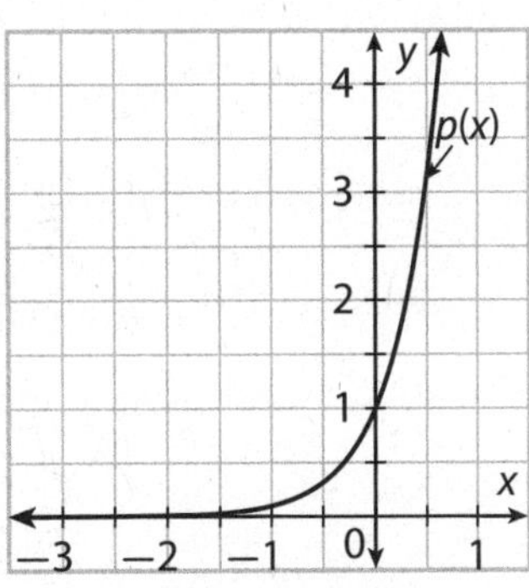

(B) The graph of $g_1(x) = -\frac{3}{4}(2^x)$ will be the graph of $f(x) = 2^x$ reflected across the ________ and vertically ________ by a factor of ____.

The graph of $g_2(x) = -5(2^x)$ will be the graph of $f(x) = 2^x$ reflected across the ________ and vertically ________ by a factor of ____.

The graph of $q_1(x) = -\frac{5}{4}(10^x)$ will be the graph of $p(x) = 10^x$ reflected across the ________ and vertically ________ by a factor of ____.

The graph of $q_2(x) = -\frac{1}{4}(10^x)$ will be the graph of $p(x) = 10^x$ reflected across the ________ and vertically ________ by a factor of ____.

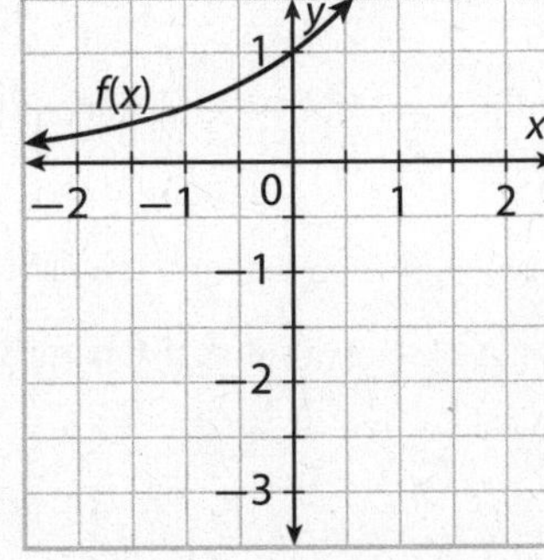

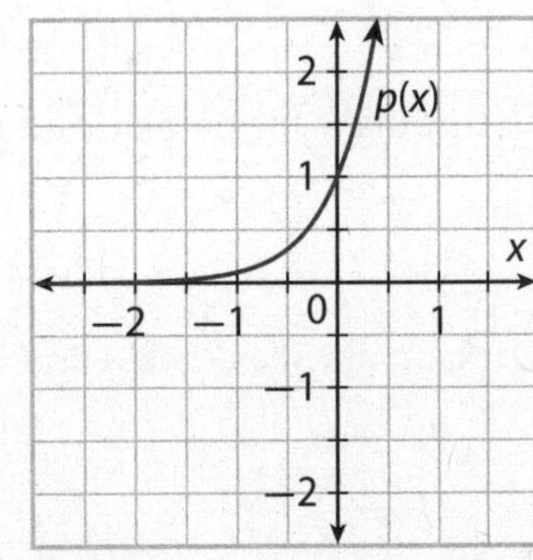

Ⓒ The graph of $g_1(x) = 2^{x+1}$ will be the graph of $f(x) = 2^x$ translated ______ unit to the ______.

The graph of $g_2(x) = 2^{x-4}$ will be the graph of $f(x) = 2^x$ translated ______ units to the ______.

The graph of $q_1(x) = 10^{x+2}$ will be the graph of $p(x) = 10^x$ translated ______ unit to the ______.

The graph of $q_2(x) = 10^{x-3}$ will be the graph of $p(x) = 10^x$ translated ______ units to the ______.

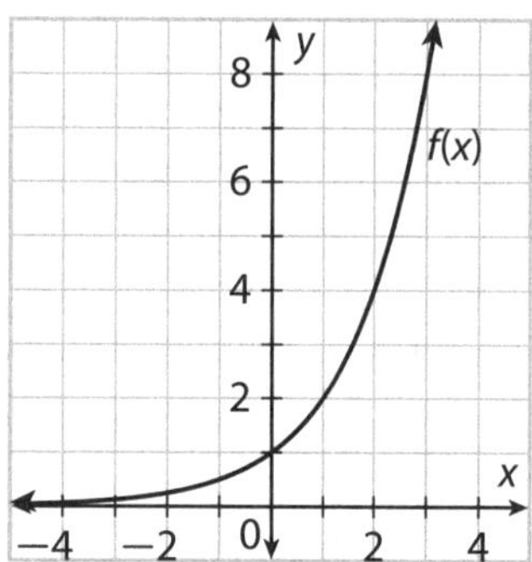

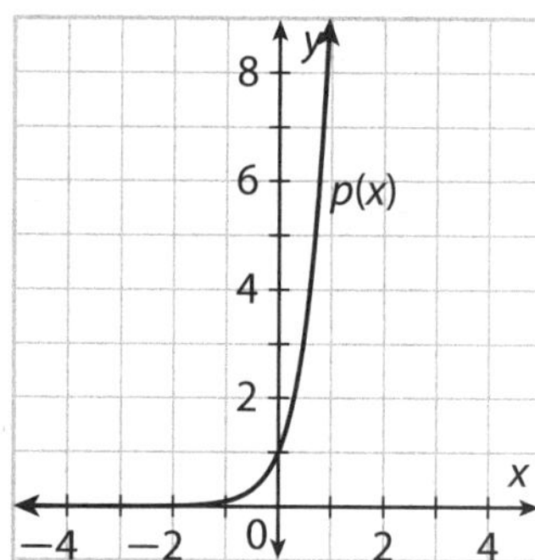

Ⓓ The graph of $g_1(x) = 2^x + 3$ will be the graph of $f(x) = 2^x$ translated ______ units ______.

The graph of $g_2(x) = 2^x - \frac{5}{2}$ will be the graph of $f(x) = 2^x$ translated ______ units ______.

The graph of $q_1(x) = 10^x + 5$ will be the graph of $p(x) = 10^x$ translated ______ units ______.

The graph of $q_2(x) = 10^x - 2$ will be the graph of $p(x) = 10^x$ translated ______ units ______.

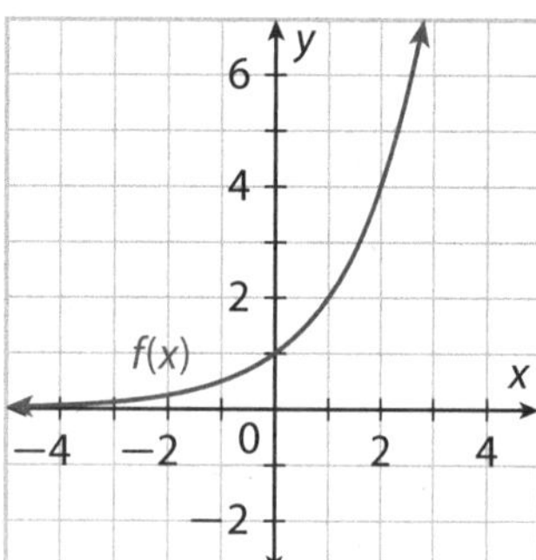

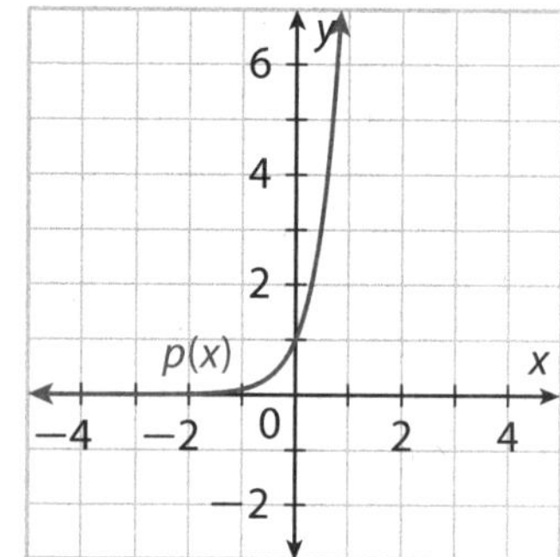

Reflect

4. **Discussion** Identify the values of a that make the domain and range of $g(x) = af(x)$ different from those of $f(x) = b^x$.

______

______

______

5. Identify the values of h that make the domain and range of $g(x) = f(x - h)$ different from those of $f(x) = b^x$.

______

6. Identify the values of k that make the domain and range of $g(x) = f(x) + k$ different from those of $f(x) = b^x$.

______

______

Explain 1 Graphing Combined Transformations of $f(x) = b^x$ Where $b > 1$

A given exponential function $g(x) = a(b^{x-h}) + k$ with base b can be graphed by recognizing the differences between the given function and its parent function, $f(x) = b^x$. These differences define the parameters of the transformation, where k represents the vertical translation, h is the horizontal translation, and a represents either the vertical stretch or compression of the exponential function and whether it is reflected across the x-axis.

You can use the parameters in $g(x) = a(b^x - h) + k$ to see what happens to two reference points during a transformation. Two points that are easily visualized on the parent exponential function are $(0, 1)$ and $(1, b)$.

In a transformation, the point $(0, 1)$ becomes $(h, a + k)$ and $(1, b)$ becomes $(1 + h, ab + k)$. The asymptote $y = 0$ for the parent function becomes $y = k$.

The graphs of $f(x) = 2^x$ and $p(x) = 10^x$ are shown below with the reference points and asymptotes labeled.

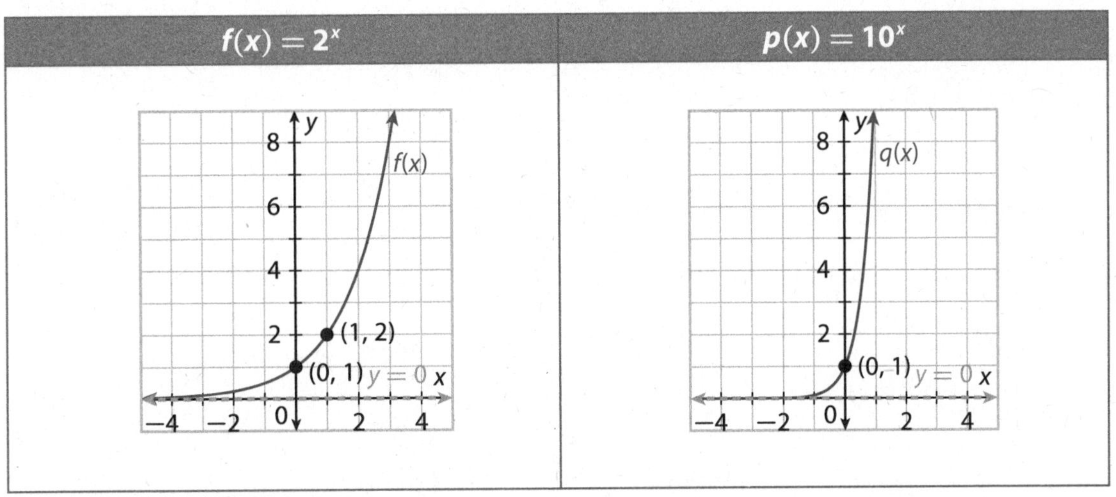

Example 1 **State the domain and range of the given function. Then identify the new values of the reference points and the asymptote. Use these values to graph the function.**

(A) $g(x) = -3(2^{x-2}) + 1$

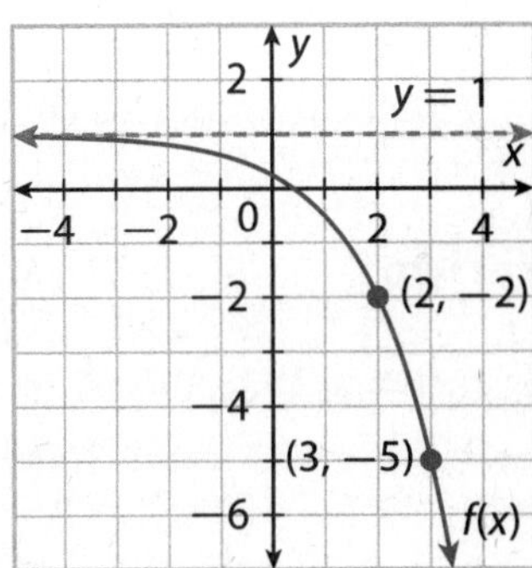

The domain of $g(x) = 3(2^{x-2}) + 1$ is $\{x| -\infty < x < \infty\}$.

The range of $g(x) = -3(2^{x-2}) + 1$ is $\{y|y < 1\}$.

Examine $g(x)$ and identify the parameters.

$a = -3$, which means that the function is reflected across the x-axis and vertically stretched by a factor of 3.

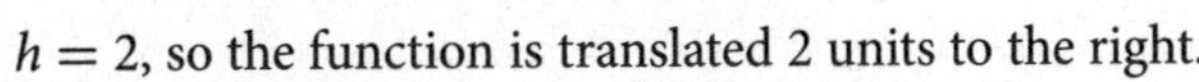

$h = 2$, so the function is translated 2 units to the right.

$k = 1$, so the function is translated 1 unit up.

The point $(0, 1)$ becomes $(h, a + k)$.

$(h, a + k) = (2, -3 + 1)$

$= (2, -2)$

$(1, b)$ becomes $(1 + h, ab + k)$.

$$(1 + h, ab + k) = (1 + 2, -3(2) + 1)$$
$$= (3, -6 + 1)$$
$$= (3, -5)$$

The asymptote becomes $y = k$.

$y = k \quad \rightarrow \quad y = 1$

Plot the transformed points and asymptote and draw the curve.

B $q(x) = 1.5(10^{x-3}) - 5$

The domain of $q(x) = 1.5(10^{x-3}) - 5$ is $\{x | \_\_\_\_\_\_\_\_\_\_ \}$.

The range of $q(x) = 1.5(10^{x-3}) - 5$ is $\{y | \_\_\_\_\_\_\_ \}$.

Examine $q(x)$ and identify the parameters.

$a =$ ☐ so the function is stretched vertically by a factor of 1.5.

$h =$ ☐ so the function is translated 3 units to the right.

$k =$ ☐ so the function is translated 5 units down.

The point $(0,1)$ becomes $(h, a + k)$.

$(h, a + k) = (3, 1.5 - 5) =$ ________

$(1, b)$ becomes $(1 + h, ab + k)$.

$(1 + h, ab + k) = (1 + 3, 1.5(10) - 5) =$ ________

The asymptote becomes $y = k$.

$y = k \quad \rightarrow \quad y =$ ☐

Plot the transformed points and asymptote and draw the curve.

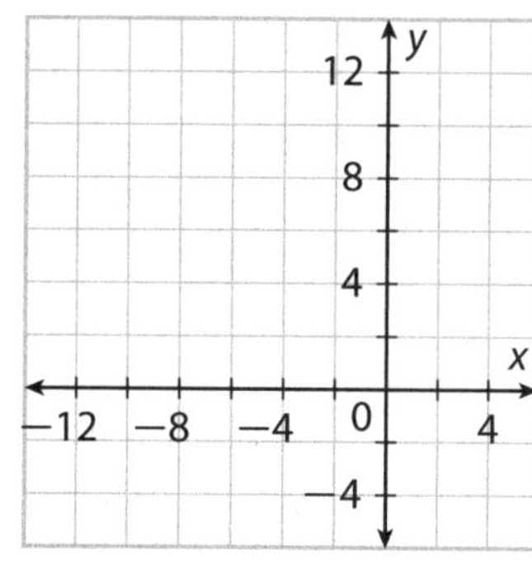

Your Turn

7. $g(x) = 4(2^{x+2}) - 6$

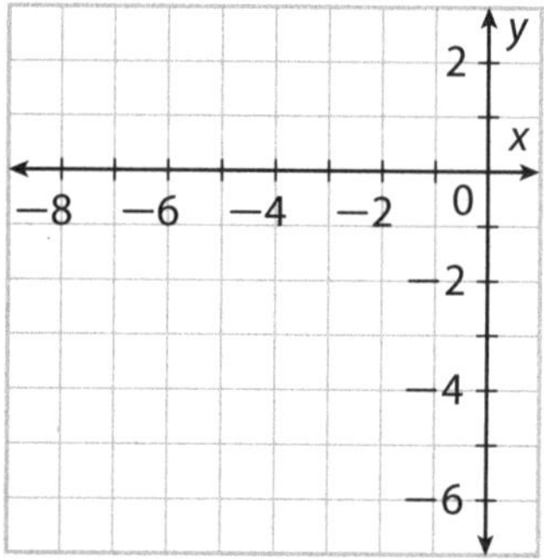

8. $q(x) = -\frac{3}{5}\left(10^{x+2}\right) + 3$

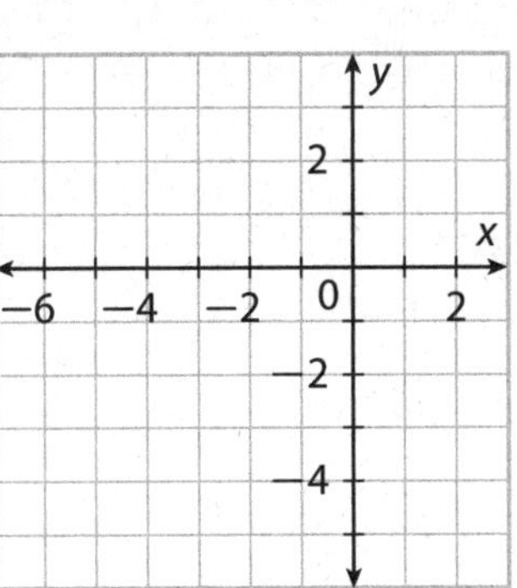

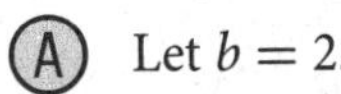

Explain 2 Writing Equations for Combined Transformations of $f(x) = b^x$ Where $b > 1$

Given the graph of an exponential function, you can use your knowledge of the transformation parameters to write the function rule for the graph. Recall that the asymptote will give the value of k and the x-coordinate of the first reference point is h. Then let y_1 be the y-coordinate of the first point and solve the equation $y_1 = a + k$ for a.

Finally, use a, h, and k to write the function in the form $g(x) = a\left(b^{x-h}\right) + k$.

Example 2 **Write the exponential function that will produce the given graph, using the specified value of b. Verify that the second reference point is on the graph of the function. Then state the domain and range of the function in set notation.**

(A) Let $b = 2$.

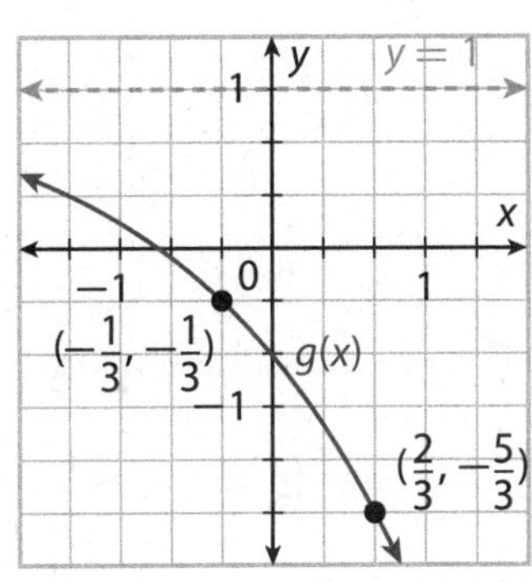

The asymptote is $y = 1$, showing that $k = 1$.

The first reference point is $\left(-\frac{1}{3}, -\frac{1}{3}\right)$. This shows that $h = -\frac{1}{3}$ and that $a + k = -\frac{1}{3}$.

Substitute $k = 1$ and solve for a.

$$a + k = -\frac{1}{3}$$

$$a + 1 = -\frac{1}{3}$$

$$a = -\frac{4}{3}$$

$$h = -\frac{1}{3}$$

$$k = 1$$

Substitute these values into $g(x) = a\left(b^{x-h}\right) + k$ to find $g(x)$.

$$g(x) = a\left(b^{x-h}\right) + k$$

$$= -\frac{4}{3}\left(2^{x+\frac{1}{3}}\right) + 1$$

Verify that $g\left(\frac{2}{3}\right) = -\frac{5}{3}$.

$$g\left(\frac{2}{3}\right) = -\frac{4}{3}\left(2^{\frac{2}{3}+\frac{1}{3}}\right) + 1$$
$$= -\frac{4}{3}\left(2^{1}\right) + 1$$
$$= -\frac{4}{3}(2) + 1$$
$$= \frac{3}{3} - \frac{8}{3}$$
$$= -\frac{5}{3}$$

The domain of $g(x)$ is $\left\{x \mid -\infty < x < +\infty\right\}$.

The range of $g(x)$ is $\left\{y \mid y < 1\right\}$.

B Let $b = 10$.

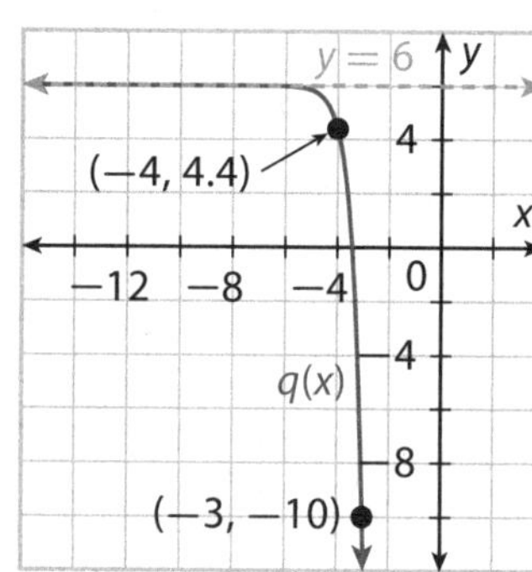

The asymptote is $y = \square$, showing that $k = \square$.

The first reference point is $(-4, 4.4)$. This shows that $h = \square$ and that $a + k = \square$. Substitute for k and solve for a.

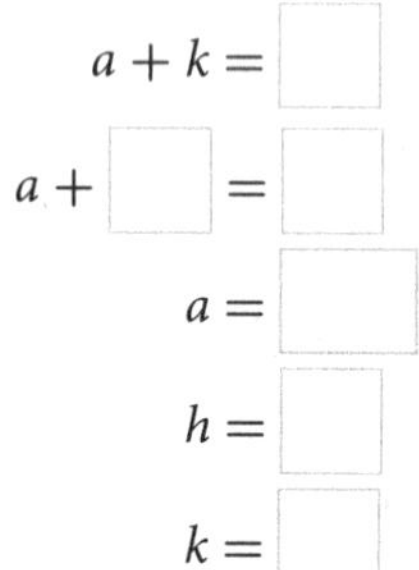

$a + k = \square$

$a + \square = \square$

$a = \square$

$h = \square$

$k = \square$

Substitute these values into $q(x) = a\left(b^{x-h}\right) + k$ to find $q(x)$.

$q(x) = a\left(b^{x-h}\right) + k = \square\left(10^{x-\square}\right) + \square$

Verify that $q(-3) = -10$.

$q(-3) = \square\left(10^{-3-\square}\right) + \square$

$= \square\left(10^{\square}\right) + \square$

$= \square + \square$

$= \square$

The domain of $q(x)$ is ____________________.

The range of $q(x)$ is __________.

Your Turn

Write the exponential function that will produce the given graph, using the specified value of *b*. Verify that the second reference point is on the graph of the function. Then state the domain and range of the function in set notation.

9. $b = 2$

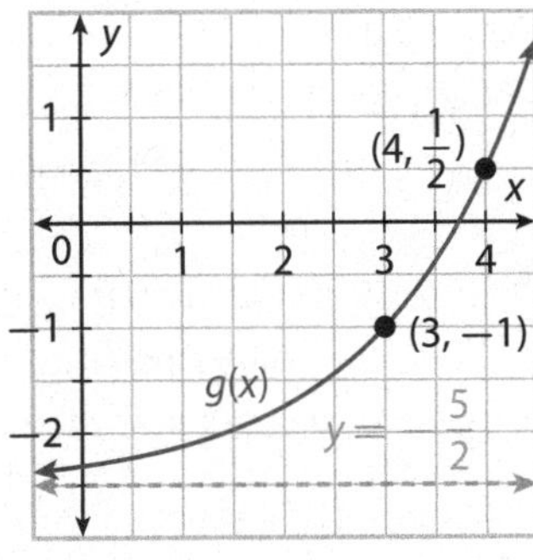

10. $b = 10$

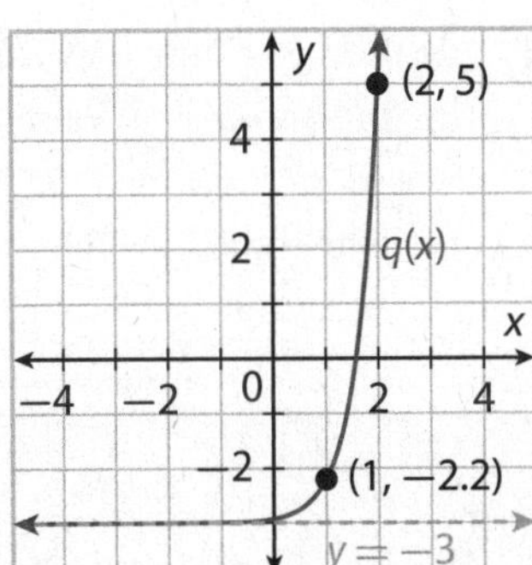

Explain 3 Modeling with Exponential Growth Functions

An **exponential growth function** has the form $f(t) = a(1 + r)^t$ where $a > 0$ and r is a constant percent increase (expressed as a decimal) for each unit increase in time t. That is, since $f(t + 1) = (1 + r) \cdot f(t) = f(t) + r \cdot f(t)$, the value of the function increases by $r \cdot f(t)$ on the interval $[t, t + 1]$. The base $1 + r$ of an exponential growth function is called the **growth factor**, and the constant percent increase r, in decimal form, is called the **growth rate**.

Example 3 **Find the function that corresponds with the given situation. Then use the graph of the function to make a prediction.**

(A) Tony purchased a rare guitar in 2000 for \$12,000. Experts estimate that its value will increase by 14% per year. Use a graph to find the number of years it will take for the value of the guitar to be \$60,000.

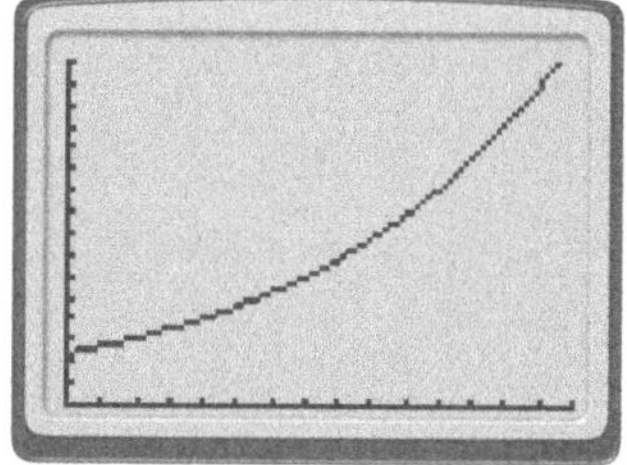

Write a function to model the growth in value for the guitar.

$$f(t) = a(1 + r)^t$$

$$= 12,000(1 + 0.14)^t$$

$$= 12,000(1.14)^t$$

Use a graphing calculator to graph the function.

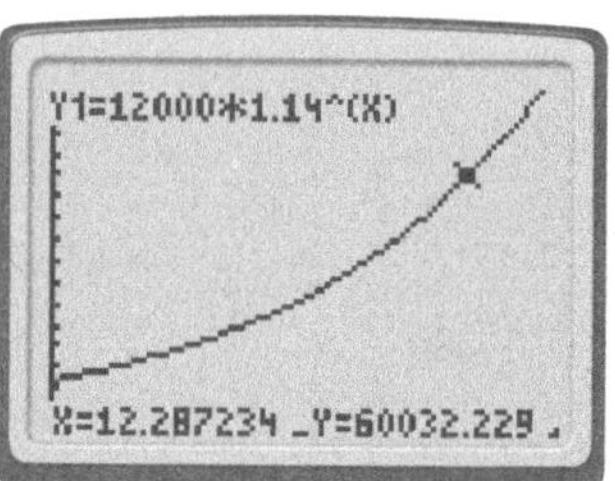

Use the graph to predict when the guitar will be worth \$60,000.

Use the TRACE feature to find the t-value where $f(t) \approx 60,000$.

So, the guitar will be worth \$60,000 approximately 12.29 years after it was purchased.

(B) At the same time that Tony bought the \$12,000 guitar, he also considered buying another rare guitar for \$15,000. Experts estimated that this guitar would increase in value by 9% per year. Determine after how many years the two guitars will be worth the same amount.

Write a function to model the growth in value for the second guitar.

$$g(t) = a(1 + r)^t$$

$$= \boxed{\quad}\left(1 + \boxed{\quad}\right)^t$$

$$= \boxed{\quad}\left(\boxed{\quad}\right)^t$$

Use a graphing calculator to graph the two functions.

Use the graph to predict when the two guitars will be worth the same amount.

Use the intersection feature to find the t-value where $g(t) = \boxed{\quad}$.

So, the two guitars will be worth the same amount ________ years after 2000.

Reflect

11. In part A, find the average rates of change over the intervals $(0, 4)$, $(4, 8)$, and $(8, 12)$. Do the rates increase, decrease, or stay the same?

Your Turn

Find the function that corresponds with the given situation. Then graph the function on a calculator and use the graph to make a prediction.

12. John researches a baseball card and finds that it is currently worth \$3.25. However, it is supposed to increase in value 11% per year. In how many years will the card be worth \$26?

Elaborate

13. How are reference points helpful when graphing transformations of $f(x) = b^x$ or when writing equations for transformed graphs?

14. Give the general form of an exponential growth function and describe its parameters.

15. **Essential Question Check-In** Which transformations of $f(x) = b^x$ change the function's end behavior? Which transformations change the function's y-intercept?

Evaluate: Homework and Practice

- Online Homework
- Hints and Help
- Extra Practice

Describe the effect of each transformation on the parent function. Graph the parent function and its transformation. Then determine the domain, range, and *y*-intercept of each function.

1. $f(x) = 2^x$ and $g(x) = 2(2^x)$

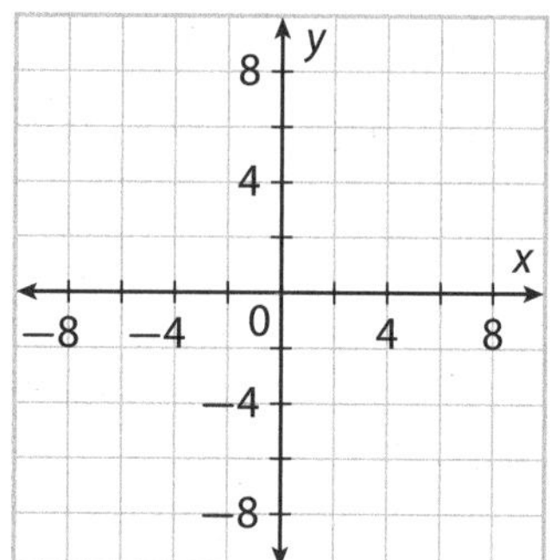

2. $f(x) = 2^x$ and $g(x) = -5(2^x)$

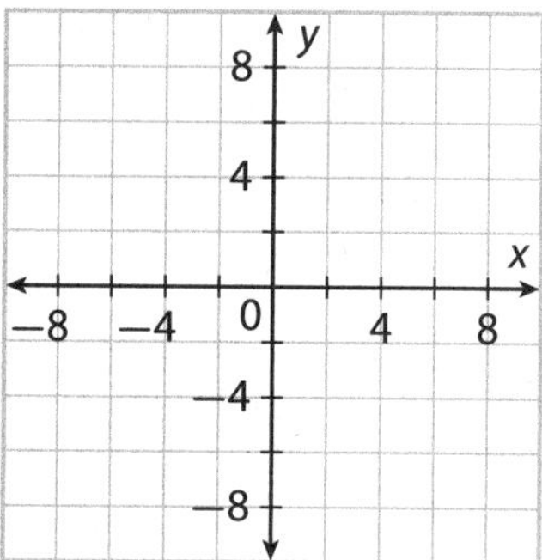

3. $f(x) = 2^x$ and $g(x) = 2^{x+2}$

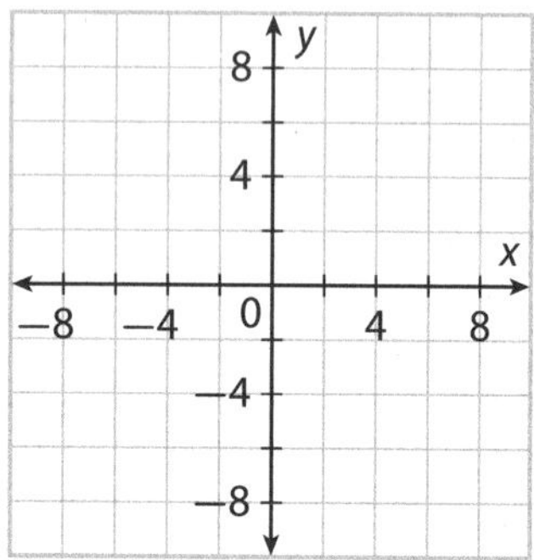

4. $f(x) = 2^x$ and $g(x) = 2^x + 5$

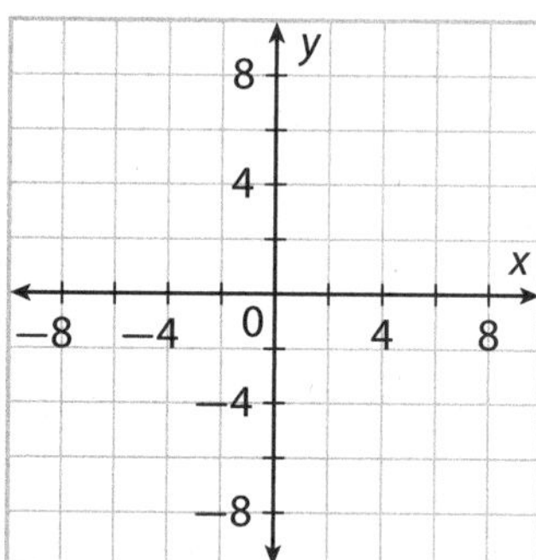

5. $f(x) = 10^x$ and $g(x) = 2(10^x)$

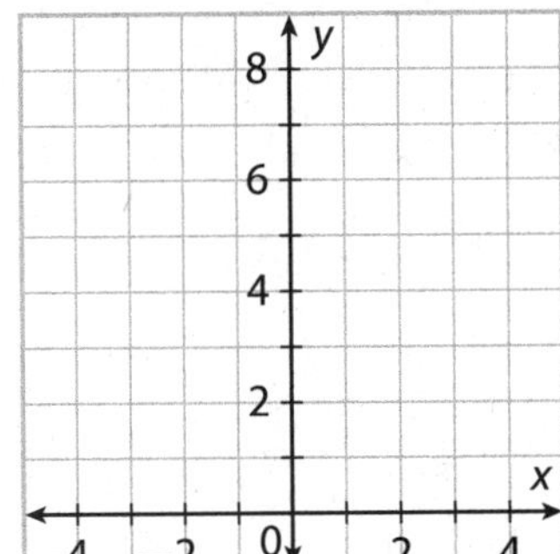

6. $f(x) = 10^x$ and $g(x) = -4(10^x)$

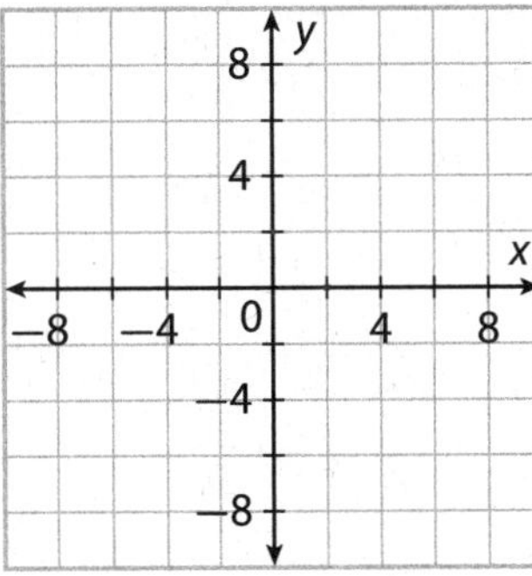

7. $f(x) = 10^x$ and $g(x) = 10^{x-2}$

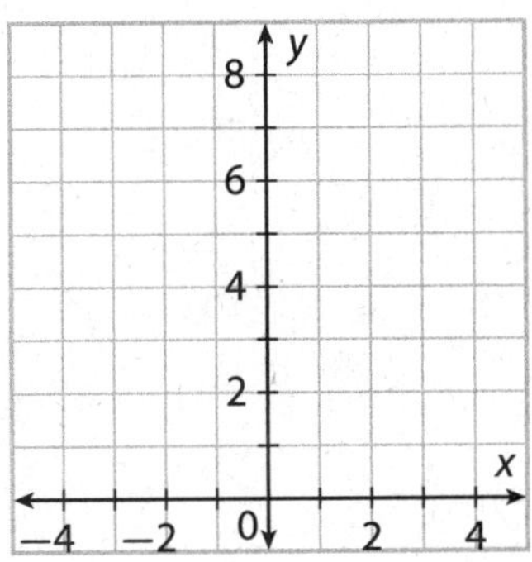

8. $f(x) = 10^x$ and $g(x) = 10^x - 6$

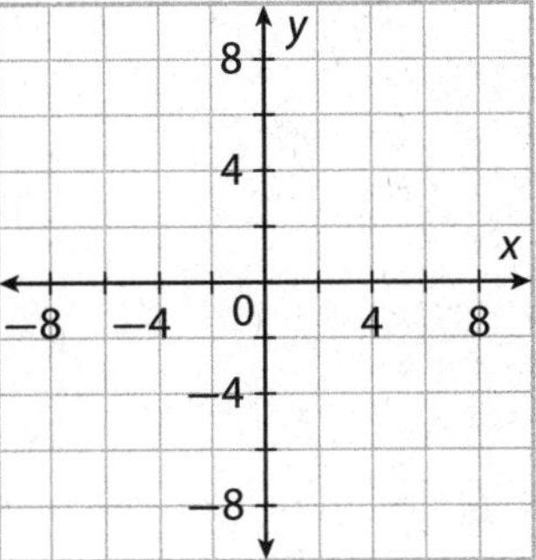

9. Describe the graph of $g(x) = 2^{(x-3)} - 4$ in terms of $f(x) = 2^x$.

10. Describe the graph of $g(x) = 10^{(x+7)} + 6$ in terms of $f(x) = 10^x$.

State the domain and range of the given function. Then identify the new values of the reference points and the asymptote. Use these values to graph the function.

11. $h(x) = 2(3^{x+2}) - 1$

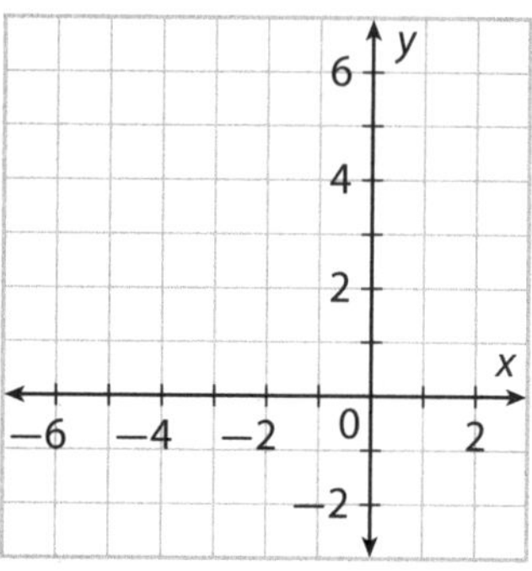

12. $k(x) = -0.5(4^{x-1}) + 2$

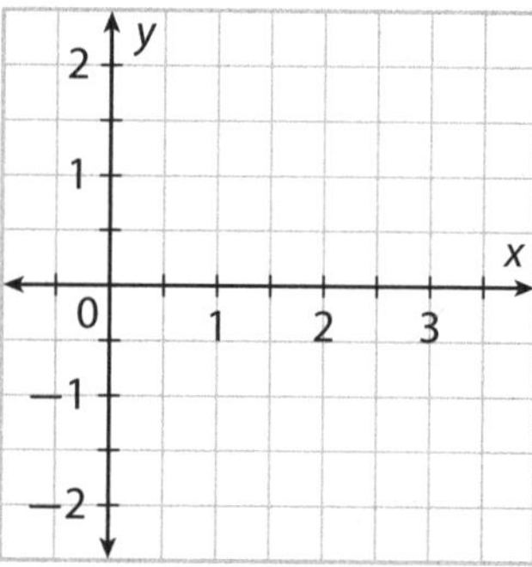

13. $f(x) = 3(6^{x-7}) - 8$

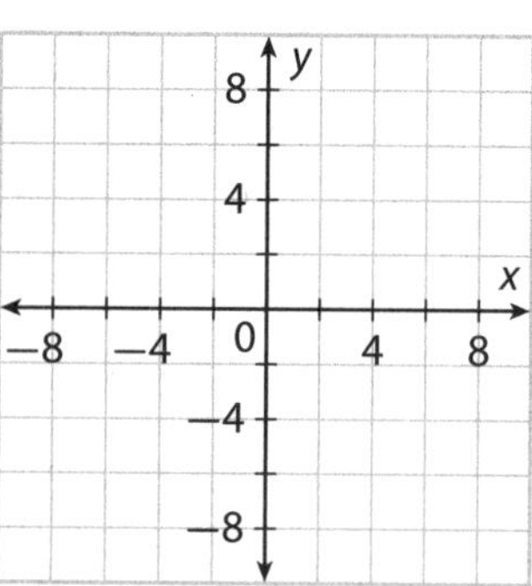

14. $f(x) = -3\left(2^{x+1}\right) + 3$

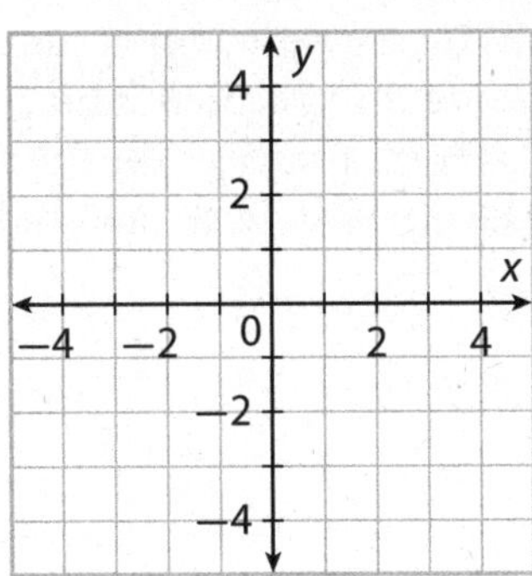

15. $h(x) = -\frac{1}{4}\left(5^{x+1}\right) - \frac{3}{4}$

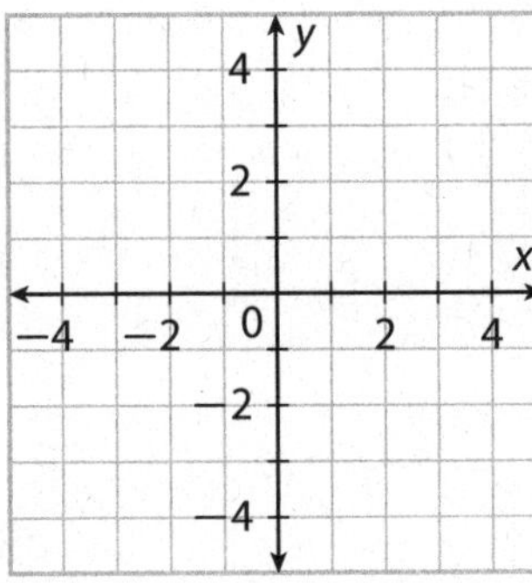

16. $p(x) = 2\left(4^{x-3}\right) - 5$

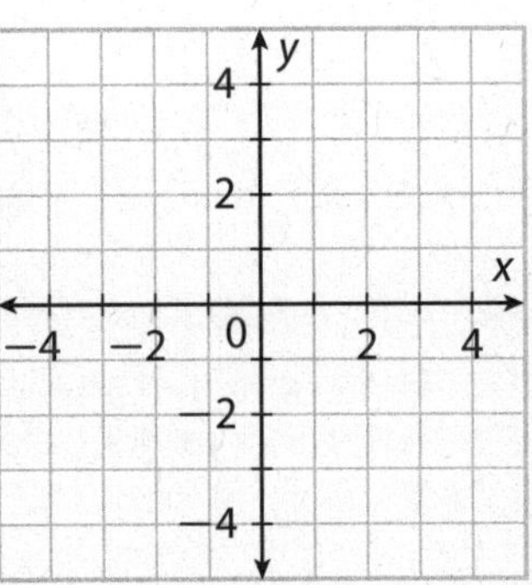

Write the exponential function that will produce the given graph, using the specified value of *b*. Verify that the second reference point is on the graph of the function. Then state the domain and range of the function in set notation.

17. $b = 2$

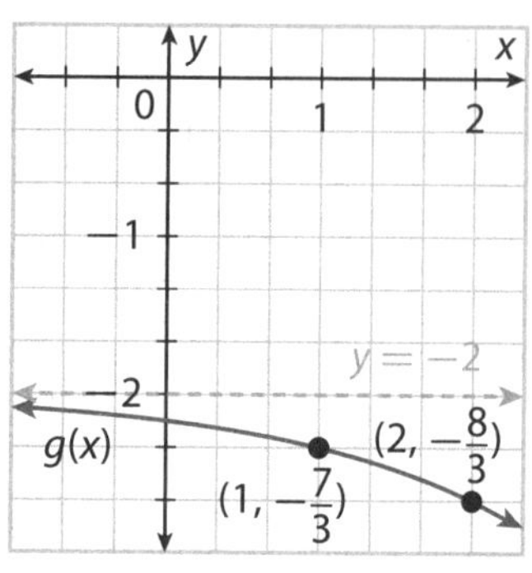

18. $b = 2$

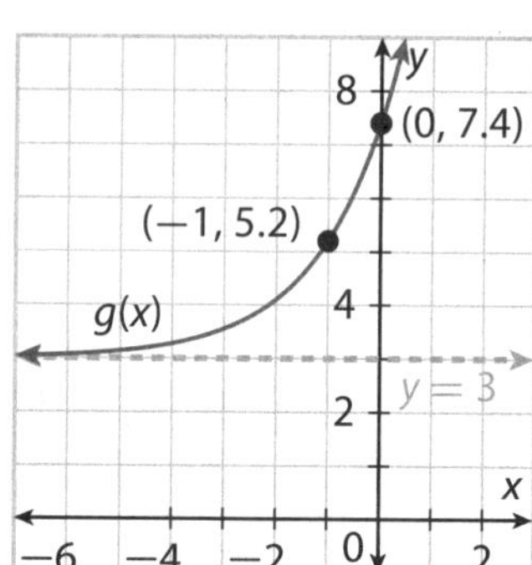

19. $b = 10$

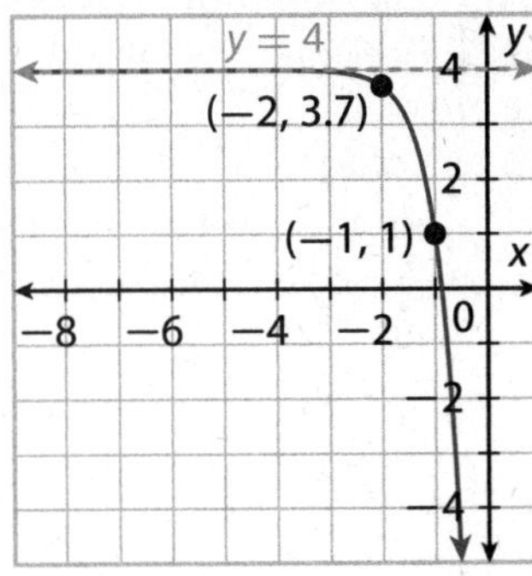

20. $b = 10$

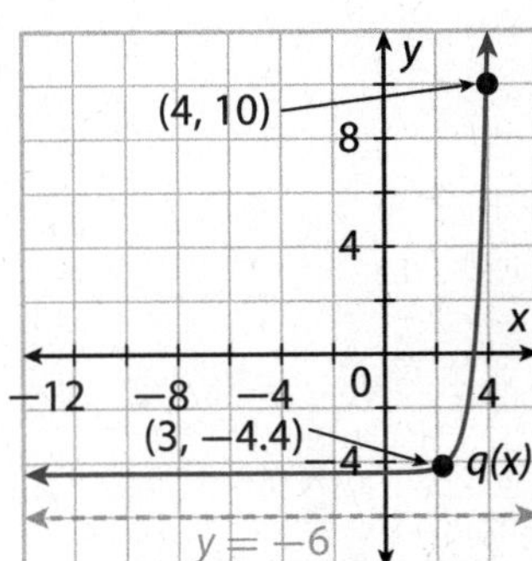

Find the function that corresponds with the given situation. Then graph the function on a calculator and use the graph to make a prediction.

21. A certain stock opens with a price of \$0.59. Over the first three days, the value of the stock increases on average by 50% per day. If this trend continues, how many days will it take for the stock to be worth \$6?

22. Sue has a lamp from her great-grandmother. She has it appraised and finds it is worth \$1000. She wants to sell it, but the appraiser tells her that the value is appreciating by 8% per year. In how many years will the value of the lamp be \$2000?

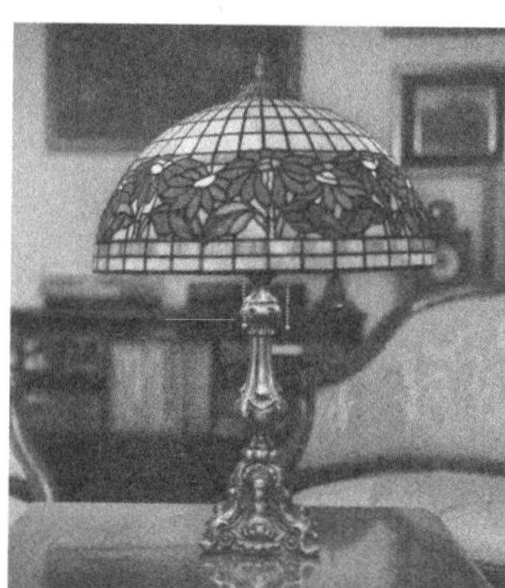

23. The population of a small town is 15,000. If the population is growing by 5% per year, how long will it take for the population to reach 25,000?

24. Bill invests \$3000 in a bond fund with an interest rate of 9% per year. If Bill does not withdraw any of the money, in how many years will his bond fund be worth \$5000?

H.O.T. Focus on Higher Order Thinking

25. **Analyze Relationships** Compare the end behavior of $g(x) = 2^x$ and $f(x) = x^2$. How are the graphs of the functions similar? How are they different?

26. **Explain the Error** A student has a baseball card that is worth \$6.35. He looks up the appreciation rate and finds it to be 2.5% per year. He wants to find how much it will be worth after 3 years. He writes the function $f(t) = 6.35(2.5)^t$ and uses the graph of that function to find the value of the card in 3 years.

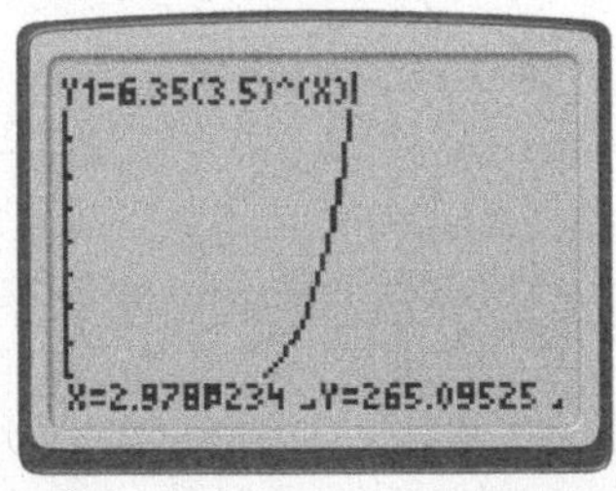

According to his graph, his card will be worth about \$265.10 in 3 years. What did the student do wrong? What is the correct answer?

Lesson Performance Task

Like all collectables, the price of an item is determined by what the buyer is willing to pay and the seller is willing to accept. The estimated value of a 1948 Tucker 48 automobile in excellent condition has risen at an approximately exponential rate from about \$500,000 in December 2006 to about \$1,400,000 in December 2013.

a. Find an equation in the form $V(t) = V_0(1 + r)^t$, where V_0 is the value of the car in dollars in December 2006, r is the average annual growth rate, t is the time in years since December 2006, and $V(t)$ is the value of the car in dollars at time t. (Hint: Substitute the known values and solve for r.)

b. What is the meaning of the value of r?

c. If this trend continues, what would be the value of the car in December 2017?

Name________________________ Class______________ Date__________

13.3 Exponential Decay Functions

Essential Question: How is the graph of $g(x) = ab^{x-h} + k$ where $0 < b < 1$ related to the graph of $f(x) = b^x$?

Resource Locker

TEKS A2.5.B Formulate exponential … equations that model real-world situations…Also A2.2.A, A2.5.A, A2.5.D, A2.7.I

Explore 1 Graphing and Analyzing $f(x) = \left(\frac{1}{2}\right)^x$ and $f(x) = \left(\frac{1}{10}\right)^x$

Exponential decay functions are exponential functions with bases between 0 and 1 assuming a positive leading coefficient. These functions can be transformed in a manner similar to exponential growth functions. Begin by plotting the parent functions of two of the more commonly used bases: $\frac{1}{2}$ and $\frac{1}{10}$.

(A) To begin, fill in the table in order to find points along the function $f(x) = \left(\frac{1}{2}\right)^x$. You may need to review the rules of the properties of exponents, including negative exponents.

x	$f(x) = \left(\frac{1}{2}\right)^x$
−3	8
−2	
−1	
0	
1	
2	
3	

(B) What does the end behavior of this function appear to be as x increases?

____

(C) Plot the points on the graph and draw a smooth curve through them.

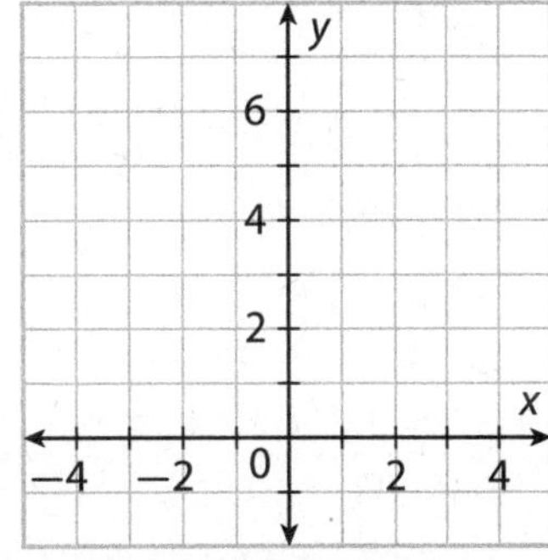

(D) Complete the table for $f(x) = \left(\frac{1}{10}\right)^x$.

x	$f(x) = \left(\frac{1}{10}\right)^x$
−3	1000
−2	
−1	
0	
1	
2	
3	

(E) Plot the points on the graph and draw a smooth curve through them.

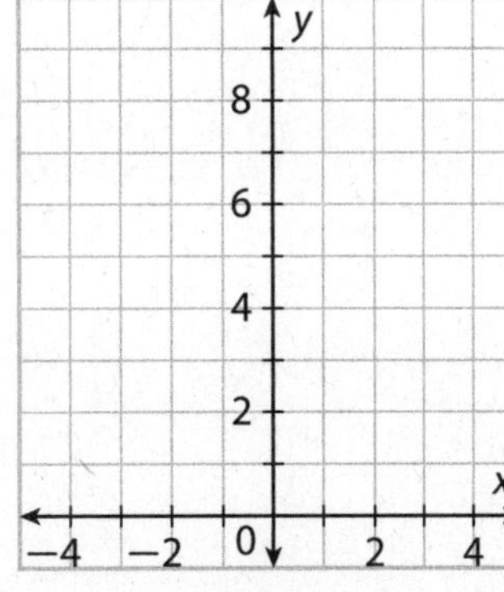

(F) Fill in the following table of properties:

	$f(x) = \left(\frac{1}{2}\right)^x$	$f(x) = \left(\frac{1}{10}\right)^x$
Domain	$\{x \mid -\infty < x < \infty\}$	$\{x \mid \square\}$
Range	$\{y \mid \square\}$	$\{y \mid \square\}$
End behavior as $x \to \infty$	$f(x) \to \square$	$f(x) \to \square$
End behavior as $x \to -\infty$	$f(x) \to \square$	$f(x) \to \square$
y-intercept	$(\square, \square)$	$(\square, \square)$

(G) Both of these functions [decrease/increase] throughout the domain.

(H) Of the two functions, $f(x) = \left(\frac{1}{\square}\right)^x$ decreases faster.

Reflect

1. **Make a Conjecture** Look at the table of properties for the functions. What do you notice? Make a conjecture about these properties for exponential decay functions of the form $f(x) = \left(\frac{1}{n}\right)^x$, where n is a constant.

____

____

Explore 2 Predicting Transformations of the Graphs of $f(x) = \left(\frac{1}{2}\right)^x$ and $f(x) = \left(\frac{1}{10}\right)^x$

Based on your experience with transforming the parent function $f(x)$ in previous lessons, make predictions about the effect of varying the parameters in $g(x) = af(x-c) + d$. Confirm your predictions using a graphing calculator.

(A) The graph of $g_1(x) = 3\left(\frac{1}{2}\right)^x$ will be the graph of $f(x) = \left(\frac{1}{2}\right)^x$ vertically ____________ by a factor of ___.

The graph of $g_2(x) = \frac{1}{4}\left(\frac{1}{2}\right)^x$ will be the graph of $f(x) = \left(\frac{1}{2}\right)^x$ vertically ____________ by a factor of ___.

(B) The graph of $g_1(x) = -\frac{3}{4}\left(\frac{1}{2}\right)^x$ will be the graph of $f(x) = \left(\frac{1}{2}\right)^x$ reflected across the ________

and vertically ____________ by a factor of __.

The graph of $g_2(x) = -5\left(\frac{1}{2}\right)^x$ will be the graph of $f(x) = \left(\frac{1}{2}\right)^x$ reflected across the ________

and vertically _________ by a factor of __.

The graph of $q_1(x) = -\frac{5}{4}\left(\frac{1}{10}\right)^x$ will be the graph of $f(x) = \left(\frac{1}{10}\right)^x$ reflected across

the ________ and vertically _________ by a factor of __.

The graph of $q_2(x) = -\frac{1}{4}\left(\frac{1}{10}\right)^x$ will be the graph of $f(x) = \left(\frac{1}{10}\right)^x$ reflected across

the ________ and vertically ____________ by a factor of __.

(C) The graph of $g_1(x) = \left(\frac{1}{2}\right)^{x+1}$ will be the graph of $f(x) = \left(\frac{1}{2}\right)^x$ translated __ unit to the _____.

The graph of $g_2(x) = \left(\frac{1}{2}\right)^{x-4}$ will be the graph of $f(x) = \left(\frac{1}{2}\right)^x$ translated __ units to

the _____.

The graph of $q_1(x) = \left(\frac{1}{10}\right)^{x+2}$ will be the graph of $f(x) = \left(\frac{1}{10}\right)^x$ translated __ units to

the _____.

The graph of $q_2(x) = \left(\frac{1}{10}\right)^{x-3}$ will be the graph of $f(x) = \left(\frac{1}{10}\right)^x$ translated __ units to

the _______.

(D) The graph of $g_1(x) = \left(\frac{1}{2}\right)^x + 3$ will be the graph of $f(x) = \left(\frac{1}{2}\right)^x$ translated __ units _____.

The graph of $g_2(x) = \left(\frac{1}{2}\right)^x - \frac{5}{2}$ will be the graph of $f(x) = \left(\frac{1}{2}\right)^x$ translated ____ units _______.

The graph of $q_1(x) = \left(\frac{1}{10}\right)^x + 5$ will be the graph of $f(x) = \left(\frac{1}{10}\right)^x$ translated __ units _____.

The graph of $q_2(x) = \left(\frac{1}{10}\right)^x - 2$ will be the graph of $f(x) = \left(\frac{1}{10}\right)^x$ translated __ units _______.

Reflect

2. Which parameters make the domain and range of $g(x)$ differ from those of the parent function? Write the transformed domain and range for $g(x)$ in set notation.

____

____

____

____

Explain 1 Graphing Combined Transformations of $f(x) = b^x$ Where $0 < b < 1$

When graphing transformations of $f(x) = b^x$ where $0 < b < 1$, it is helpful to consider the effect of the transformation on two reference points, $(0, 1)$ and $\left(-1, \frac{1}{b}\right)$, vas well as the effect on the asymptote, $y = 0$. The table shows these reference points and the asymptote $y = 0$ for $f(x) = b^x$ and the corresponding points and asymptote for the transformed function, $g(x) = ab^{x-h} + k$.

	$f(x) = b^x$	$g(x) = ab^{x-h} + k$
First reference point	$(0, 1)$	$(h, a + k)$
Second reference point	$\left(-1, \frac{1}{b}\right)$	$\left(h - 1, \frac{a}{b} + k\right)$
Asymptote	$y = 0$	$y = k$

Example 1 **For each of the transformed functions, use the reference points and the asymptote to draw the transformed function on the grid with the parent function. Then describe the domain and range of the transformed function using set notation.**

(A) $g(x) = 3\left(\frac{1}{2}\right)^{x-2} - 2$

Identify parameters: $a = 3$ $b = \frac{1}{2}$ $h = 2$ $k = -2$

Find reference points:

$$(h, a + k) = (2, 3 - 2) = (2, 1)$$

$$\left(h - 1, \frac{a}{b} + k\right) = \left(2 - 1, \frac{3}{\frac{1}{2}} - 2\right) = (1, 4)$$

Find the asymptote: $y = -2$

Plot the points and draw the asymptote. Then connect the points with a smooth curve that approaches the asymptote without crossing it.

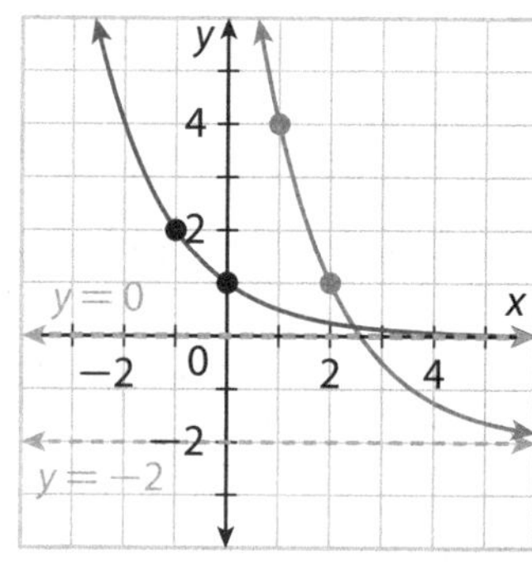

Domain: $\{x \mid -\infty < x < \infty\}$

Range: $\{y \mid y > -2\}$

(B) $g(x) = -\left(\frac{1}{10}\right)^{x+2} + 8$

Identify parameters:

$a = \square$ $b = \square$ $h = \square$ $k = \square$

Find reference points:

$\left(h, \square\right) = (-2, -1 + 8)\ (-2, 7)$

$\left(h - 1, \frac{a}{b} + k\right) = \left(-2 - 1, \frac{\square}{\square} + 8\right) = \left(\square, \square\right)$

Find the asymptote:

$y = \square$

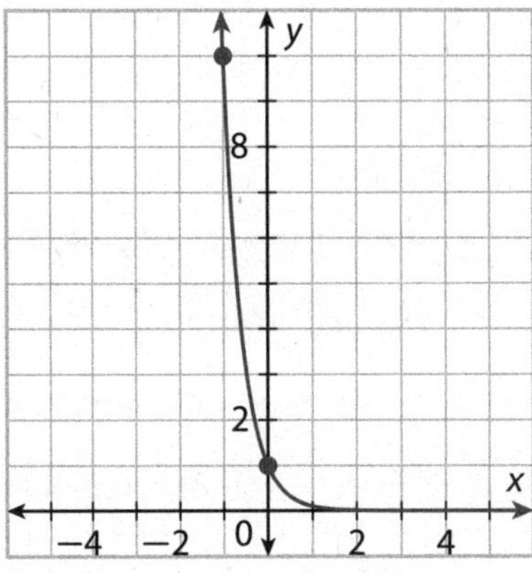

Plot the points and draw the asymptote. Then connect the points with a smooth curve that approaches the asymptote without crossing it.

Domain: $\left\{x \mid \square\right\}$

Range: $\left\{y \mid \square\right\}$

Your Turn

For the transformed function, use the reference points and the asymptote to draw the transformed function on the grid with the parent function. Then describe the domain and range of the transformed function using set notation.

3. $g(x) = 3\left(\frac{1}{3}\right)^{x+2} - 4$

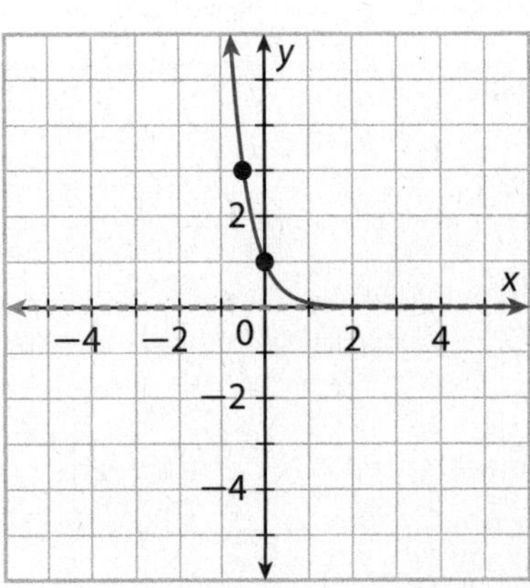

Explain 2 Writing Equations for Combined Transformations of $f(x) = b^x$ where $0 < b < 1$

Given a graph of an exponential decay function, $g(x) = ab^{x-h} + k$, the reference points and the asymptote can be used to identify the transformation parameters in order to write the function rule.

Example 2 **Write the function represented by this graph and state the domain and range using set notation.**

Ⓐ

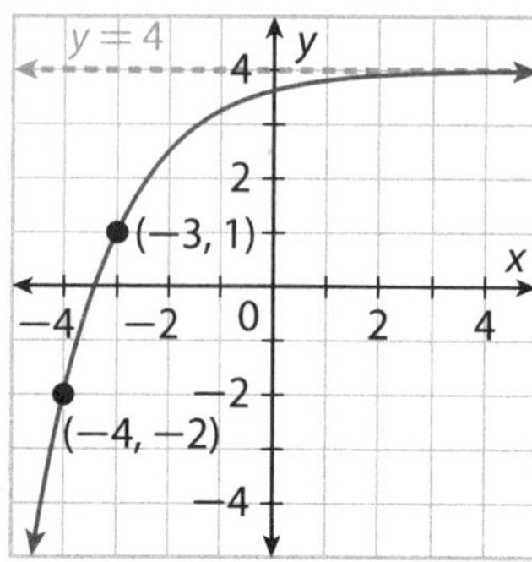

Find k from the asymptote: $k = 4$.

The first reference point is at $(-3,1)$.

Equate point value with parameters-based expression. $\quad (-3, 1) = (h, a + k)$

Use the x-coordinate to solve for h. $\quad h = -3$

Use the y-coordinate to solve for a. $\quad a = 1 - k$

$$= -3$$

The second reference point is at $(-4, -2)$.

Equate point value with parameters-based expression. $\quad (-4, -2) = \left(h - 1, \frac{a}{b} + k\right)$

Equate y-coordinate with parameters. $\quad \frac{-3}{b} + 4 = -2$

Solve for b.

$$\frac{-3}{b} = -6$$

$$b = \frac{-3}{-6}$$

$$= \frac{1}{2}$$

$$g(x) = -3\left(\frac{1}{2}\right)^{x+3} + 4$$

Domain: $\{x \mid -\infty < x < \infty\}$

Range: $\{y \mid y < 4\}$

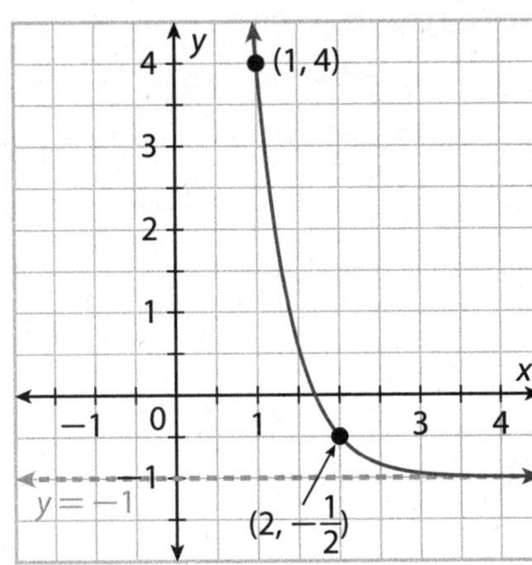

Find k from the asymptote: $k = \square$.

The first reference point is at $(\square, \square)$, so $\left(\square, -\frac{1}{2}\right) = (h, \square)$

$h = \square$ $\qquad a = \square - k$

$= \square$

The second reference point is at $(\square, \square)$, so $(\square, 4) = (h - 1, \square)$

$$\frac{\square}{b} - 1 = \square$$

$$\frac{\frac{1}{2}}{b} = \square$$

$$b = \frac{\frac{1}{2}}{5}$$

$$= \square$$

$$g(x) = \square \left(\frac{1}{10}\right)^{x-\square} + \square$$

Domain: $\{x \mid -\infty < x < \infty\}$

Range: $\{y \mid y \square -1\}$

Reflect

4. Compare the y-intercept and the asymptote of the function shown in this table to the function plotted in Example 2A.

x	-5	-4	-3	-2	-1	0	1	2
$g(x)$	-10	-4	-4	$\frac{1}{2}$	$1\frac{1}{4}$	$1\frac{5}{8}$	$1\frac{13}{16}$	$1\frac{29}{32}$

5. Compare the y-intercept and the asymptote of the function shown in this table to the function plotted in Example 2B.

x	-3	-2	-1	0	1	2
$g(x)$	49	4	-0.5	-0.95	-0.995	-0.9995

Your Turn

Write the function represented by this graph and state the domain and range using set notation.

6.

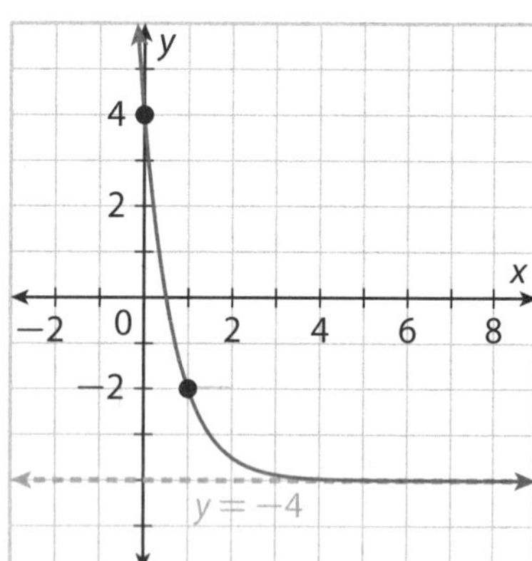

Explain 3 Modeling with Exponential Decay Functions

Exponential decay functions can be applied to situations in which a quantity decreases by a constant percentage for each unit increase in time.

$$f(t) = a(1 - r)^t$$

In this form of the decay function, r (which must be expressed as a decimal or a fraction rather than a percentage) is called the **decay rate**. The term $(1 - r)$ is known as the **decay factor**. The vertical stretch parameter, a, is also the value of the decay function at the start (when $t = 0$).

Example 3 **Given the description of the decay terms, write the exponential decay function in the form $f(t) = a(1 - r)^t$ and graph it with a graphing calculator.**

(A) The value of a truck purchased new for \$28,000 decreases by 9.5% each year. Write an exponential function for this situation and graph it using a calculator. Use the graph to predict after how many years the value of the truck will be \$5000.

"Purchased new for \$28,000..." $a = 28{,}000$

"...decreases by 9.5% each year." $r = 0.095$

Substitute parameter values. $V_T(t) = 28{,}000(1 - 0.095)^t$

Simplify. $V_T(t) = 28{,}000(0.905)^t$

Graph the function with a graphing calculator. Use WINDOW to adjust the graph settings so that you can see the function and the function values that are important.

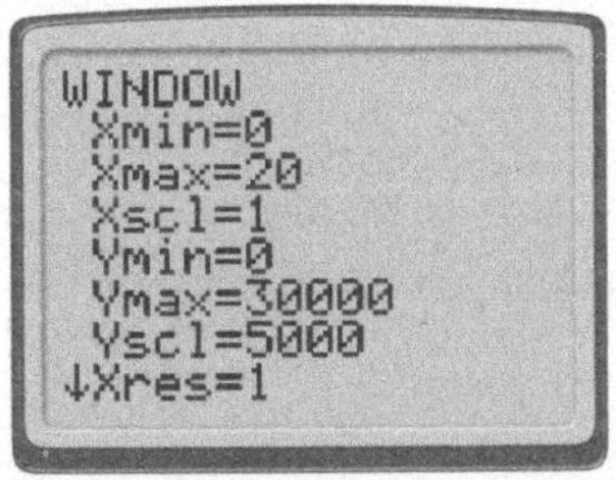

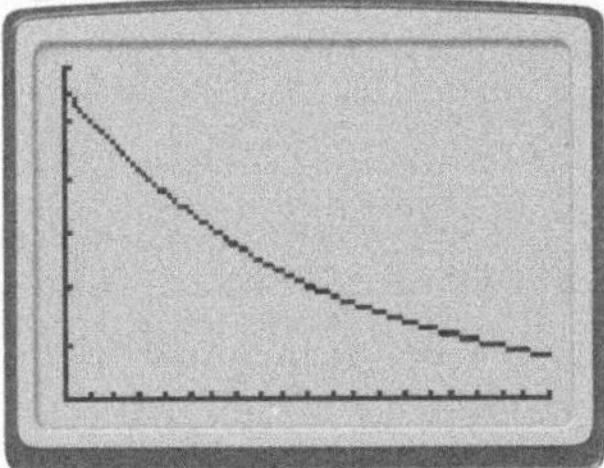

Find when the value reaches \$5000 by finding the intersection between $V_T(t) = 28{,}000(0.905)^t$ and $V_T(t) = 5000$ on the calculator.

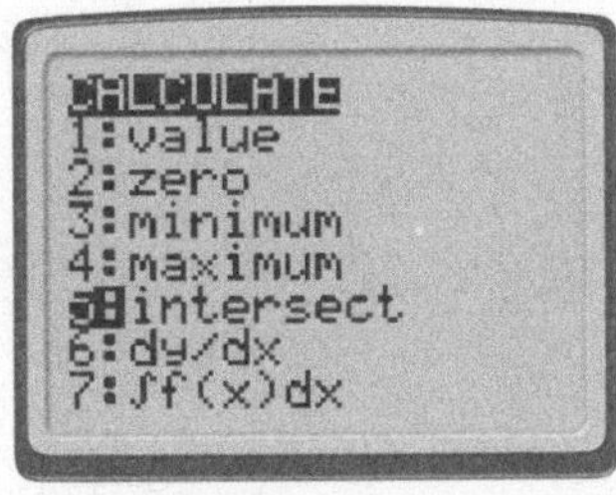

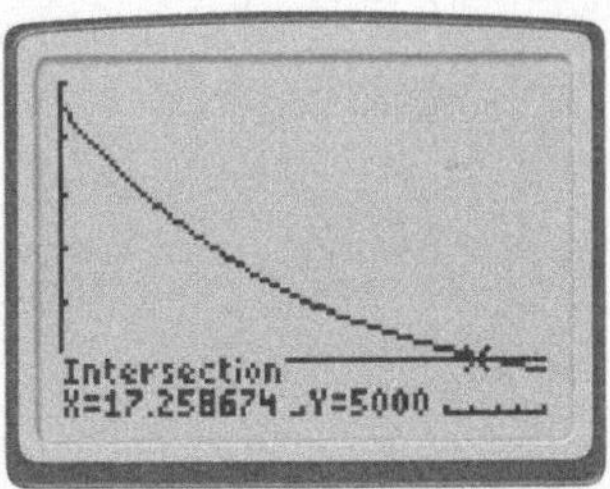

The intersection is at the point $(17.26, 5000)$, which means after 17.26 years, the truck will have a value of \$5000.

(B) The value of a sports car purchased new for \$45,000 decreases by 15% each year. Write an exponential function for the depreciation of the sports car, and plot it along with the previous example. After how many years will the two vehicles have the same value if they are purchased at the same time?

"Purchased new for \$45,000..." $\square = 45{,}000$

"...decreases by 15% each year." $r = \square$

Substitute parameter values. $V_c(t) = \square\left(1 - \square\right)^t$

Simplify. $V_c(t) = 45{,}000\left(\square\right)^t$

Add this plot to the graph for the truck value from Example A and find the intersection of the two functions to determine when the values are the same.

The intersection point is $\left(\square, \square\right)$.

After $\square$ years, the values of both vehicles will be

\$ $\square$.

Reflect

7. What reference points could you use if you plotted the value function for the sports car on graph paper? Confirm that the graph passes through them using the calculate feature on a graphing calculator.

8. Using the sports car from example B, calculate the average rate of change over the course of the first year and the second year of ownership. What happens to the absolute value of the rate of change from the first interval to the second? What does this mean in this situation?

Your Turn

9. On federal income tax returns, self-employed people can depreciate the value of business equipment. Suppose a computer valued at \$2765 depreciates at a rate of 30% per year. Use a graphing calculator to determine the number of years it will take for the computer's value to be \$350.

Elaborate

10. Which transformations of $f(x) = \left(\frac{1}{2}\right)^x$ or $f(x) = \left(\frac{1}{10}\right)^x$ change the function's end behavior?

11. Which transformations change the location of the graph's y-intercept?

12. **Discussion** How are reference points and asymptotes helpful when graphing transformations of $f(x) = \left(\frac{1}{2}\right)^x$ or $f(x) = \left(\frac{1}{10}\right)^x$ or when writing equations for transformed graphs?

13. Give the general form of an exponential decay function based on a known decay rate and describe its parameters.

14. **Essential Question Check-In** How is the graph of $f(x) = b^x$ used to help graph the function $g(x) = ab^{x-h} + k$?

Evaluate: Homework and Practice

- Online Homework
- Hints and Help
- Extra Practice

1. Graph the function $f(x) = \left(\frac{1}{3}\right)^x$ by plotting points with integer x-values from −2 to 2.

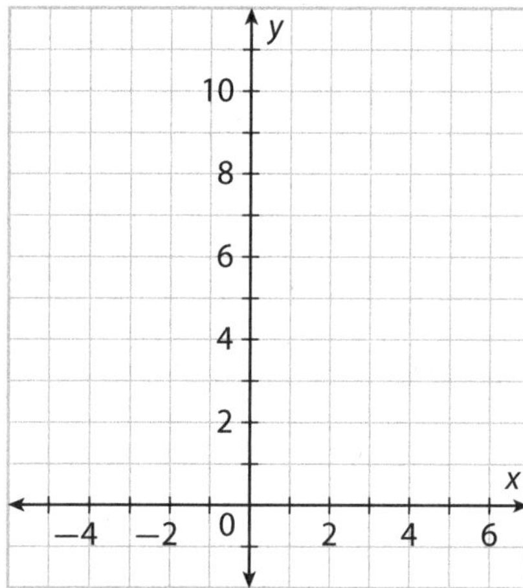

Describe the transformation(s) from each parent function and give the domain and range of each function.

2. $g(x) = \left(\frac{1}{2}\right)^x + 3$

3. $g(x) = \left(\frac{1}{10}\right)^{x+4}$

4. $g(x) = -\left(\frac{1}{10}\right)^{x-1} + 2$

5. $g(x) = 3\left(\frac{1}{2}\right)^{x+3} - 6$

For each of the transformed functions, use the reference points and the asymptote to draw the transformed function on the grid. Then describe the domain and range of the transformed function using set notation.

6. $g(x) = -2\left(\frac{1}{2}\right)^{x-1} + 2$

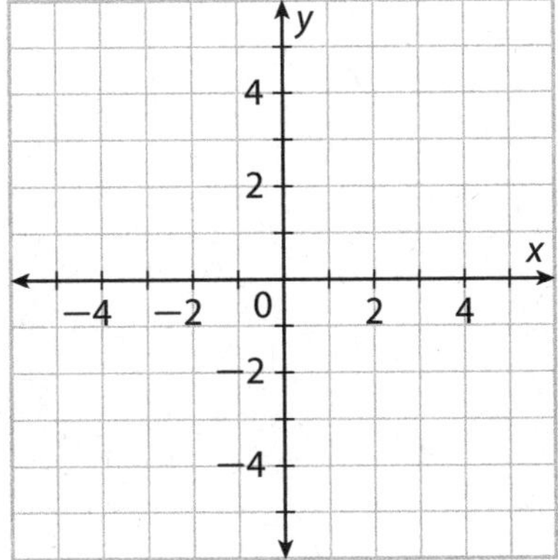

7. $g(x) = \left(\frac{1}{4}\right)^{x+2} + 3$

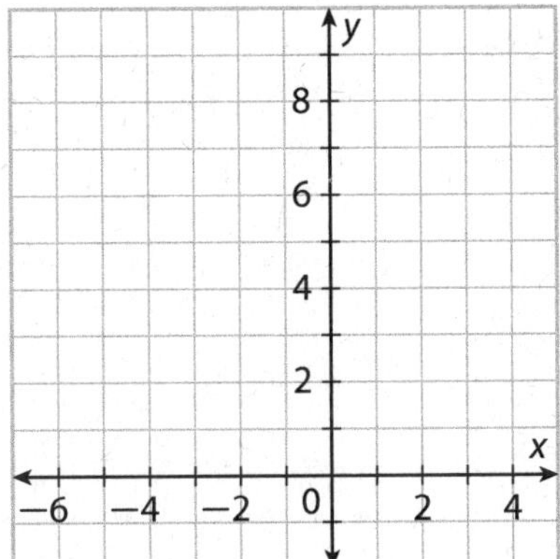

8. $g(x) = \frac{1}{2}\left(\frac{1}{3}\right)^{x-\frac{1}{2}} + 2$

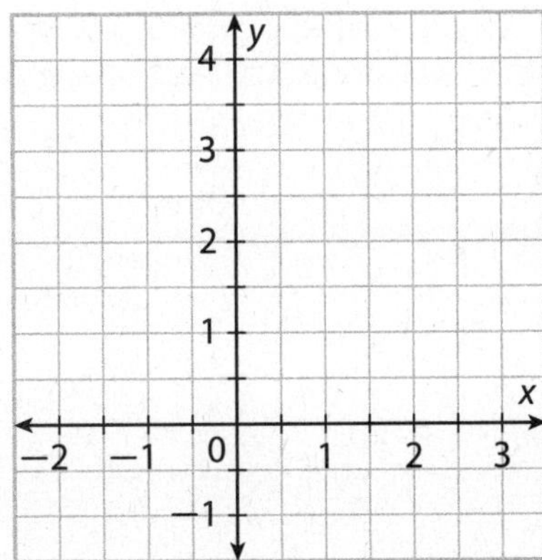

9. $g(x) = 2\left(\frac{1}{4}\right)^{x-2} - 3$

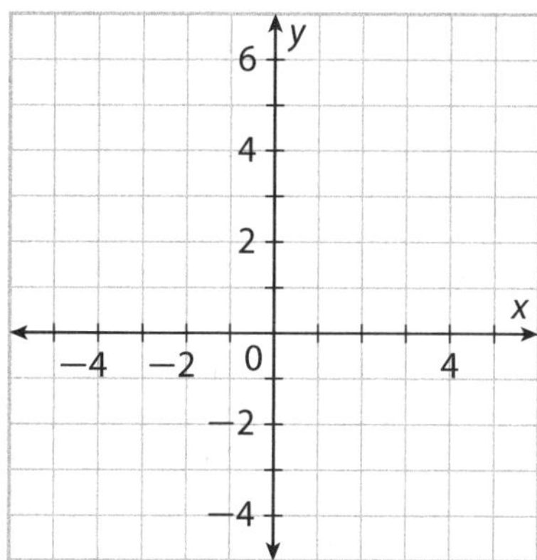

10. $g(x) = -3\left(\frac{1}{2}\right)^{x+2} + 7$

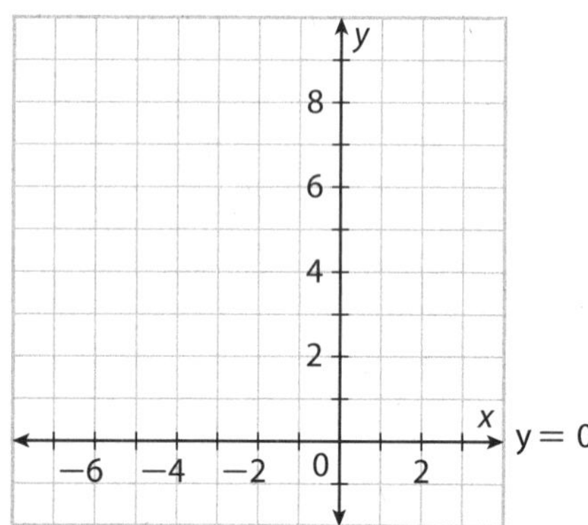

11. $g(x) = -\left(\frac{2}{3}\right)^{x+1} + \frac{1}{2}$

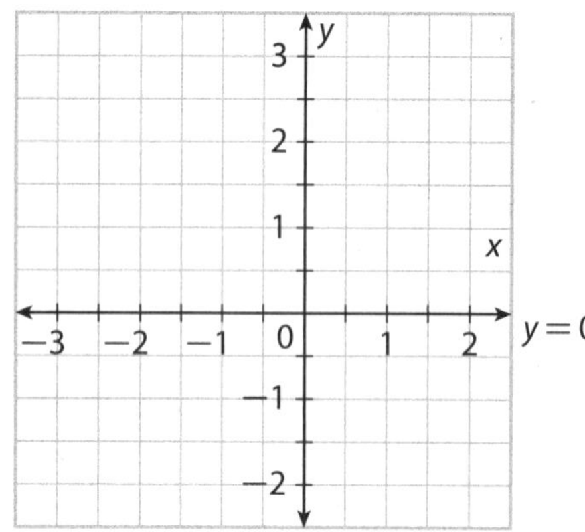

Write the function represented by each graph and state the domain and range using set notation.

12.

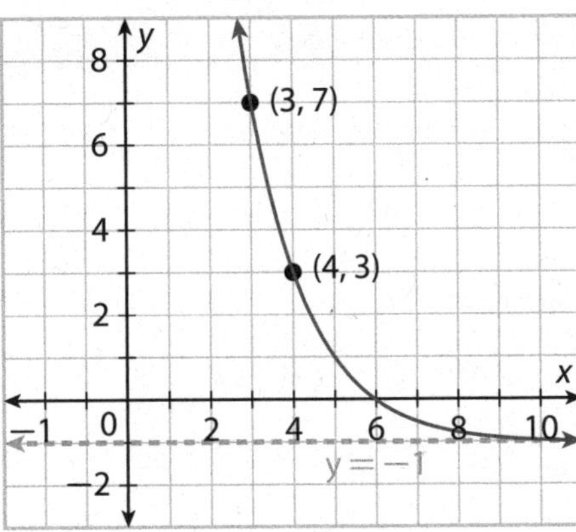

13.

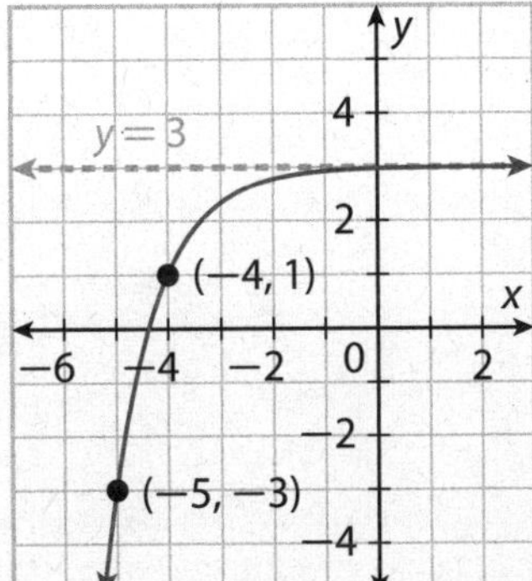

Write the exponential decay function described in the situation and use a graphing calculator to answer each question asked.

14. **Medicine** A quantity of insulin used to regulate sugar in the bloodstream breaks down by about 5% each minute after the injection. A bodyweight-adjusted dose is generally 10 units. How long does it take for the remaining insulin to be half of the original injection?

15. **Paleontology** Carbon-14 is a radioactive isotope of carbon that is used to date fossils. There are about 1.5 atoms of carbon-14 for every trillion atoms of carbon in the atmosphere, which known as 1.5 ppt (parts per trillion). Carbon in a living organism has the same concentration as carbon-14. When an organism dies, the carbon-14 content decays at a rate of 11.4% per millennium (1000 years). Write the equation for carbon-14 concentration (in ppt) as a function of time (in millennia) and determine how old a fossil must be that has a measured concentration of 0.2 ppt.

16. **Music** Stringed instruments like guitars and pianos create a note when a string vibrates back and forth. The distance that the middle of the string moves from the center is called the amplitude (a), and for a guitar, it starts at 0.75 mm when a note is first struck. Amplitude decays at a rate that depends on the individual instrument and the note, but a decay rate of about 25% per second is typical. Calculate the time it takes for an amplitude of 0.75 mm to reach 0.1 mm.

H.O.T. Focus on Higher Order Thinking

17. **Analyze Relationships** Compare the graphs of $f(x) = \left(\frac{1}{2}\right)^x$ and $g(x) = x^{\frac{1}{2}}$. Which of the following properties are the same? Explain.

    a. Domain

    b. Range

    c. End behavior as x increases

    d. End behavior as x decreases

18. **Communicate Mathematical Ideas** A quantity is reduced to half of its original amount during each given time period. Another quantity is reduced to one quarter of its original amount during the same given time period. Determine each decay rate, state which is greater, and explain your results.

19. **Multiple Representations** Exponential decay functions are written as transformations of the function $f(x) = b^x$, where $0 < b < 1$. However, it also possible to use negative exponents as the basis of an exponential decay function. Use the properties of exponents to show why the function $f(x) = 2^{-x}$ is an exponential decay function.

20. **Represent Real-World Problems** You buy a video game console for \$500 and sell it 5 years later for \$100. The resale value decays exponentially over time. Write a function that represents the resale value, R, in dollars, over the time, t, in years. Explain how you determined your function.

Lesson Performance Task

Sodium-24 is a radioactive isotope of sodium used as a diagnostic aid in medicine. It undergoes radioactive decay to form the stable isotope magnesium-24 and has a half-life of about 15 hours. This means that, in this time, half the amount of a sample mass of sodium-24 decays to magnesium-24. Suppose we start with an initial mass of of 100 grams sodium-24.

a. Use the half-life of sodium-24 to write an exponential decay function of the form $m_{Na}(t) = m_0(1 - r)^t$, where m_0 is the initial mass of sodium-24, r is the decay rate, t is the time in hours, and $m_{Na}(t)$ is the mass of sodium-24 at time t. What is the meaning of r?

b. The combined amounts of sodium-24 and magnesium-24 must equal m_0, or 100, for all possible values of t. Show how to write a function for $m_{Mg}(t)$, the mass of magnesium-24 as a function of t.

c. Use a graphing calculator to graph $m_{Na}(t)$ and $m_{Mg}(t)$. Describe the graph of $m_{Mg}(t)$ as a series of transformations of $m_{Na}(t)$. What does the intersection of the graphs represent?

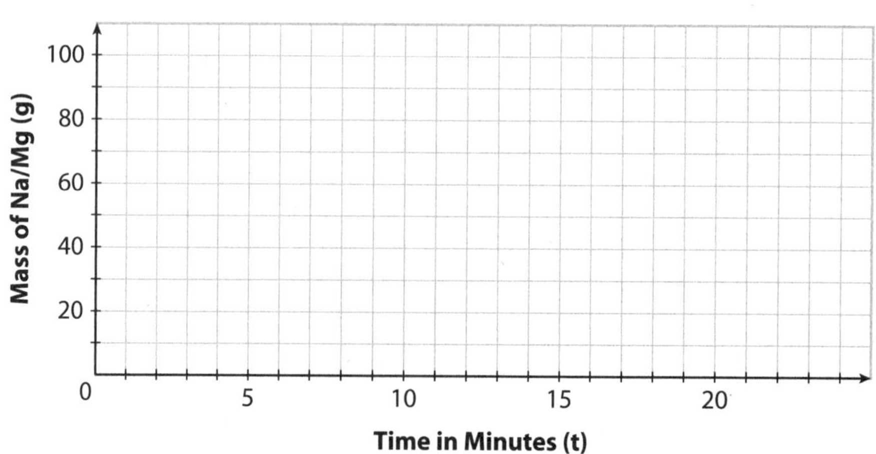

Name _______________ Class _______________ Date _______________

13.4 The Base e

Essential Question: How is the graph of $g(x) = ae^{x-h} + k$ related to the graph of $f(x) = e^x$?

A2.5.A Determine the effects on the key attributes on the graphs of $f(x) = bx$...where b is...e when $f(x)$ is replaced by $af(x)$, $f(x) + d$, and $f(x - c)$ for specific positive and negative real values of a, c, and d. Also A2.2.A, A2.5.B, A2.5.D, A2.7.I/

Explore 1 Graphing and Analyzing $f(x) = e^x$

The following table represents the function $f(x) = \left(1 + \frac{1}{x}\right)^x$ for several values of x.

x	1	10	100	1000	...
$f(x)$	2	2.5937...	2.7048...	2.7169...	...

As the value of x increases without bound, the value of $f(x)$ approaches a number whose decimal value is 2.718... This number is irrational and is called e. You can write this in symbols as $f(x) \to e$ as $x \to +\infty$.

If you graph $f(x)$ and the horizontal line $y = e$, you can see that $y = e$ is the horizontal asymptote of $f(x)$.

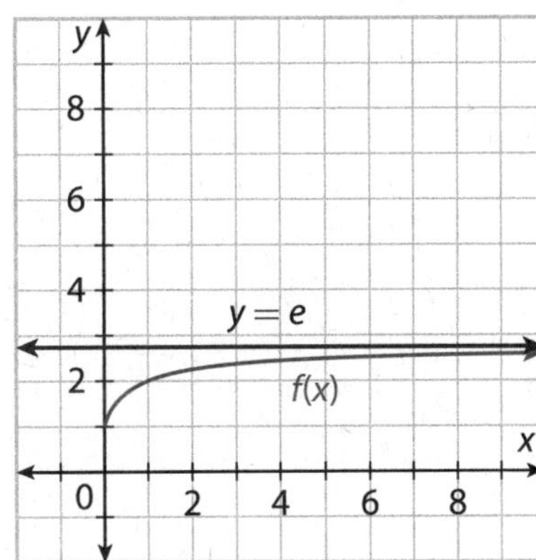

Even though e is an irrational number, it can be used as the base of an exponential function. The number e is sometimes called the natural base of an exponential function and is used extensively in scientific and other applications involving exponential growth and decay.

(A) Fill out the table of values below for the function $f(x) = e^x$. Use decimal approximations.

x	−10	−1	−0.5	0	0.5	1	1.5	2
$f(x) = e^x$	4.54×10^{-5}	$\frac{1}{e} = 0.367...$	0.606...		$\sqrt{e} =$			

(B) Plot the points on a graph.

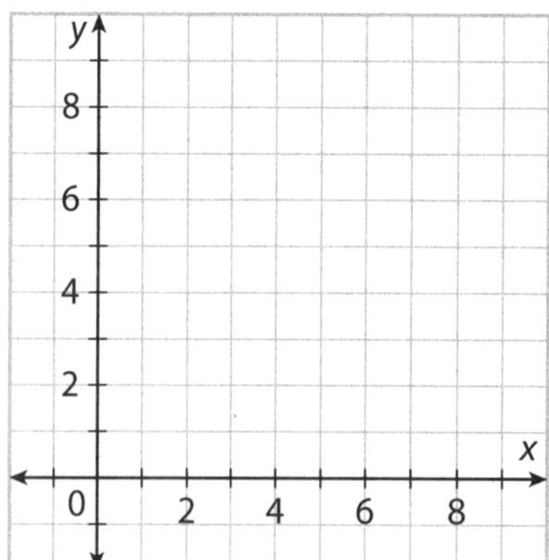

(C) The domain of $f(x) = e^x$ is $\{x| \quad \}$.

The range of $f(x) = e^x$ is $\{y| \quad \}$.

(D) Is the function increasing or decreasing? For what values of x is it increasing/decreasing?

(E) The function's y-intercept is $(0, \quad)$ because $f(0) = e^0 = \quad$ and $x = 0$ is in the domain of the function.

(F) Another point on the graph that can be used as a reference point is $(1, \quad)$.

(G) Identify the end behavior.

$f(x) \to \quad$ as $x \to \infty$

$f(x) \to \quad$ as $x \to -\infty$

There is a horizontal asymptote at $y = \quad$.

Reflect

1. What is the relationship between the graphs of $f(x) = e^x$, $g(x) = 2^x$, and $h(x) = 3^x$? (Hint: Sketch the graphs on your own paper.)

Explore 2 Predicting Transformations of the Graph of $f(x) = e^x$

The parent function, $f(x) = e^x$, can be transformed into a different exponential function with base e depending on the value and sign of the constant parameters h, k, and a. As in previous transformation of graphs, the effect of h on the graph of $g(x) = f(x - h)$, the effect of k on the graph of $g(x) = f(x) + k$, and the effect of a on the graph of $g(x) = af(x)$ can all be predicted from the value and sign of the parameters. Predict the effect of each transformation on the following graphs and then use a graphing calculator to confirm your prediction.

(A) Transform $f(x) = e^x$ into $g(x) = e^{x-1}$.

The function $g(x) = e^{x-1}$ is of the form $g(x) =$ ☐, so $g(x) = e^{x-1}$ represents a ________ translation by ☐ units(s) to the ______.

(B) Transform $f(x) = e^x$ into $g(x) = e^{x+1}$.

The function $g(x) = e^{x+1}$ is of the form $g(x) =$ ☐, so $g(x) = e^{x+1}$ represents a ________ translation by ☐ units(s) to the ______.

(C) Transform $f(x) = e^x$ into $g(x) = e^x + 2$.

The function $g(x) = e^x + 2$ is of the form $g(x) =$ ☐, so $g(x) = e^x + 2$ represents a ________ translation by ☐ units(s) ______.

(D) $f(x) = e^x$ into $g(x) = e^x - 2$.

The function $g(x) = e^x - 2$ is of the form $g(x) =$ ☐, so $g(x) = e^x - 2$ represents a ________ translation by ☐ units(s) ______.

(E) Transform $f(x) = e^x$ into $g(x) = 2e^x$.

The function $g(x) = 2e^x$ is of the form $g(x) =$ ☐, so $g(x) = 2e^x$ represents a vertical ________ by a factor of ☐.

(F) Transform $f(x) = e^x$ into $g(x) = \frac{1}{2}e^x$.

The function $g(x) = \frac{1}{2}e^x$ is of the form $g(x) =$ ☐, so $g(x) = \frac{1}{2}e^x$ represents a vertical ________ by a factor of ☐.

(G) Transform $f(x) = e^x$ into $g(x) = -2e^x$.

The function $g(x) = -2e^x$ is of the form $g(x) =$ ☐, so $g(x) = -2e^x$ represents a vertical ________ by a factor of ☐ and a reflection across the ☐ -axis.

(H) Transform $f(x) = e^x$ into $g(x) = -\frac{1}{2}e^x$.

The function $g(x) = -\frac{1}{2}e^x$ is of the form $g(x) =$ ☐, so $g(x) = -\frac{1}{2}e^x$ represents a vertical ________ by a factor of ☐ and a reflection across the ☐ -axis.

Reflect

2. **Discussion** Describe the effects of the parameters h, k, and a on the domain, range, and asymptote of $g(x)$ in regards to the domain, range, and asymptote of the parent function $f(x)$.

Explain 1 Graphing Combined Transformations of $f(x) = e^x$

When graphing combined transformations of $f(x) = e^x$ that result in the function $g(x) = a \cdot e^{x-h} + k$, it helps to focus on two reference points on the graph of $f(x)$, $(0, 1)$ and $(1, e)$, as well as on the asymptote $y = 0$. The table shows these reference points and the asymptote $y = 0$ for $f(x) = e^x$ and the corresponding points and asymptote for the transformed function, $g(x) = a \cdot e^{x-h} + k$.

	$f(x) = e^x$	$g(x) = a \cdot e^{x-h} + k$
First reference point	$(0, 1)$	$(h, a + k)$
Second reference point	$(1, e)$	$(h + 1, ae + k)$
Asymptote	$y = 0$	$y = k$

Example 1 **Given a function of the form $g(x) = a \cdot e^{x-h} + k$, identify the reference points and use them to draw the graph. State the transformations that compose the combined transformation, the asymptote, the domain, and range. Write the domain and range using set notation.**

(A) $g(x) = 3 \cdot e^{x+1} + 4$

Compare $g(x) = 3 \cdot e^{x+1} + 4$ to the general form $g(x) = a \cdot e^{x-h} + k$ to find that $h = -1$, $k = 4$, and $a = 3$.

Find the reference points of $f(x) = 3 \cdot e^{x+1} + 4$.

$(0, 1) \rightarrow (h, a + k) = (-1, 3 + 4) = (-1, 7)$

$(1, e) \rightarrow (h + 1, ae + k) = (-1 + 1, 3e + 4) = (0, 3e + 4)$

State the transformations that compose the combined transformation.

$h = -1$, so the graph is translated 1 unit to the left.

$k = 4$, so the graph is translated 4 units up.

$a = 3$, so the graph is vertically stretched by a factor of 3.

a is positive, so the graph is not reflected across the x-axis.

The asymptote is vertically shifted to $y = k$, so $y = 4$.

The domain is $\{x | -\infty < x < \infty\}$.

The range is $\{y | y > 4\}$.

Use the information to graph the function $g(x) = 3 \cdot e^{x+1} + 4$.

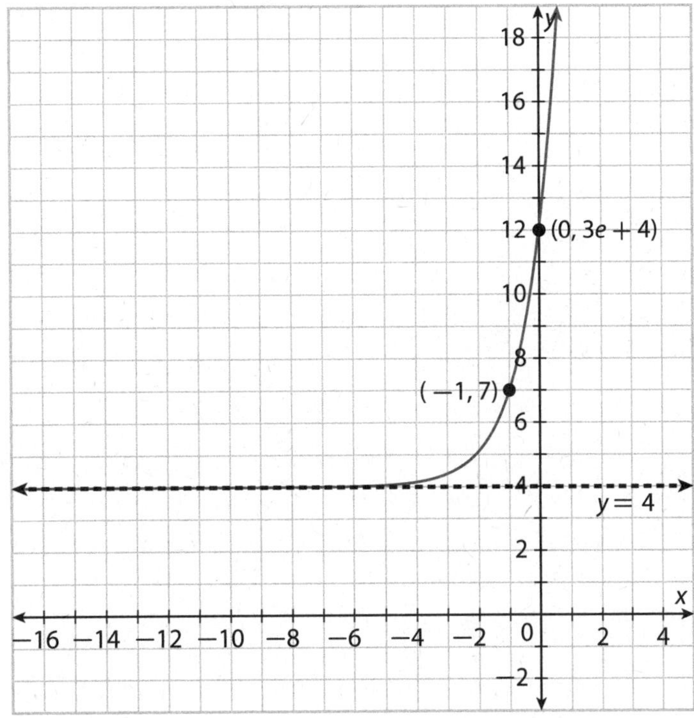

(B) $g(x) = -0.5 \cdot e^{x-2} - 1$

Compare $g(x) = -0.5 \cdot e^{x-2} - 1$ to the general form $g(x) = a \cdot e^{x-h} + k$ to find that $h =$ ☐, $k =$ ☐, and $a =$ ☐.

Find the reference points of $g(x) = -0.5 \cdot e^{x-2} - 1$.

$(0, 1) \rightarrow (h, a + k) = (\square, \square + \square) = (\square, \square)$

$(1, e) \rightarrow (h + 1, ae + k) = (\square + 1, \square e + \square) = (\square, \square)$

State the transformations that compose the combined transformation.

$h =$ ☐, so the graph is translated ☐ units to the ________.

$k =$ ☐, so the graph is translated ☐ unit ________.

$a =$ ☐, so the graph is vertically ________ by a factor of ☐.

a is negative, so the graph is reflected across the ☐-axis.

The asymptote is vertically shifted to $y = k$, so $y = \square$.

The domain is $\left\{x \mid \square \right\}$.

The range is $\left\{y \mid \square \right\}$.

Use the information to graph the function $g(x) = -0.5 \cdot e^{x-2} - 1$.

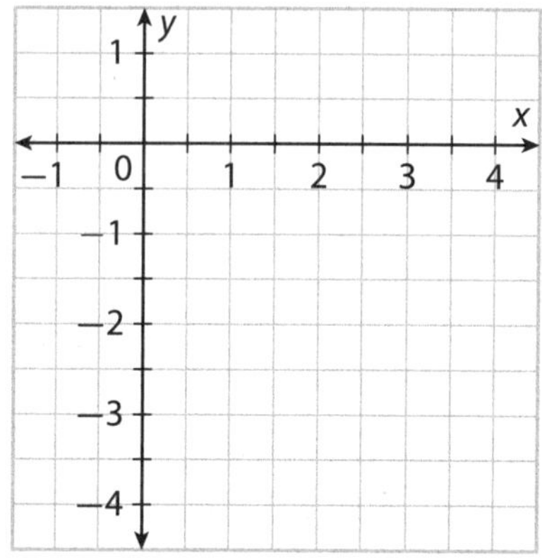

Your Turn

Given a function of the form $g(x) = a \cdot e^{x-h} + k$, identify the reference points and use them to draw the graph. State the asymptote, domain, and range. Write the domain and range using set notation.

3. $g(x) = (-1) \cdot e^{x+2} - 3$

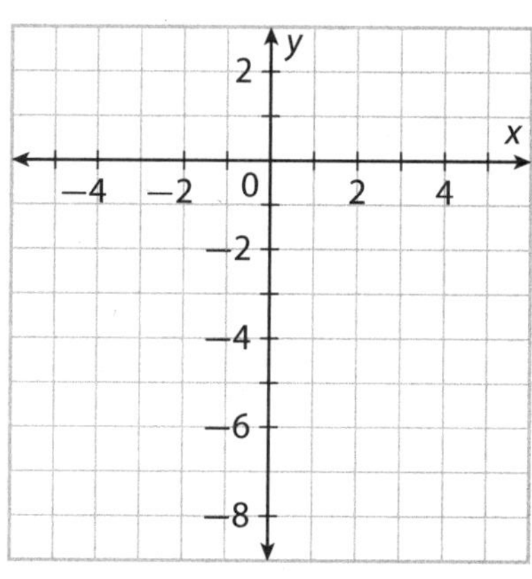

4. $g(x) = 2 \cdot e^{x-1} + 1$

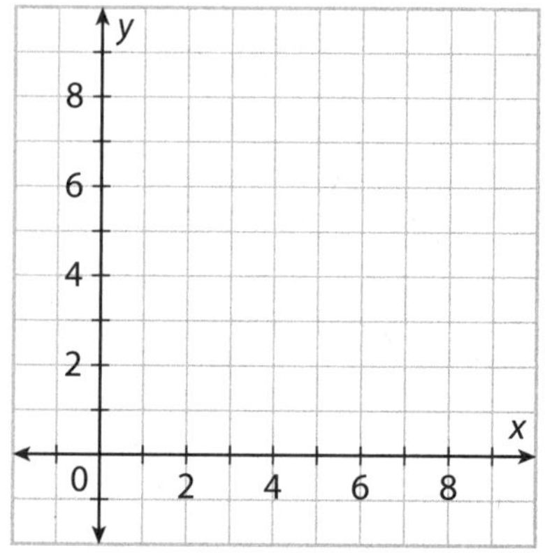

Explain 2 Writing Equations for Combined Transformations of $f(x) = e^x$

If you are given the transformed graph $g(x) = a \cdot e^{x-h} + k$, it is possible to write the equation of the transformed graph by using the reference points $(h, a + k)$ and $(1 + h, ae + k)$.

Example 2 **Write the function whose graph is shown. State the domain and range in set notation.**

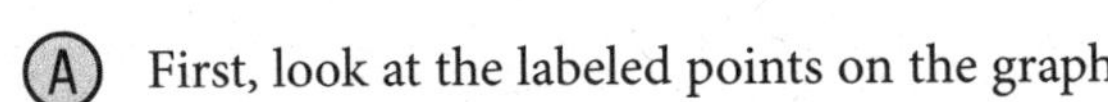

(A) First, look at the labeled points on the graph.

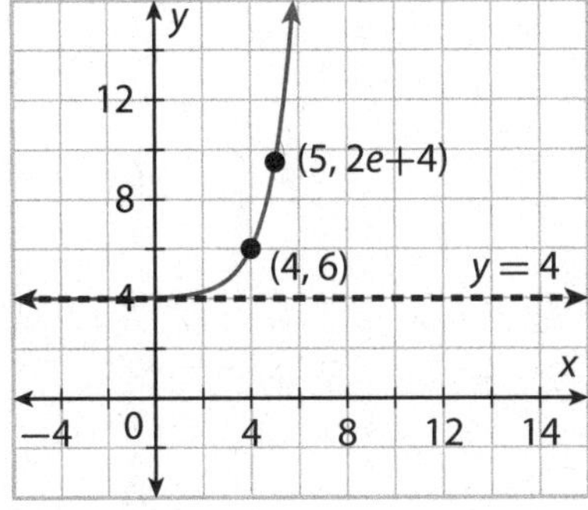

$(h, a + k) = (4, 6)$

$(1 + h, ae + k) = (5, 2e + 4)$

Find a, h, and k.

$(h, a + k) = (4, 6)$, so $h = 4$.

$(1 + h, ae + k) = (5, 2e + 4)$, so $ae + k = 2e + 4$.
Therefore, $a = 2$ and $k = 4$.

Write the equation by substituting the values of a, h, and k into the function $g(x) = a \cdot e^{x-h} + k$.

$g(x) = 2e^{x-4} + 4$

State the domain and range.

Domain: $\{x \mid -\infty < x < \infty\}$

Range: $\{y \mid y > 4\}$

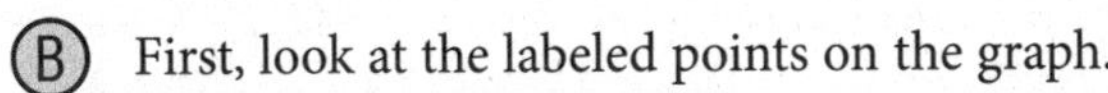

(B) First, look at the labeled points on the graph.

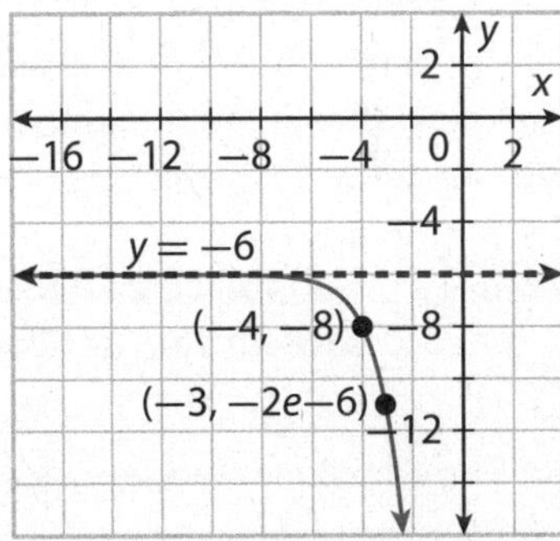

$(h, a + k) = (\square, \square)$

$(1 + h, ae + k) = (\square, \square)$

Find a, h, and k.

$(h, a + k) = (-4, -8)$, so $h = \square$.

$(1 + h, ae + k) = (-3, -2e - 6)$, so $ae + k = \square$.

Therefore, $a = \square$ and $k = \square$.

Write the equation by substituting the values of a, h, and k into the function $g(x) = a \cdot e^{x-h} + k$.

$g(x) = \square$

State the domain and range.

Domain: $\{x \mid \square\}$

Range: $\{y \mid \square\}$

Your Turn

Write the function whose graph is shown. State the domain and range in set notation.

5.

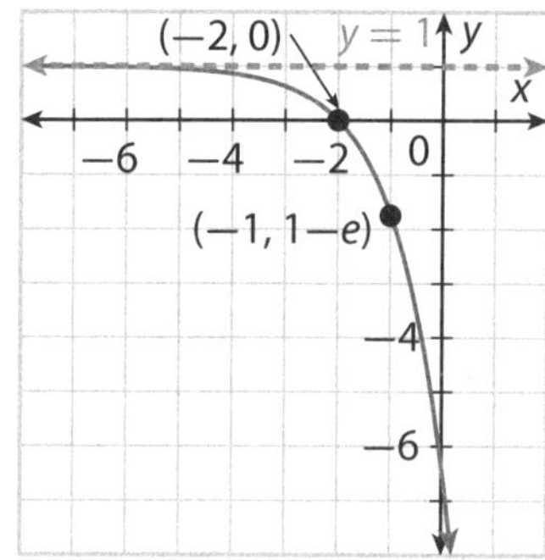

6.

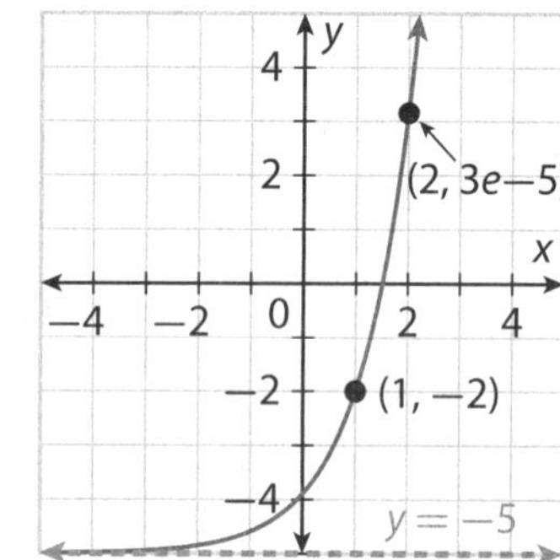

Explain 3 Modeling with Exponential Functions Having Base *e*

Although the function $f(x) = e^x$ has base $e \approx 2.718$, the function $g(x) = e^{cx}$ can have any positive base (other than 1) by choosing an appropriate positive or negative value of the constant c. This is because you can write $g(x)$ as $(e^c)^x$ by using the Power of a Power Property of Exponents.

Example 3 **Solve each problem using a graphing calculator. Then determine the growth rate or decay rate of the function.**

Ⓐ The Dow Jones index is a stock market index for the New York Stock Exchange. The Dow Jones index for the period 1980-2000 can be modeled by $V_{DJ}(t) = 878e^{0.121t}$, where t is the number of years after 1980. Determine how many years after 1980 the Dow Jones index will reach 3000.

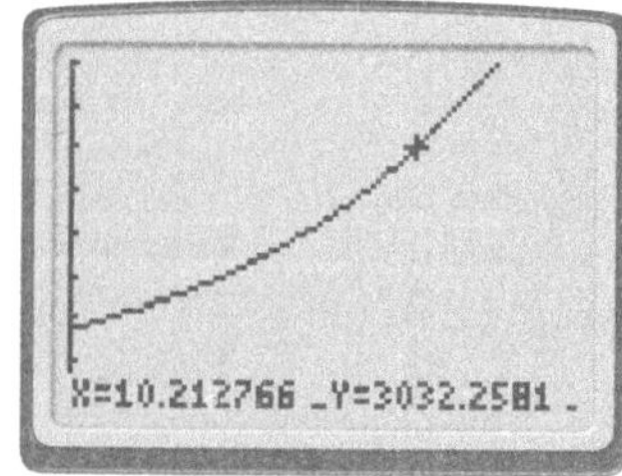

Use a graphing calculator to graph the function.

The value of the function is about 3000 when $x \approx 10.2$. So, the Dow Jones index will reach 3000 after 10.2 years, or after the year 1990.

In an exponential growth model of the form $f(x) = ae^{cx}$, the growth factor $1 + r$ is equal to e^c.

To find r, first rewrite the function in the form $f(x) = a(e^c)^x$.

$$V_{DJ}(t) = 878e^{0.121t}$$
$$= 878\left(e^{0.121}\right)^t$$

Find r by using $1 + r = e^c$.

$$1 + r = e^c$$
$$1 + r = e^{0.121}$$
$$r = e^{0.121} - 1 \approx 0.13$$

So, the growth rate is about 13%.

(B) The Nikkei 225 index is a stock market index for the Tokyo Stock Exchange. The Nikkei 225 index for the period 1990-2010 can be modeled by $V_{N225}(t) = 23{,}500e^{-0.0381t}$, where t is the number of years after 1990. Determine how many years after 1990 the Nikkei 225 index will reach 15,000.

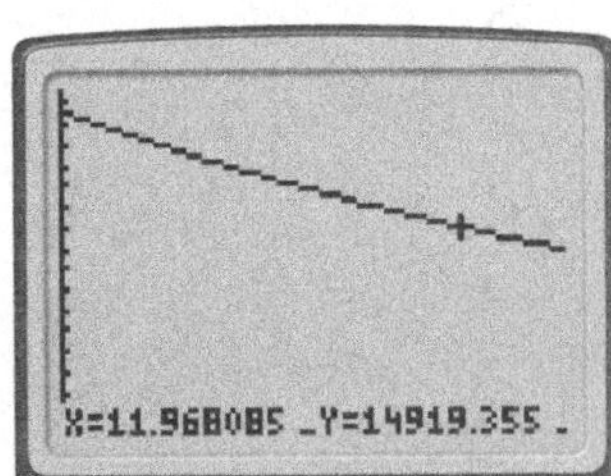

Use a graphing calculator to graph the function.

The value of the function is about 15,000 when $x \approx$ ☐. So, the Nikkei 225 index will reach 15,000 after ☐ years, or after the year ☐.

In an exponential decay model of the form $f(x) = ae^{cx}$, the decay factor ☐ is equal to e^c.

To find r, first rewrite the function in the form $f(x) = a(e^c)^x$.

$$V_{N225}(t) = 23{,}500e^{-0.0381t}$$
$$= 23{,}500\left(\square\right)^t$$

Find r by using $1 - r = e^c$.

$$1 - r = e^c$$
$$1 - r = \square$$
$$r = \square \approx \square$$

So, the growth rate is ☐%.

Your Turn

7. A paleontologist uncovers a fossil of a saber-toothed cat in California. The paleontologist analyzes the fossil and concludes that the specimen contains 15% of its original carbon-14. The percent of original carbon-14 in a specimen after t years can be modeled by $N(t) = 100e^{-0.00012t}$, where t is the number of years after the specimen died. Use a graphing calculator to determine the age of the fossil. Then determine the decay rate of the function.

Elaborate

8. Which transformations of $f(x) = e^x$ change the function's end behavior?

9. Which transformations change the location of the graph's y-intercept?

10. Why can the function $f(x) = ae^{cx}$ be used as an exponential growth model and as an exponential decay model? How can you tell if the function represents growth or decay?

11. **Essential Question Check-In** How are reference points helpful when graphing transformations of $f(x) = e^x$ or when writing equations for transformed graphs?

Evaluate: Homework and Practice

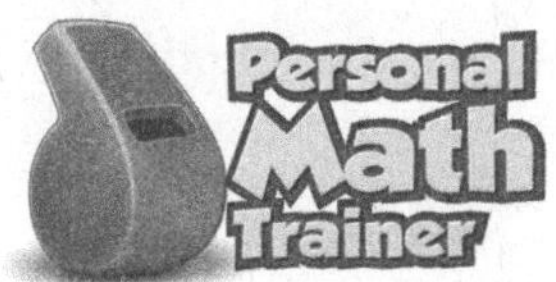

- Online Homework
- Hints and Help
- Extra Practice

1. What is the greatest value of $f(x) = \left(1 + \frac{1}{x}\right)^x$ for any positive value of x?

2. Identify the key attributes of $f(x) = e^x$, including the domain and range in set notation, the end behavior, and all intercepts.

Predict the effect of the parameters h, k, or a on the graph of the parent function $f(x) = e^x$. Identify any changes of domain, range, or end behavior.

3. $g(x) = f\left(x - \frac{1}{2}\right)$

4. $g(x) = f(x) - \frac{5}{2}$

5. $g(x) = -\frac{1}{4}f(x)$

6. $g(x) = \frac{27}{2}f(x)$

7. The graph of $f(x) = ce^x$ crosses the y-axis at $(0, c)$, where c is some constant. Where does the graph of $g(x) = f(x) - d$ cross the y-axis?

Given the function of the form $g(x) = a \cdot e^{x-h} + k$, identify the reference points and use them to draw the graph. State the domain and range in set notation.

8. $g(x) = e^{x-1} + 2$

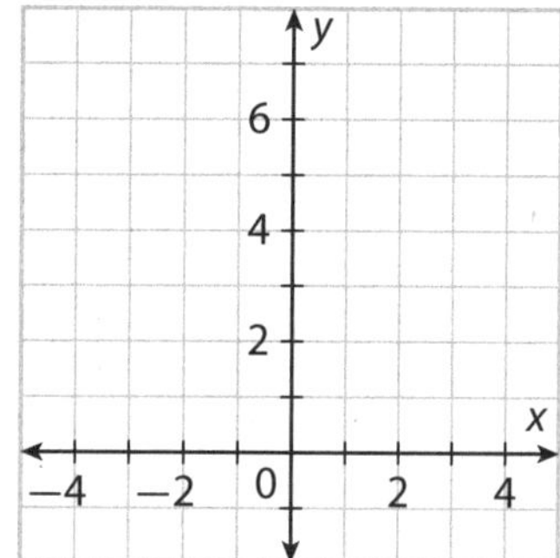

9. $g(x) = -e^{x+1} - 1$

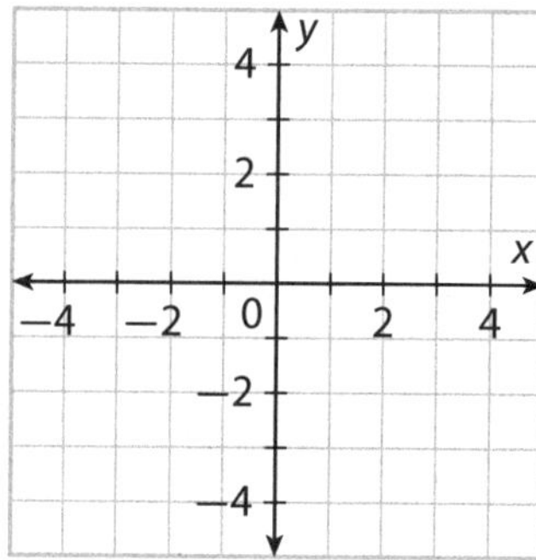

10. $g(x) = \frac{1}{2} e^{x+3} + 2$

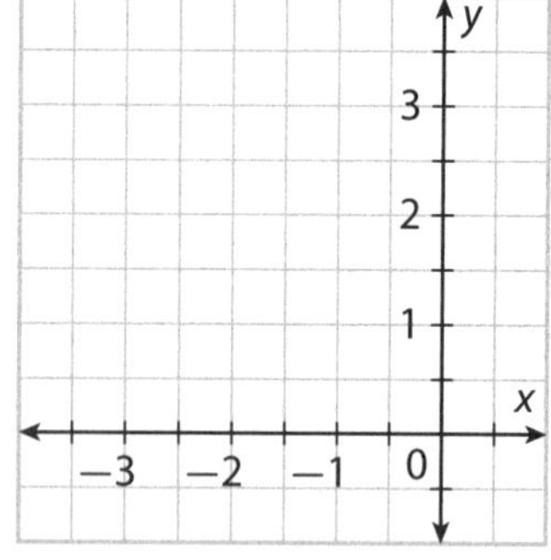

11. $g(x) = -\frac{3}{4} e^{x+2} - 4$

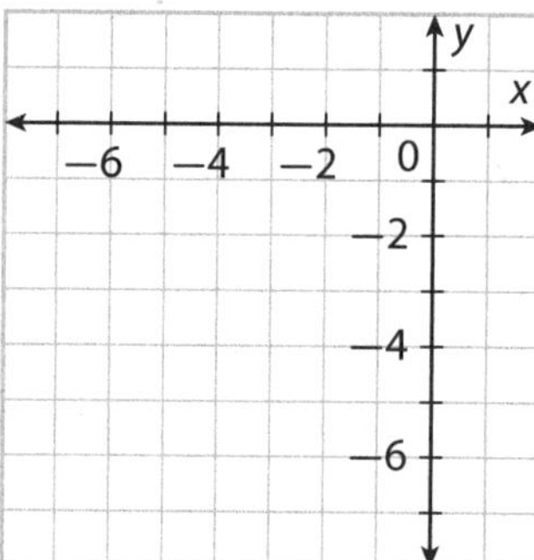

12. $g(x) = \frac{3}{2}e^{x-1} - 3$

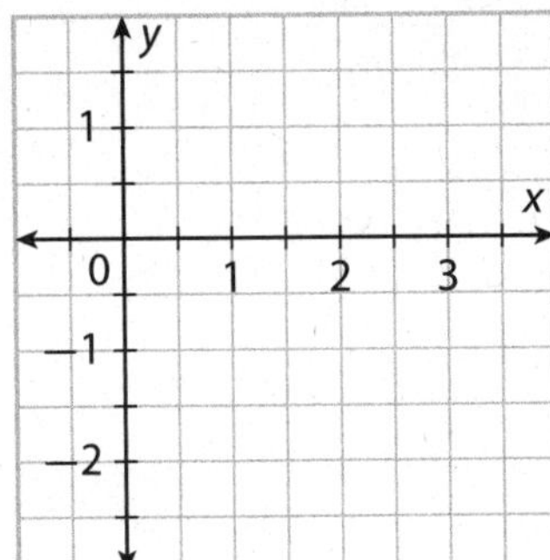

13. $g(x) = -\frac{5}{3}e^{x-4} + 2$

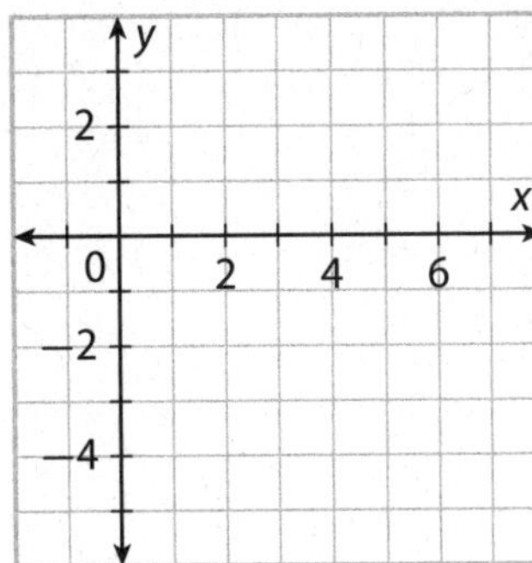

Write the function whose graph is shown. State the domain and range in set notation.

14.

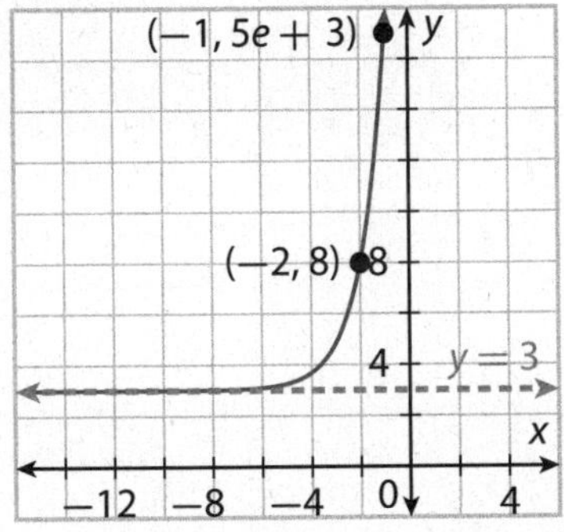

15.

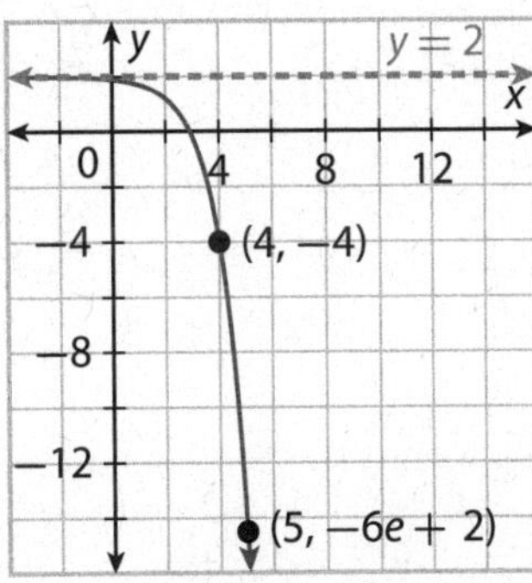

Solve each problem using a graphing calculator. Then determine the growth rate or decay rate of the function.

16. **Medicine** Technetium-99m, a radioisotope used to image the skeleton and the heart muscle, has a half-life of about 6 hours. Use the decay function $N(t) = N_0 e^{-0.1155t}$, where N_0 is the initial amount and t is the time on hours, to determine how many hours it takes for a 250 milligram dose to decay to 16 milligrams.

17. **Ecology** The George River herd of caribou in Canada was estimated to be about 4700 in 1954 and grew at an exponential rate to about 472,000 in 1984. Use the exponential growth function $P(t) = P_0 e^{0.154t}$, where P_0 is the initial population, t is the time in years after 1954, and $P(t)$ is the population at time t, to determine after how many years the herd will be 25 million.

18. Chemistry Radioactive plutonium (Pu-239) has a half-life about 24,110 years. Use the function $N(t) = N_0 e^{-0.000029t}$ to find how many years it will take for 20 grams of Pu-239 to decay to 1 gram. N_0 represents the initial amount of Pu-239 and t is the time in years.

19. Population The population of a town was estimated to be about 7200 in 1990 and grew at an exponential rate to about 40,000 in 2010. Use the exponential growth function $P(t) = P_0 e^{0.086t}$, where P_0 is the initial population, t is the time in years after 1990, and $P(t)$ is the population at time t, to determine after how many years the population will be 50,000.

H.O.T. Focus on Higher Order Thinking

20. Explain the Error A classmate claims that the function $g(x) = -4e^{x-5} + 6$ is the parent function $f(x) = e^x$ reflected across the y-axis, vertically compressed by a factor of 4, translated to the left 5 units, and translated up 6 units. Explain what the classmate described incorrectly and describe $g(x)$ as a series of transformations of $f(x)$.

21. Multi-Step Newton's law of cooling states that the temperature of an object decreases exponentially as a function of time, according to $T = T_s + (T_0 - T_s)e^{-kt}$, where T_0 is the initial temperature of the liquid, T_s is the surrounding temperature, and k is a constant. For a time in minutes, the constant for coffee is approximately 0.283. The corner coffee shop has an air temperature of 70°F and serves coffee at 206°F. Coffee experts say coffee tastes best at 140°F.

a. How long does it take for the coffee to reach its best temperature?

b. The air temperature on the patio outside the coffee shop is 86 °F. How long does it take for coffee to reach its best temperature there?

c. Find the time it takes for the coffee to cool to 71°F in both the coffee shop and the patio. Explain how you found your answer.

22. Analyze Relationships The graphing calculator screen shows the graphs of the functions $f(x) = 2^x$, $f(x) = 10^x$, and $f(x) = e^x$ on the same coordinate grid. Identify the common attributes and common point(s) of the three graphs. Explain why the point(s) is(are) common to all three graphs.

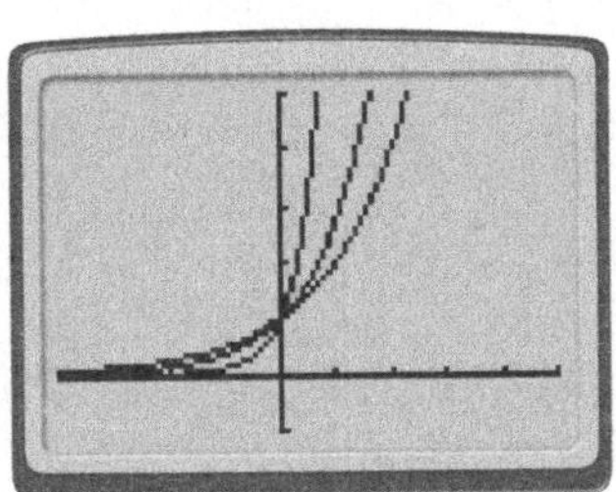

Lesson Performance Task

The ever-increasing amount of carbon dioxide in Earth's atmosphere is an area of concern for many scientists. In order to more accurately predict what the future consequences of this could be, scientists make mathematic models to extrapolate past increases into the future. A model developed to predict the annual mean carbon dioxide level L in Earth's atmosphere in parts per million t years after 1960 is $L(t) = 36.9 \cdot e^{0.0223t} + 280$.

a. Use the function $L(t)$ to describe the graph of $L(t)$ as a series of transformations of $f(t) = e^t$.

b. Find and interpret $L(80)$, the carbon dioxide level predicted for the year 2040. How does it compare to the carbon dioxide level in 2015?

c. Can $L(t)$ be used as a model for all positive values of t? Explain.

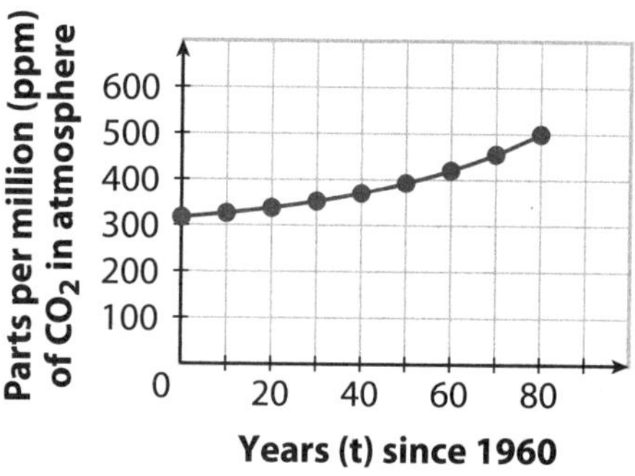

STUDY GUIDE REVIEW

Exponential Functions

MODULE 13

Essential Question: How can you use exponential functions to solve real-world problems?

Key Vocabulary
exponential decay
(decremento exponencial)
exponential function
(función exponencial)
exponential growth
(crecimiento exponencial)
geometric sequence
(sucesión geométrica)

KEY EXAMPLE *(Lesson 13.1)*

Find the 12th term of the geometric sequence 5, 15, 45,. . .

$r = \frac{15}{5} = 3$ Find the common ratio of the sequence.

$a_n = a_1 r^{n-1}$ Write the formula for a geometric sequence.

$a_{12} = 5(3)^{12-1}$ Substitute in a_1, r, and n.

$a_{12} = 5(177{,}147)$ Use a calculator to solve for a_{12}.

$a_{12} = 885{,}735$ Simplify.

KEY EXAMPLE *(Lesson 13.3)*

$g(x) = 3^{x+1}$ is a transformation of the function $f(x) = 3^x$. Sketch a graph of $g(x) = 3^{x+1}$.

$g(x) = 3^{x+1} = f(x + 1)$

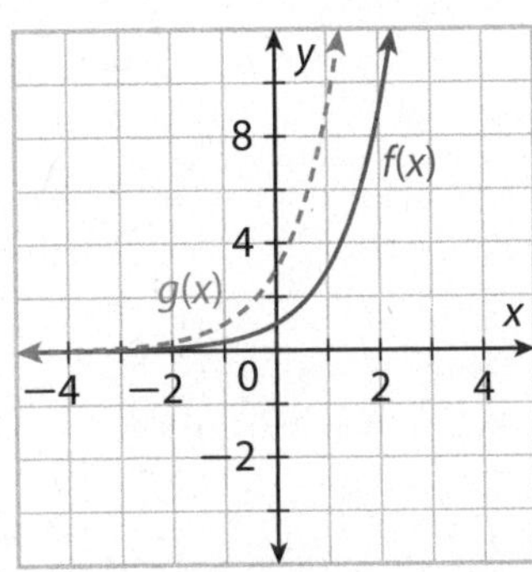

Because $g(x) = 3^{x+1} = f(x + 1)$, the graph of g can be obtained by shifting the graph of f one unit to the left, as shown.

The solid line represents $f(x) = 3^x$.
The dashed line represents $g(x) = 3^{x+1}$.

EXERCISES

1. If the first three terms of a geometric sequence are 3, 12, and 48, what is the seventh term? *(Lesson 13.1)*

Sketch the graphs of the following transformations. *(Lessons 13.2, 13.3, 13.4)*

2. $g(x) = -2(0.5)^x$

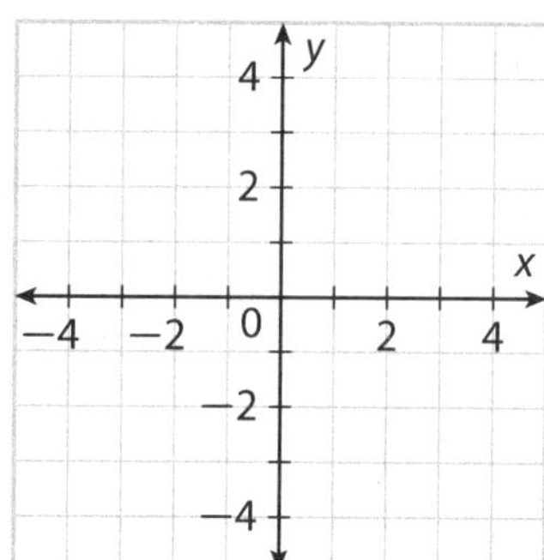

3. $f(x) = \left(\frac{1}{2}\right)^{-x}$

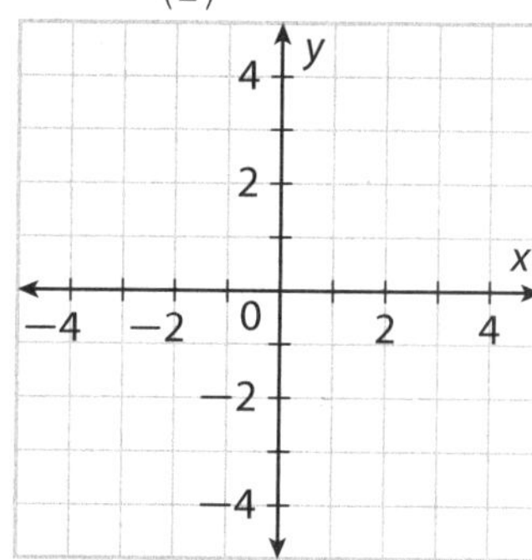

4. $f(x) = 2e^{x-2} + 1$

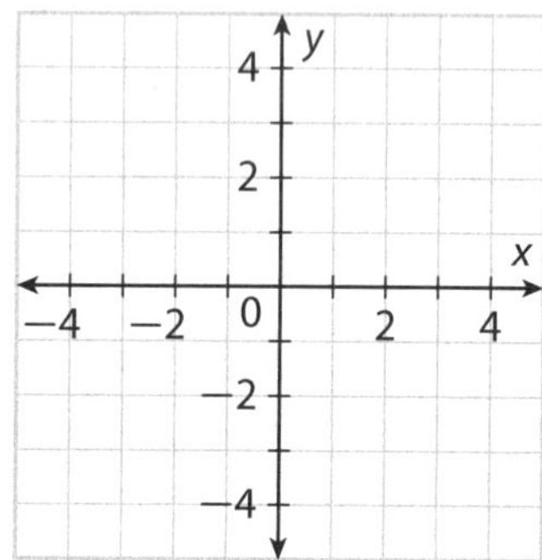

5. $g(x) = \left(\frac{3}{5}\right)^x$

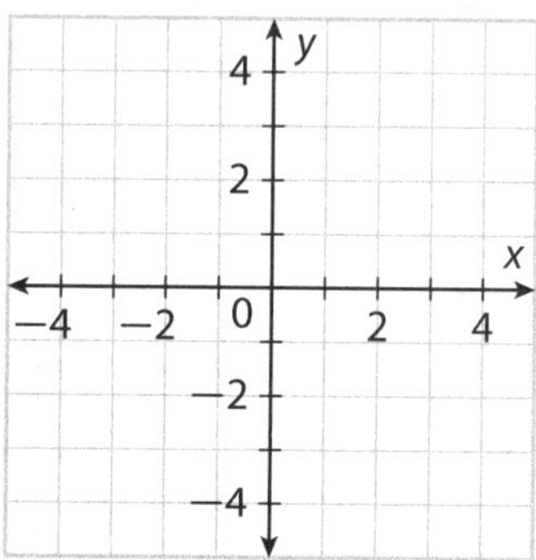

MODULE PERFORMANCE TASK

That's The Way the Ball Bounces

Kingston is a chemical engineer who is testing the "bounciness" of two novel materials. He formed each material into two equal-sized spheres. Kingston then dropped each sphere on a hard surface and measured the heights the spheres reached after each bounce. The results are shown in the table.

	Heights (cm)					
	h_0	h_1	h_2	h_3	h_4	h_5
Material A	90	63.0	44.1	30.9	21.6	15.1
Material B	90	49.5	27.2	15.0	8.2	4.5

Use the data to create mathematical models for the heights of the bounces. Compare the two materials. One of the two materials will be used for the tip of a pogo stick. Which material should Kingston recommend and why?

Use your own paper to complete the task. Be sure to write down all your data and assumptions. Then use graphs, numbers, words, or algebra to explain how you reached your conclusion.

Ready to Go On?

13.1–13.4 Exponential Functions

- Online Homework
- Hints and Help
- Extra Practice

Write a recursive rule and an explicit rule for each geometric sequence. *(Lesson 13.1)*

1. 9, 27, 81, 243,...

2. 5, −5, 5, −5,...

State the domain and range for the graphs of the following functions. *(Lessons 13.2, 13.3)*

3. $y = \left(\frac{1}{4}\right)^x$

4. $y = \left(\frac{1}{3}\right)^{x-2} + 2$

5. $y = -3 \cdot 2^{x+2}$

6. $y = 3^{x-2} - 1$

ESSENTIAL QUESTION

7. How can you tell whether an exponential function models exponential growth or exponential decay?

MODULE 13
MIXED REVIEW

Assessment Readiness

1. Which of the following is a geometric sequence?

A. 10, 15, 20, 25,...

B. 5, 15, 45, 135,...

C. 1, 3, 5, 7,...

D. $\frac{1}{2}, \frac{3}{4}, 1, 1\frac{1}{4}, \ldots$

2. When the base in an exponential function is between 0 and 1, the function shows

A. exponential decay

B. exponential growth

C. the natural base *e*

D. a geometric sequence

3. What is the asymptote of the graph of $y = \left(\frac{1}{2}\right)^{x-2} + 3$?

A. $y = -3$

B. $y = -2$

C. $y = 2$

D. $y = 3$

4. Solve $\frac{3}{4}|x + 3| - 8 = 4$ for x.

A. $x = 13$ or $x = -19$

B. $x = 8$ or $x = -3$

C. $x = -3$

D. $x = 5$ or $x = -11$

5. The graphs of $f(x) = 2^x$, $f(x) = 10^x$ and $f(x) = e^x$ all pass through a common point. Explain why the point $(0, 1)$ is common to all three functions.

Modeling with Exponential and Other Functions

MODULE 14

TEKS

Essential Question: How can modeling with exponential and other functions help you to solve real-world problems?

LESSON 14.1
Fitting Exponential Functions to Data
TEKS A2.8.A, A2.8.B, A2.8.C

LESSON 14.2
Choosing Among Linear, Quadratic, and Exponential Models
TEKS A2.8.A, A2.8.B, A2.8.C

REAL WORLD VIDEO
Most people have to save up money for a major purchase like a car or new home. Check out some of the factors to consider when investing for long-term goals.

MODULE PERFORMANCE TASK PREVIEW
Double Your Money!

If you had some money to invest, how would you pick the investment option that would let your money grow fastest? The return on an investment depends on factors such as the length of time of the investment and the return rate. How can you use an exponential model to find out when an investment will double in value? Let's find out!

Are YOU Ready?

Complete these exercises to review skills you will need for this module.

Personal Math Trainer

- Online Homework
- Hints and Help
- Extra Practice

Writing Linear Equations

Example 1 Write an equation for the line that passes through the points $(2, 3)$ and $(4, -1)$.

$\frac{-1-3}{4-2} = \frac{-4}{2} = -2$ Find the slope.

So $y = -2x + b$ Substitute slope for m in $y = mx + b$.

$(3) = -2(2) + b$ Substitute $(2, 3)$ for x- and y-values.

$b = 7$ Solve for b.

The equation is $y = -2x + 7$.

Write an equation for the line that passes through the given points.

1. $(2, -5), (6, -3)$

2. $(4, -3), (-2, 15)$

3. $(4, 7), (-2, -2)$

Transforming Linear Functions

Example 2 Write the equation of $y = 9x - 2$ after a reflection across the x-axis followed by a reflection across the y-axis.

$y = -1(9x - 2) \rightarrow y = -9x + 2$ Reflection across the x-axis

$y = -9(-x) + 2 \rightarrow y = 9x + 2$ Reflection across the y-axis

So $y = 9x - 2$ reflected across both axes is $y = 9x + 2$.

Write the equation of each function after a reflection across both axes.

4. $y = 2x + 1$

5. $y = -3x - 4$

6. $y = -0.2x + 6$

Equations Involving Exponents

Example 3 Solve $x^{\frac{2}{3}} = 16$ for x.

$\left(x^{\frac{2}{3}}\right)^{\frac{3}{2}} = \pm (16)^{\frac{3}{2}}$ Raise both sides to the same power.

$x = \pm \left(16^{\frac{1}{2}}\right)^3 = \pm (4)^3 = \pm 64$ Evaluate the right side.

So, $x = \pm 64$.

Solve for x.

7. $x^6 = 4096$

8. $x^{\frac{3}{2}} = 27$

9. $\frac{1}{3}x^{\frac{2}{5}} = 3$

Name ______________________ Class ____________ Date ________

14.1 Fitting Exponential Functions to Data

Resource Locker

Essential Question: What are ways to model data using an exponential function of the form $f(x) = ab^x$?

TEKS A2.8.B Use regression methods available through technology to write ... an exponential function from a given set of data. Also A2.8.A, A2.8.C

Explore Identifying Exponential Functions from Tables of Values

Notice for an exponential function $f(x) = ab^x$ that $f(x + 1) = ab^{x+1}$. By the product of powers property, $ab^{x+1} = a(b^x \cdot b^1) = ab^x \cdot b = f(x) \cdot b$. So, $f(x + 1) = f(x) \cdot b$. This means that increasing the value of x by 1 multiplies the value of $f(x)$ by b. In other words, for successive integer values of x, each value of $f(x)$ is b times the value before it, or, equivalently, the ratio between successive values of $f(x)$ is b. This gives you a test to apply to a given set of data to see whether it represents exponential growth or decay.

Each table gives function values for successive integer values of x. Find the ratio of successive values of $f(x)$ to determine whether each set of data can be modeled by an exponential function.

x	0	1	2	3	4
f(x)	1	4	16	64	256

$\frac{f(1)}{f(0)} = \square$; $\frac{f(2)}{f(1)} = \square$; $\frac{f(3)}{f(2)} = \square$; $\frac{f(4)}{f(3)} = \square$

The data [are/are not] exponential.

x	0	1	2	3	4
f(x)	1	7	13	19	25

$\frac{f(1)}{f(0)} = \square$; $\frac{f(2)}{f(1)} = \square$; $\frac{f(3)}{f(2)} = \square$; $\frac{f(4)}{f(3)} = \square$

The data [are/are not] exponential.

x	0	1	2	3	4
f(x)	1	4	13	28	49

$\frac{f(1)}{f(0)} = \square$; $\frac{f(2)}{f(1)} = \square$; $\frac{f(3)}{f(2)} = \square$; $\frac{f(4)}{f(3)} = \square$

The data [are/are not] exponential.

(D)

x	0	1	2	3	4
$f(x)$	1	0.25	0.0625	0.015625	0.00390625

$\frac{f(1)}{f(0)} = \square$; $\frac{f(2)}{f(1)} = \square$; $\frac{f(3)}{f(2)} = \square$; $\frac{f(4)}{f(3)} = \square$

The data [are/are not] exponential.

Reflect

1. In which step(s) does the table show exponential growth? Which show(s) exponential decay? What is the base of the growth or decay?

2. In which step are the data modeled by the exponential function $f(x) = 4^{-x}$?

3. What type of function model would be appropriate in each step not modeled by an exponential function? Explain your reasoning.

4. **Discussion** In the introduction to this Explore, you saw that the ratio between successive terms of $f(x) = ab^x$ is b. Find and simplify an expression for $f(x + c)$ where c is a constant. Then explain how this gives you a more general test to determine whether a set of data can be modeled by an exponential function.

Explain 1 Roughly Fitting an Exponential Function to Data

As the answer to the last Reflect question above indicates, if the ratios of successive values of the dependent variable in a data set for equally-spaced values of the independent variable are equal, an exponential function model fits. In the real world, sets of data rarely fit a model perfectly, but if the ratios are approximately equal, an exponential function can still be a good model.

Example 1

(A) **Population Statistics** The table gives the official population of the United States for the years 1790 to 1890.

Year	Total Population
1790	3,929,214
1800	5,308,483
1810	7,239,881
1820	9,638,453
1830	12,860,702
1840	17,063,353
1850	23,191,876
1860	31,443,321
1870	38,558,371
1880	50,189,209
1890	62,979,766

Create an approximate exponential model for the data set. Then graph your function with a scatter plot of the data and assess its fit.

It appears that the ratio of the population in each decade to the population of the decade before it is pretty close to one and one third, so an exponential model should be reasonable.

For a model of the form $f(x) = ab^x$, $f(0) = a$. So, if x is the number of decades after 1790, the value when $x = 0$ is a, the initial population in 1790, or 3,929,214.

One way to estimate the growth factor, b, is to find the population ratios from decade to decade and average them:

$$\frac{1.35 + 1.36 + 1.33 + 1.33 + 1.33 + 1.36 + 1.36 + 1.23 + 1.30 + 1.26}{10} \approx 1.32$$

An approximate model is $f(x) = 3.93(1.32)^x$, where $f(x)$ is in millions.

The graph is shown.

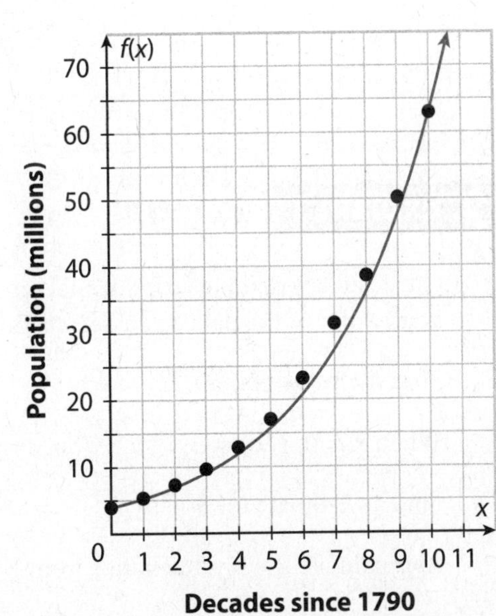

The graph looks like a good fit for the data. All of the points lie on, or close to, the curve.

Another way to estimate b is to choose a point other than $(0, a)$ from the scatter plot that appears would lie on, or very close to, the best-fitting exponential curve. Substitute the coordinates in the general formula and solve for b. For the plot shown, the point (8, 38.56) looks like a good choice.

$$38.56 = 3.93 \cdot b^8$$

$$(9.81)^{\frac{1}{8}} \approx \left(b^8\right)^{\frac{1}{8}}$$

$$1.33 \approx b$$

An approximate model is $f(x) = 3.93(1.33)^x$. The graph is shown.

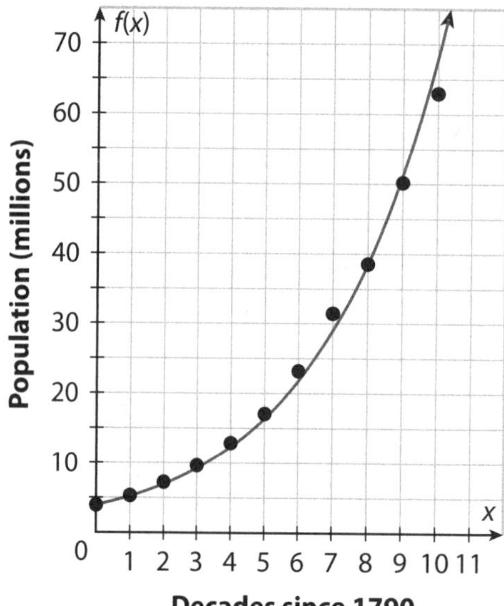

Notice that this function model appears to be even better than the first one for the period from 1790 to 1870, but the first model better represents the data at the end of the 100-year period.

(B) **Movies** The table shows the decline in weekly box office revenue from its peak for one of 2013's top-grossing summer movies.

Week	Revenue (in Millions of Dollars)
0	95.3
1	55.9
2	23.7
3	16.4
4	8.8
5	4.8
6	3.3
7	1.9
8	1.1
9	0.6

Create an approximate exponential model for the data set. Then graph your function with a scatter plot of the data and assess its fit.

Find the value of a in $f(x) = ab^x$.

When $x = 0$, $f(x) =$ ________. So, $a =$ ________.

Find an estimate for b.

Approximate the revenue ratios from week to week and average them:

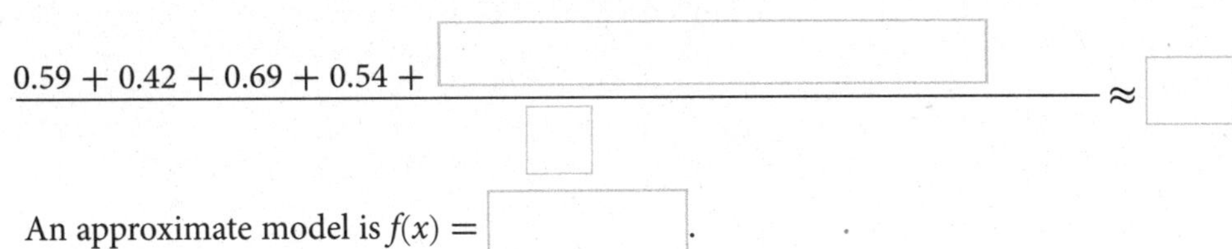

An approximate model is $f(x) =$ ______.

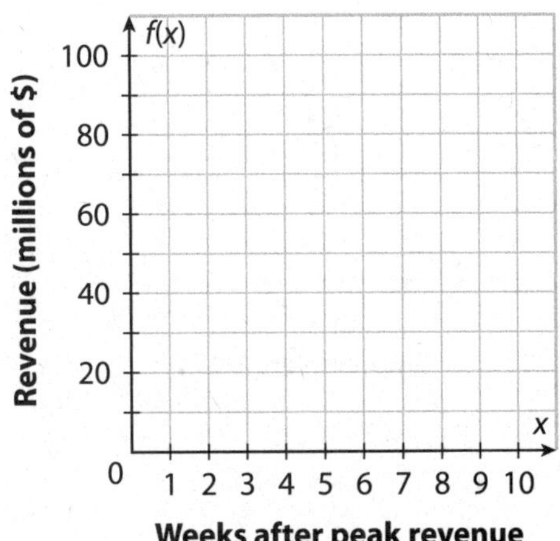

The graph looks like a very good fit for the data. All of the points except one (2 weeks after peak revenue) lie on, or very close to, the curve.

Your Turn

5. Fisheries The total catch in tons for Iceland's fisheries from 2002 to 2010 is shown in the table.

Year	Total Catch (Millions of Tons)
2002	2.145
2003	2.002
2004	1.750
2005	1.661
2006	1.345
2007	1.421
2008	1.307
2009	1.164
2010	1.063

Create an approximate exponential model for the data set. Then graph your function with a scatter plot of the data and assess its fit.

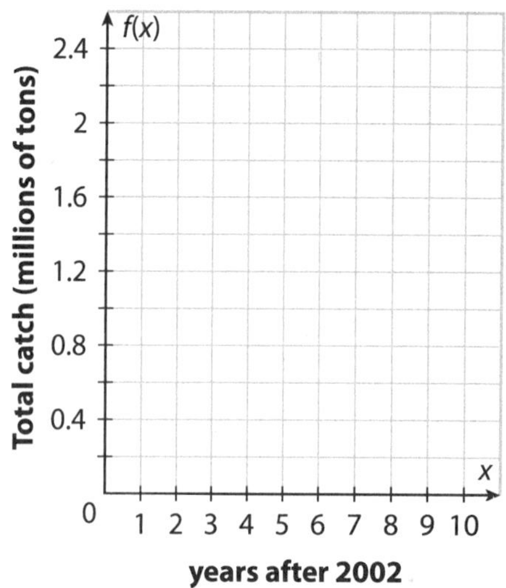

Explain 2 Fitting an Exponential Function to Data Using Technology

Previously you have used a graphing calculator to find a linear regression model of the form $y = ax + b$ to model data, and have also found quadratic regression models of the form $y = ax^2 + bx + c$. Similarly, you can use a graphing calculator to perform exponential regression to produce a model of the form $f(x) = ab^x$.

Example 2

Ⓐ **Population Statistics** Use the data from Example 1 Part A and a graphing calculator to find the exponential regression model for the data, and show the graph of the model with the scatter plot.

Using the STAT menu, enter the number of decades since 1790 in List1 and the population to the nearest tenth of a million in List2.

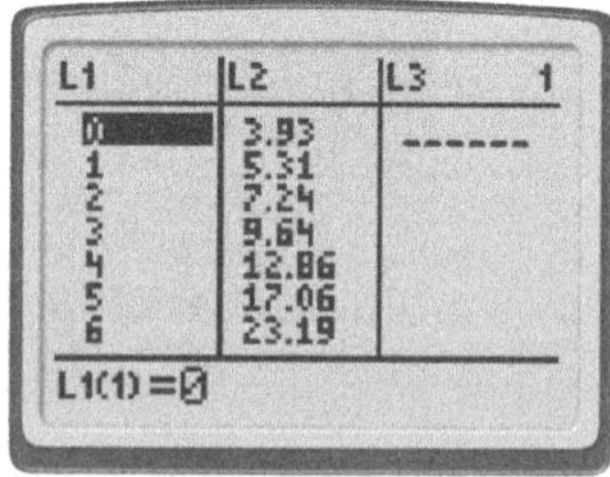

Using the STAT CALC menu, choose "ExpReg" and press ENTER until you see this screen:

An approximate model is $f(x) = 4.116(1.323)^x$.

Making sure that STATPLOT is turned "On," enter the model into the Y = menu either directly or using the VARS menu and choosing "Statistics," "EQ," and "RegEQ." The graphs are shown using the ZoomStat window:

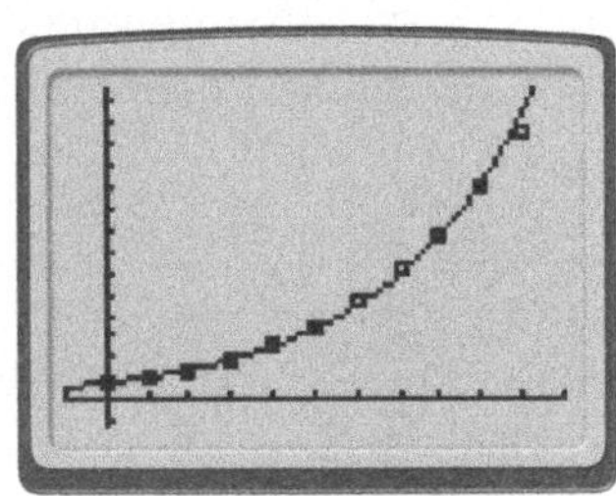

Plotted with the second graph from Example 1 (shown dotted), you can see that the graphs are nearly identical.

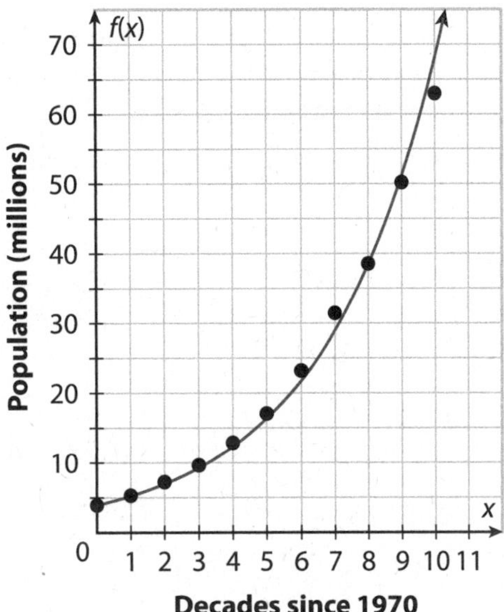

Ⓑ **Movies** Use the data from Example 1 Part B and a graphing calculator to find the exponential regression model for the data. Graph the regression model on the calculator, then graph the model from Example 1 on the same screen using a dashed curve. How do the graphs of the models compare? What can you say about the actual decline in revenue from one week after the peak to two weeks after the peak compared to what the regression model indicates?

Enter the data and perform exponential regression.

The model (using 3 digits of precision) is $f(x) =$ ______.

Reflect

6. **Discussion** The U.S. population in 2014 was close to 320 million people. What does the regression model in Part A predict for the population in 2014? What does this tell you about extrapolating far into the future using an exponential model? How does the graph of the scatter plot with the regression model support this conclusion? (Note: The decade-to-decade U.S. growth dropped below 30% to stay after 1880, and below 20% to stay after 1910. From 2000 to 2010, the rate was below 10%.)

Your Turn

7. **Fisheries** Use the data from Your Turn 5 and a graphing calculator to find the exponential regression model for the data. Graph the regression model on the calculator, then graph the model from your answer to YourTurn5 on the same screen. How do the graphs of the models compare?

Explain 3 Solving a Real-World Problem Using a Fitted Exponential Function

In the real world, the purpose of finding a mathematical model is to help identify trends or patterns, and to use them to make generalizations, predictions, or decisions based on the data.

Example 3

(A) The Texas population increased from 20.85 million to 25.15 million from 2000 to 2010.

a. Assuming exponential growth during the period, write a model where $x = 0$ represents the year 2000 and $x = 1$ represents the year 2010. What was the growth rate over the decade?

b. Use the power of a power property of exponents to rewrite the model so that b is the yearly growth factor instead of the growth factor for the decade. What is the yearly growth rate for this model? Verify that the model gives the correct population for 2010.

c. The Texas population was about 26.45 million in 2013. How does this compare with the prediction made by the model?

d. Find the model's prediction for the Texas population in 2035. Do you think it is reasonable to use this model to guide decisions about future needs for water, energy generation, and transportation needs. Explain your reasoning.

Solution:

a. For a model of the form $f(x) = ab^x$, $a = f(0)$, so $a = 20.85$. To find an estimate for b, substitute $(x, f(x)) = (1, 25.5)$ and solve for b.

$$f(x) = a \cdot b^x$$

$$25.15 = 20.85 \cdot b^1$$

$$\frac{25.15}{20.85} = b$$

$$1.206 \approx b$$

An approximate model is $f(x) = 20.85(1.206)^x$. The growth rate was about 20.6%.

b. Because there are 10 years in a decade, the 10th power of the new b must give 1.206, the growth factor for the decade. So, $b^{10} = 1.206$, or $b = 1.206^{\frac{1}{10}}$. Use the power of a power property:

$$f(x) = 20.85(1.206)^x = 20.85\left(1.206^{\frac{1}{10}}\right)^{10x} \approx 20.85(1.019)^x$$

Because x is decades after 2000, this is equivalent to $f(x) = 20.85(1.019)^x$ where x is years after 2000.

The model gives a 2010 population of $f(x) = 20.85(1.019)^{10} \approx 25.17$. This agrees with the actual population within a rounding error.

c. Substitute $x = 13$ into the model $f(x) = 20.85(1.019)^x$:

$$f(13) = 20.85(1.019)^{13} \approx 26.63$$

The prediction is just a little bit higher than the actual population.

d. For 2035, $x = 35$: $f(35) = 20.85(1.019)^{35} \approx 40.3$. The model predicts a Texas population of about 40 million in 2035. Possible answer: Because it is very difficult to maintain a high growth rate with an already very large population, and with overall population growth slowing, it seems unreasonable that the population would increase from 25 to 40 million so quickly. But because using the model to project even to 2020 gives a population of over 30 million, it seems reasonable to make plans for the population to grow by several million people over a relatively short period.

(B) The average revenue per theater for the movie in Part B of the previous Examples is shown in the graph. (Note that for this graph, Week 0 corresponds to Week 2 of the graphs from the previous Examples.) The regression model is $y = 5.65(0.896)^x$.

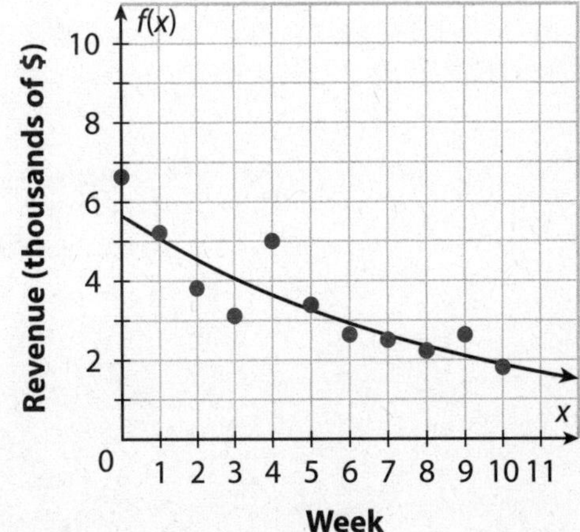

a. From Week 3 to Week 4, there is a jump of over 60% in the average weekly revenue per theater, but the total revenue for the movie for the corresponding week fell by over 30%. What must have occurred for this to be true?

b. A new theater complex manager showing a similar summer movie in a single theater worries about quickly dropping revenue the first few weeks, and wants to stop showing the movie. Suppose you are advising the manager. Knowing that the model shown reflects the long-term trend well for such movies, what advice would you give the manager?

Reflect

8. **Discussion** Consider the situation in Example 3B about deciding when to stop showing the movie. How does an understanding of what other theater managers might do affect your decision?

Your Turn

9. Using a graphing calculator, graph the regression model for the catch in Icelandic fisheries, $f(x) = 2.119(0.9174^x)$, to find when the model predicts the total catches to drop below 0.5 million tons (remember that $x = 0$ corresponds to 2002). Should the model be used to project actual catch into the future? Why or why not? What are some considerations that the model raises about the fishery?

Elaborate

10. How can you tell whether a given set of data can reasonably be modeled using an exponential function?

11. What are some ways that an exponential growth or decay model can be used to guide decisions, preparations, or judgments about the future?

12. **Essential Question Check-In** What are some ways to find an approximate exponential model for a set of data without using a graphing calculator?

Evaluate: Homework and Practice

- Online Homework
- Hints and Help
- Extra Practice

Determine whether each set of data can be modeled by an exponential function. If it can, tell whether it represents exponential growth or exponential decay. If it can't, tell whether a linear or quadratic model is instead appropriate.

1.

x	0	1	2	3	4
$f(x)$	2	6	18	54	162

2.

x	1	2	3	4	5	6
$f(x)$	1	2	3	5	8	13

3.

x	0	1	2	3	4
$f(x)$	2	8	18	32	50

4.

x	5	10	15	20	25
$f(x)$	76.2	66.2	59.1	50.9	44.6

Three students, Anja, Ben, and Celia, are asked to find an approximate exponential model for the data shown. Use the data and scatter plot for Exercises 5–7.

x	0	1	2	3	4	5	6	7	8	9	10
$f(x)$	10	6.0	5.4	3.9	3.7	2.3	1.4	1.0	0.9	0.8	0.5

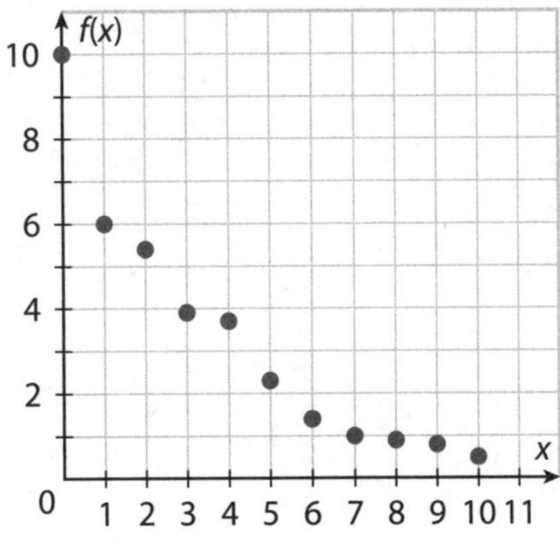

5. To find an approximate exponential model, Anja uses the first data point to find a, and then estimates b by finding the ratio of the first two function values. What is her model?

6. To find his model, Ben uses the first and last data points. What is his model?

7. Celia thinks that because the drop between the first two points is so large, the best model might actually have a y-intercept a little below 10. She uses $(0, 9.5)$ to estimate a in her model. To estimate b, she finds the average of the ratios of successive data values. What is her model? (Use two digits of precision for all quantities.)

8. **Classic Cars** The data give the estimated value in dollars of a model of classic car over several years.

15,300	16,100	17,300	18,400	19,600	20,700	22,000

a. Find an approximate exponential model for the car's value by averaging the successive ratios of the value. Then make a scatter plot of the data, graph your model with the scatter plot, and assess its fit to the data.

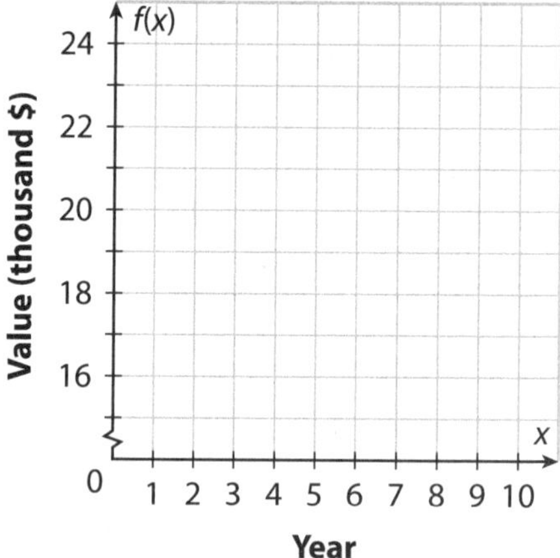

b. In the last year of the data, a car enthusiast spends $15,100 on a car of the given model that is in need of some work. The owner then spends $8300 restoring it. Use your model to create a table of values with a graphing calculator. How long does the function model predict the owner should keep the car before it can be sold for a profit of at least $5000?

9. **Movies** The table shows the average price of a movie ticket in the United States from 2001 to 2010.

Year	2001	2002	2003	2004	2005	2006	2007	2008	2009	2010
Price ($)	5.66	5.81	6.03	6.21	6.41	6.55	6.88	7.18	7.50	7.89

a. Make a scatter plot of the data. Then use the first point and another point on the plot to find an approximate exponential model for the average ticket price. Then graph the model with your scatter plot and assess its fit to the data.

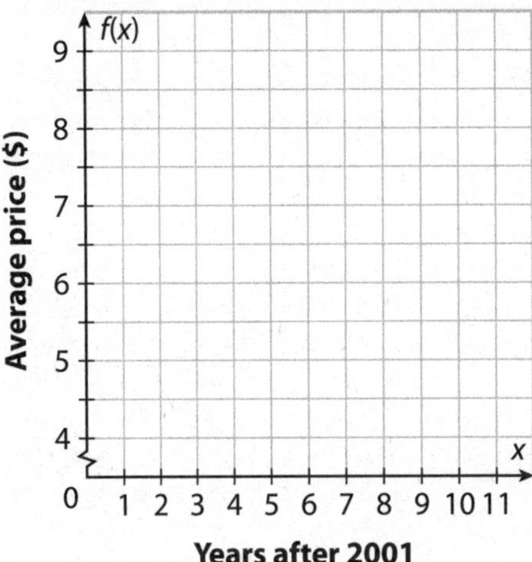

b. Use a graphing calculator to find a regression model for the data, and graph the model with the scatter plot. How does this model compare to your previous model?

c. What does the regression model predict for the average cost in 2014? How does this compare with the actual 2014 cost of about $8.35? A theater owner uses the model in 2010 to project income for 2014 assuming average sales of 490 tickets per day at the predicted price. If the actual price is instead $8.35, did the owner make a good business decision? Explain.

10. Pharmaceuticals A new medication is being studied to see how quickly it is metabolized in the body. The table shows how much of an initial dose of 15 milligrams remains in the bloodstream after different intervals of time.

Hours Since Administration	Amount Remaining (mg)
0	15
1	14.3
2	13.1
3	12.4
4	11.4
5	10.7
6	10.2
7	9.8

a. Use a graphing calculator to find a regression model. Use the calculator to graph the model with the scatter plot. How much of the drug is eliminated each hour?

b. The half-life of a drug is how long it takes for half of the drug to be broken down or eliminated from the bloodstream. Using the Table function, what is the half-life of the drug to the nearest hour?

c. Doctors want to maintain at least 7 mg of the medication in the bloodstream for maximum therapeutic effect, but do not want the amount much higher for a long period. This level is reached after 12 hours. A student suggests that this means that a 15 mg dose should be given every 12 hours. Explain whether you agree with the student. (*Hint*: Given the medicine's decay factor, how much will be in the bloodstream after the first few doses?)

11. Housing The average selling price of a unit in a high-rise condominium complex over 5 consecutive years was approximately \$184,300; \$195,600; \$204,500; \$215,300; \$228,200.

a. Find an exponential regression model where x represents years after the initial year and $f(x)$ is in thousands of dollars.

b. A couple wants to buy a unit in the complex. First, they want to save 20% of the selling price for a down payment. What is the model that represents 20% of the average selling price for a condominium?

c. At the time that the average selling price is \$228,200 (or when $x = 4$), the couple has \$20,000 saved toward a down payment. They are living with family, and saving \$1000 per month. Graph the model from Part b and a function that represents the couple's total savings on the same calculator screen. How much longer does the model predict it will take them to save enough money?

12. Business growth The growth in membership in thousands of a rapidly-growing Internet site over its first few years is modeled by $f(x) = 60(3.61)^x$ where x is in years and $x = 0$ represents the first anniversary of the site. Rewrite the model so that the growth factor represents weeks instead of years. What is the weekly growth factor? What does this model predict for the membership 20 weeks after the anniversary?

13. Which data set can be modeled by an exponential function $f(x) = ab^x$?

a. $(0, 0.1), (1, 0.5), (2, 2.5), (3, 12.5)$

b. $(0, 0.1), (1, 0.2), (2, 0.3), (3, 0.4)$

c. $(0, 1), (1, 2), (2, 4), (4, 8)$

d. $(0, 0.8), (1, 0.4), (2, 0.10), (3, 0.0125)$

H.O.T. Focus on Higher Order Thinking

14. Error analysis From the data $(2, 72.2), (3, 18.0), (4, 4.4), (5, 1.1), (6, 0.27)$, a student sees that the ratio of successive values of $f(x)$ is very close to 0.25, so that an exponential model is appropriate. From the first term, the student obtains $a = 72.2$, and writes the model $f(x) = 72.2(0.25)^x$. The student graphs the model with the data and observes that it does not fit the data well. What did the student do wrong? Correct the student's model.

15. Critical thinking For the data $(0, 5), (1, 4), (2, 3.5), (3, 3.25), (4, 3.125), (5, 3.0625)$, the ratio of consecutive y-values is not constant, so you cannot write an exponential model $f(x) = ab^x$. But the difference in the values from term to term, 1, 0.5, 0.25, 0.125, 0.0625, shows exponential decay with a decay factor of 0.5. How can you use this fact to write a model of the data that contains an exponential expression of the form ab^x?

16. Challenge Suppose that you have two data points (x_1, y_1) and (x_2, y_2) that you know are fitted by an exponential model $f(x) = ab^x$. Can you always find an equation for the model? Explain.

Lesson Performance Task

According to data from the U.S. Department of Agriculture, the number of farms in the United States has been decreasing over the past several decades. During this time, however, the average size of each farm has increased.

a. The average size in acres of a U.S. farm from 1940 to 1980 can be modeled by the function $A(t) = 174e^{0.022t}$ where t is the number of years since 1940. What was the average farm size in 1940? In 1980?

Farms in the United States

Year	Farms (Millions)
1940	6.35
1950	5.65
1960	3.96
1970	2.95
1980	2.44
1990	2.15
2000	2.17

b. The table shows the number of farms in the United States from 1940 to 2000. Find an exponential model for the data using a calculator.

c. If you were to determine the exponential model without a calculator, would the value for a be the same as the value from the calculator? Explain your answer.

d. Based on the data in the table, estimate the number of farms in the United States in 2014.

e. Using a graphing calculator, determine how many years it takes for the number of farms to decrease by 50%.

f. Using a graphing calculator, determine when the number of farms in the United States will fall below 1 million.

g. Does an exponential model seem appropriate for all of the data listed in the table? Why or why not?

Name_______________ Class_______________ Date_______________

14.2 Choosing Among Linear, Quadratic, and Exponential Models

Resource Locker

Essential Question: **How do you choose among, linear, quadratic, and exponential models for a given set of data?**

A2.8.A Analyze data to select the appropriate model from among linear, quadratic, and exponential models. Also A2.8.B, A2.8.C

Explore Developing Rules of Thumb for Visually Choosing a Model

Previously, you have found models for different function classes, including linear, quadratic, and exponential. In those cases, you were working with one kind of model. When you are working with data, you may not know ahead of time what kind of model may be appropriate. Sometimes, the choice may be relatively easy. For example, data lying along a curve that rises and then falls or falls and then rises will likely be well-fitted by a quadratic model. But sometimes it may not be as clear.

Use the scatter plots shown for the steps following.

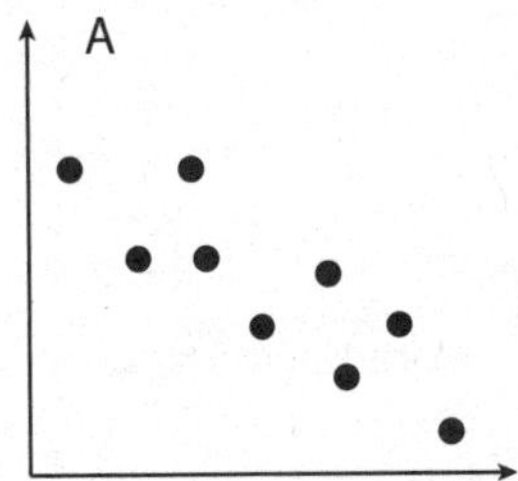

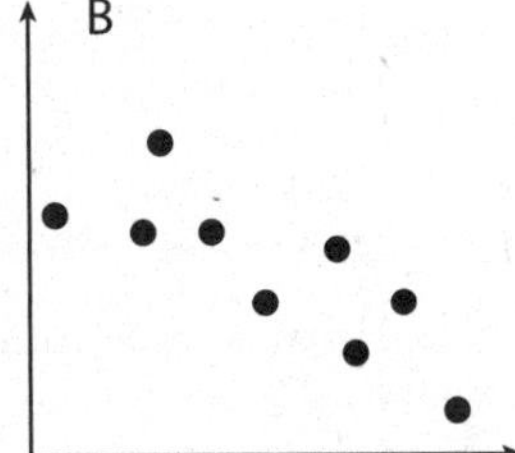

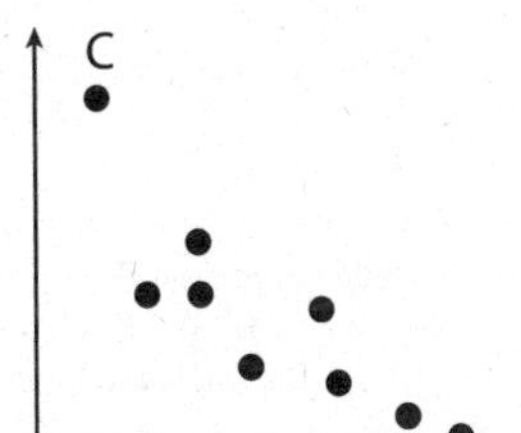

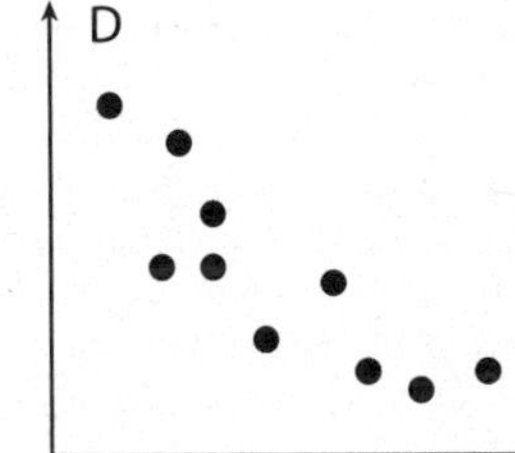

(A) Look at scatter plot A. Do you think a linear model will be appropriate? Explain your reasoning. If you think a linear model is appropriate, what do you know about the lead coefficient?

____

____

(B) Look at scatter plot B. What is different now that indicates that another kind of model might be appropriate? What characteristics would this model have?

____

____

____

(C) Look at scatter plot C. What about this plot indicates that yet another kind of model might be appropriate? What characteristics would this model have?

(D) Look at scatter plot D. This plot is very similar to plot C, but what indicates that a different model would be appropriate? What characteristics would this model have?

Reflect

1. When can it be difficult to distinguish whether a quadratic or an exponential model is most appropriate?

2. Under what circumstances might it be difficult to tell exponential or quadratic data from linear data?

3. For data that do not lie tightly along a curve or line, what is different about the last data point that can make it potentially more misleading than other points?

Explain 1 Modeling with a Linear Function

As noted in the Explore, it is not always immediately clear what kind of model best represents a data set. With experience, your ability to recognize signs and reasons for choosing one model over another will increase.

Example 1 **Examine each scatter plot. Then complete the steps below.**

Step 1: Choose the data set that appears to be best modeled by a linear function. Explain your choice, whether you think a linear model will be a close fit, and whether any other model might possibly be appropriate. What characteristics do you expect the linear model will have?

Step 2: Enter the data for your choice into your graphing calculator in two lists, and perform linear regression. Then give the model, defining your variables. What are the initial value and the rate of change of the model?

Step 3: Graph the model along with the scatter plot using your calculator, then assess how well the model appears to fit the data.

(A) **Wildlife Conservation** Data sets and scatter plots for populations over time of four endangered, threatened, or scarce species are shown.

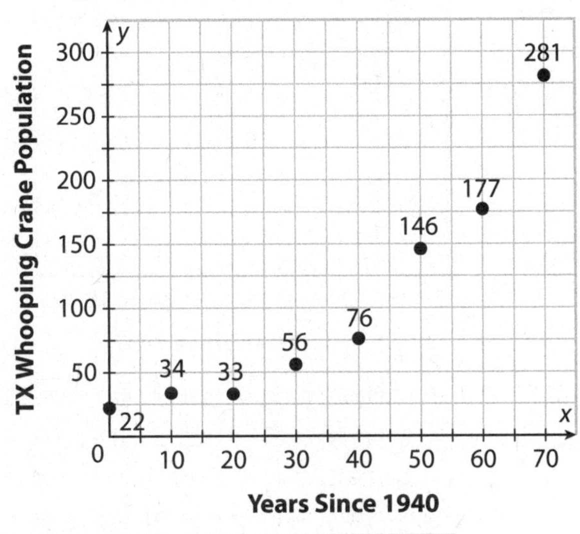

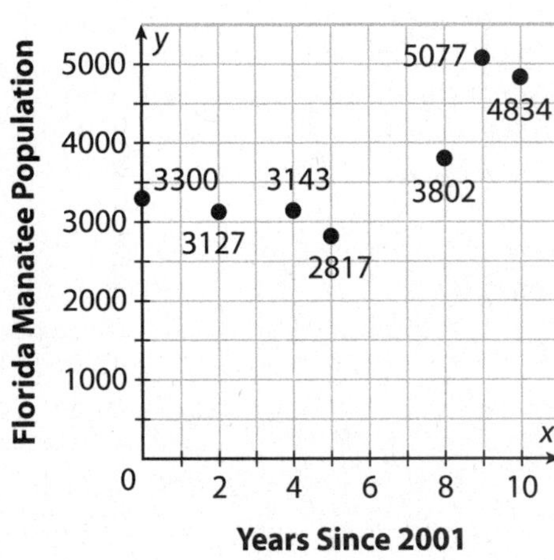

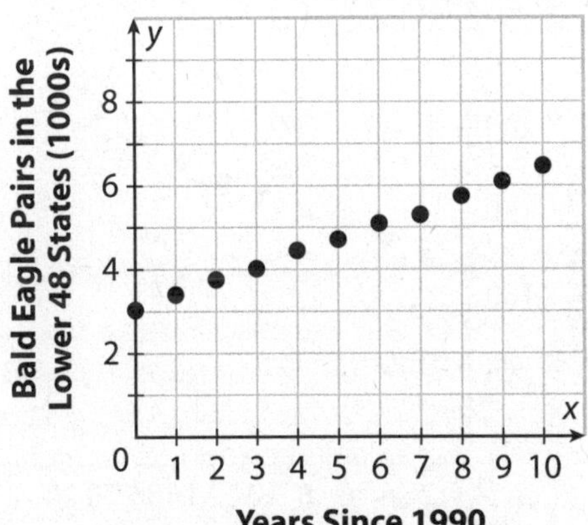

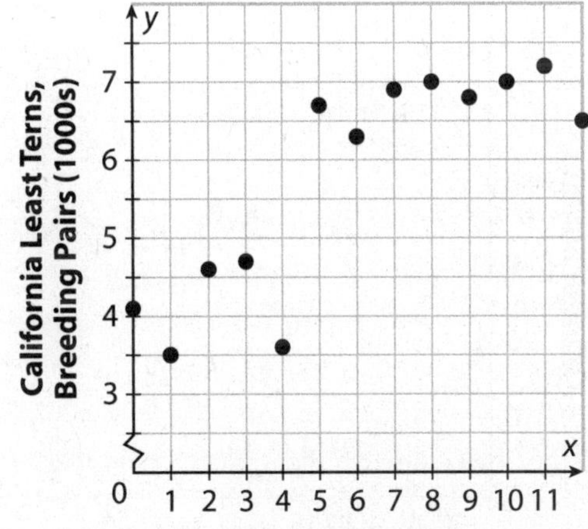

(1990, 3035) (1994, 4449) (1998, 5748)
(1991, 3399) (1995, 4712) (1999, 6104)
(1992, 3749) (1996, 5094) (2000, 6471)
(1993, 4015) (1997, 5295)

(1998, 4100) (2003, 6700) (2007, 6800)
(1999, 3500) (2004, 6300) (2008, 7000)
(2000, 4600) (2005, 6900) (2009, 7200)
(2001, 4700) (2006, 7000) (2010, 6500)
(2002, 3600)

Step 1: The bald eagle population is clearly the one best modeled by a linear function, as the increase in the number of pairs is very steady, with no apparent curving or changes in the trend that might indicate a different model. The model will have a y-intercept of about 3 (in thousands) and will have a slope very close to 0.34, since the rate of change all along the graph remains close to the average rate of change from the first point to the last.

Step 2: A regression model is $y = 338.2x + 3042$ where x is the number of years after 1990 and y is the number of breeding pairs in thousands. The initial value is 3042, and the rate of change is 338.2 pairs per year.

Step 3: The model is a very close fit to the data. It fits both the overall trend and the individual points very closely.

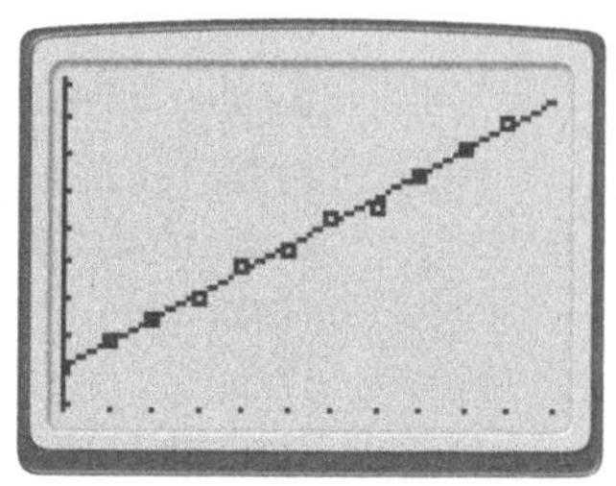

(B) **Automobiles** Data sets and scatter plots for various statistics about changes in automobiles of different model years are shown.

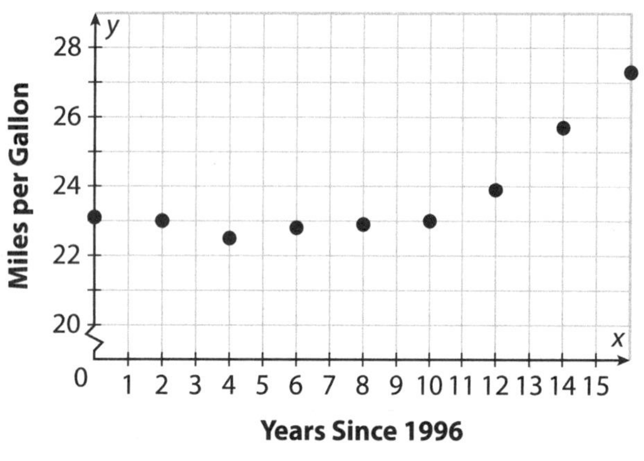

(1996, 23.1) (2002, 22.8) (2008, 23.9)
(1998, 23.0) (2004, 22.9) (2010, 25.7)
(2000, 22.5) (2006, 23.0) (2012, 27.3)

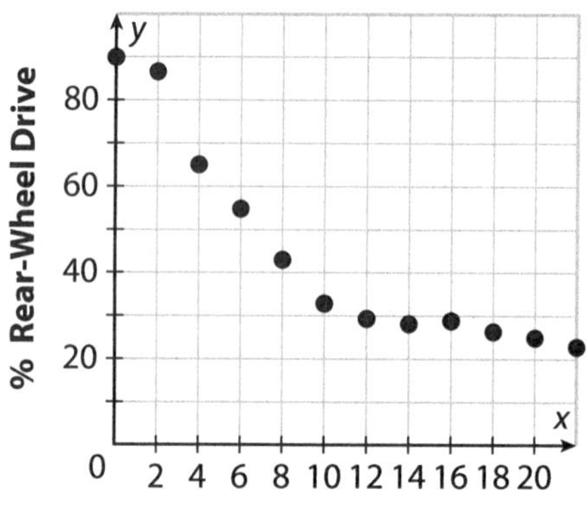

(1977, 89.8) (1985, 42.9) (1993, 28.8)
(1979, 86.5) (1987, 32.8) (1995, 26.3)
(1981, 65.0) (1989, 29.3) (1997, 24.9)
(1983, 54.8) (1991, 28.1) (1999, 22.9)

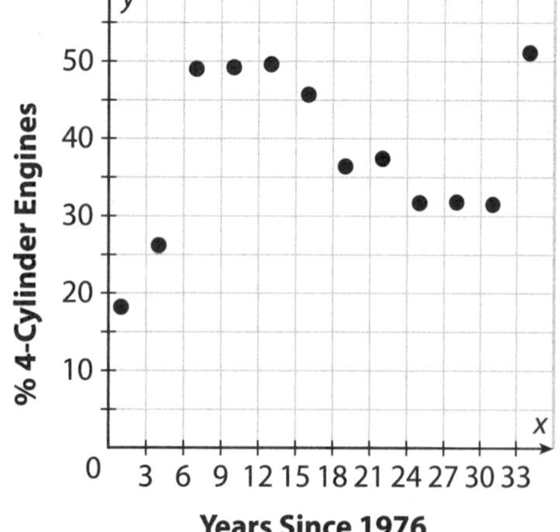

(1976, 18.2) (1988, 49.6) (2000, 31.7)
(1979, 26.2) (1991, 45.7) (2003, 31.8)
(1982, 49.0) (1994, 36.4) (2006, 31.5)
(1985, 49.2) (1997, 37.4) (2009, 51.1)

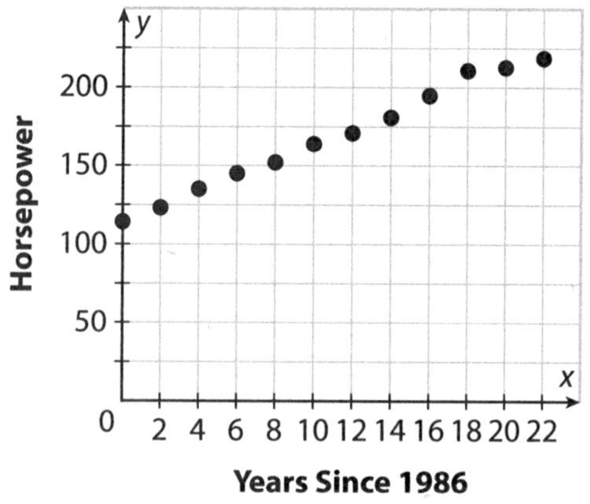

(1986, 114) (1994, 152) (2002, 195)
(1988, 123) (1996, 164) (2004, 211)
(1990, 135) (1998, 171) (2006, 213)
(1992, 145) (2000, 181) (2008, 219)

Step 1: The [ ] are the data best modeled by a linear function, as the increase in [ ] over time is very steady, with the amount of increase never varying too much from the trend. All of the other data sets show clear variation in the rate of change. The model will have a y-intercept of about [ ] and will have a positive slope that should be close to the overall average rate of change of [ ], since the rate of change all along the graph remains close to the average rate of change from the first point to the last.

Step 2: A regression model is [ ] where x is the number of years after 1986 and y is the [ ] for the model year. The initial value is about 114 horsepower, and the rate of change is about [ ].

Step 3: The model is a very close fit to the data. It fits both the overall trend and the individual points very closely.

Reflect

4. Discussion In Example 1A, the linear model is a very close fit for the data. What does the model predict for the number of pairs in 2007, the year the bald eagle was removed from the Endangered Species list? Do you think that this was a good decision given the model and the fact that the actual 2007 number of pairs was 11,040? Explain.

5. What does the model in Example 1B predict for the horsepower in 2013? Given that the actual average horsepower for new vehicles in 2013 was about 230, observing your plot and model, and thinking about changes in passenger vehicles, do you think the trend in the model will continue, or change significantly? Explain.

Your Turn

6. Demographics Data sets and scatter plots for various changes in the United States population over time are shown. Using these data, complete the three steps described at the beginning of Example 1. Also, tell whether you would expect the trend indicated by your model to continue for a time after the data shown, or whether you expect that it would soon change, and explain your answer.

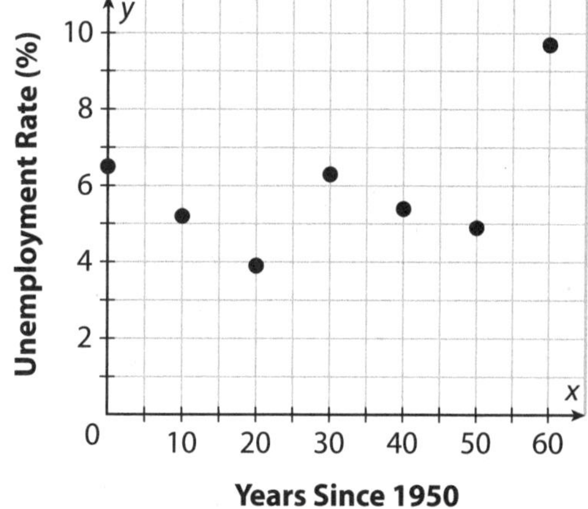

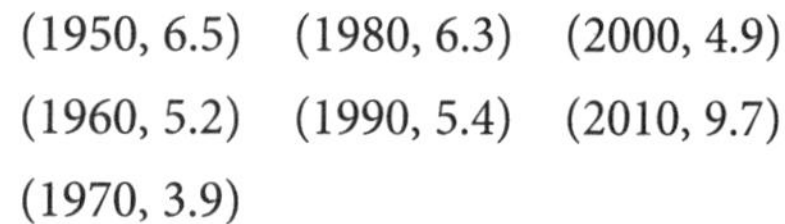
(1950, 6.5) (1980, 6.3) (2000, 4.9)
(1960, 5.2) (1990, 5.4) (2010, 9.7)
(1970, 3.9)

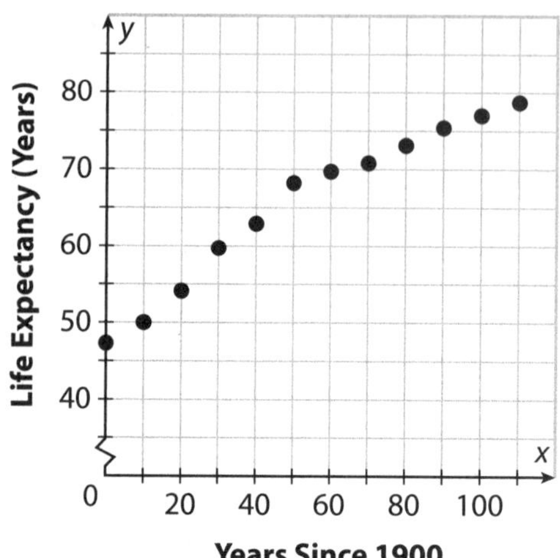

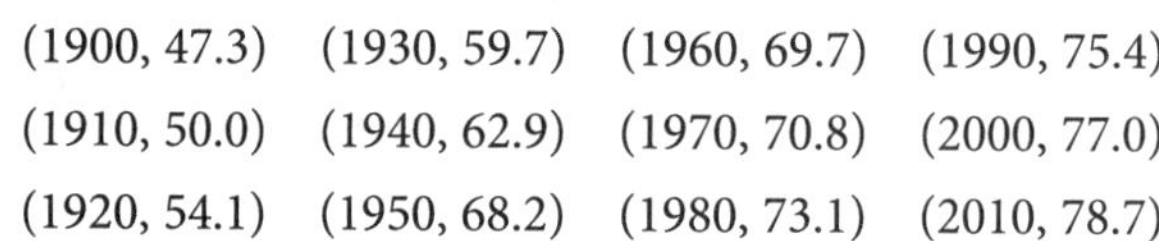
(1900, 47.3) (1930, 59.7) (1960, 69.7) (1990, 75.4)
(1910, 50.0) (1940, 62.9) (1970, 70.8) (2000, 77.0)
(1920, 54.1) (1950, 68.2) (1980, 73.1) (2010, 78.7)

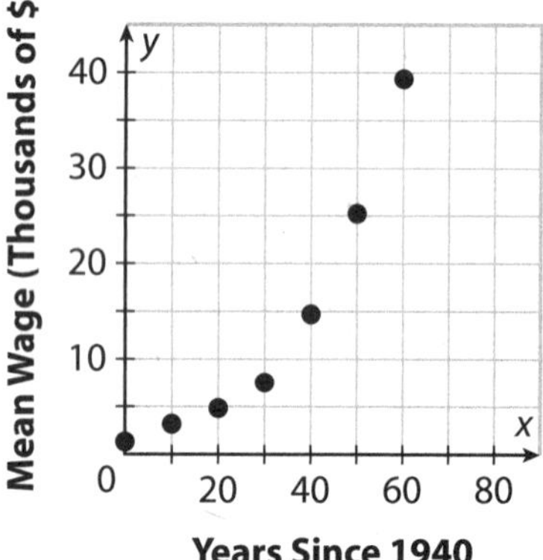

(1940, 1315) (1970, 7501) (1990, 25,236)
(1950, 3180) (1980, 14,610) (2000, 39,237)
(1960, 4816)

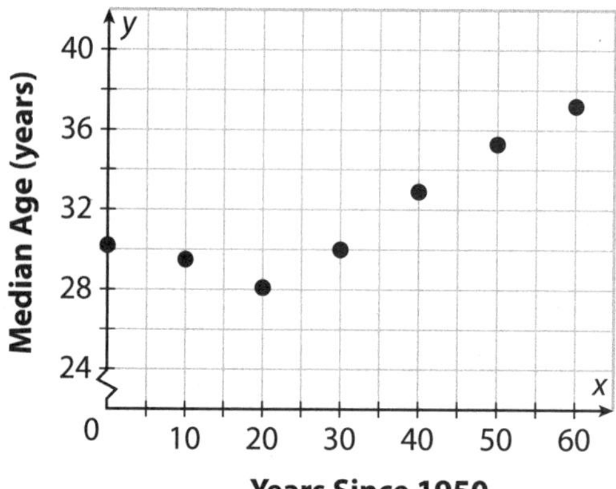

(1950, 30.2) (1980, 30.0) (2000, 35.3)
(1960, 29.5) (1990, 32.9) (2010, 37.2)
(1970, 28.1)

Explain 2 Modeling With a Quadratic Function

Example 2 **Using the groups of data sets and their scatter plots from Example 1:**

Step 1: Choose the data set that appears to be best modeled by a quadratic function. Explain your choice, whether you think a quadratic model will be a close fit, and whether any other model might possibly be appropriate. What characteristics do you expect the quadratic model will have?

Step 2: Enter the data for your choice into your graphing calculator in two lists, and perform quadratic regression. Then give the model, defining your variables.

Step 3: Graph the model along with the scatter plot using your calculator, then assess how well the model appears to fit the data.

(A) Use the data about animal populations in Example 1 Part A.

Step 1: The Florida manatee population appears to be the one best modeled by a quadratic function, as its scatter plot is the only one with a clear change in direction, and it clearly would not be well represented by a linear or exponential model. The whooping crane data might also be fit fairly well on one side of a quadratic model, but it might also be exponential. The quadratic model for the manatee population will have a positive leading coefficient since it opens upward, but it is hard to predict what the y-intercept or the vertex will be. Because the graph is not very symmetrical, the fit may not be very close.

Step 2: A regression model is $y = 48.50x^2 - 317.3x + 3411$ where x is the number of years after 2001 and y is the number of manatees

Step 3: The model is not a close fit, but it does look like an appropriate model for the overall trend during the time of the data. It misses the horizontal position for the vertex by a fairly wide margin, but otherwise is not too far from the data.

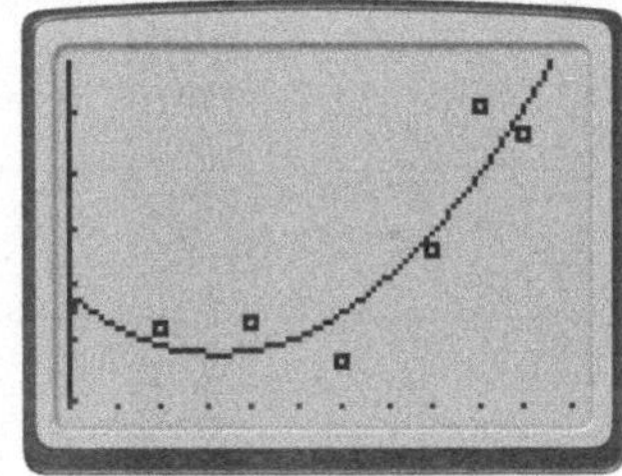

(B) Use the data about automobiles in Example 1 Part B.

Step 1:

Step 2:

Step 3:

Reflect

7. **Discussion** The Florida manatee has been under consideration for being downgraded from endangered to threatened. Do you believe the graph and model of the manatee population in Part A of this Example support this concept or argue against it? Explain your reasoning.

8. How might the model for miles per gallon affect a decision on when to purchase a car?

Your Turn

9. Using the data in Your Turn Exercise 6, complete the three steps described at the beginning of Example 2. Also, tell whether you would expect the trend indicated by your model to continue for a time after the data shown, or whether you expect that it would soon change, and explain your answer.

Explain 3 Modeling with an Exponential Function

Example 3 **Using the groups of data sets and their scatter plots from Example 1:**

Step 1: Choose the data set that appears to be best modeled by an exponential function. Explain your choice, whether you think an exponential model will be a close fit, and whether any other model might possibly be appropriate. What characteristics do you expect the exponential model will have?

Step 2: Enter the data for your choice into your graphing calculator in two lists, and perform exponential regression. Then give the model, defining your variables. What are the initial value, growth or decay factor, and growth or decay rate of the model?

Step 3: Graph the model along with the scatter plot using your calculator, then assess how well the model appears to fit the data.

(A) Use the data about animal populations in Example 1 Part A.

Step 1: The whooping crane population appears to be the one best modeled by an exponential function, as it rises increasingly quickly, but does not reflect a change in direction as a quadratic model can. Though the whooping crane plot is nearly linear in its midsection, the slow initial rise and fast later rise indicate that an exponential model is better. The California least tern data show a significant jump, but no clear pattern. An appropriate whooping crane model shows exponential growth, so the parameter b is greater than 1. Because the growth is not very large considering the time period of 70 years, however, the yearly growth factor will not be much above 1.

Step 2: A regression model is $y = 20.05(1.0374)^x$ where x is the number of years after 1940 and y is the population. The initial value for the model is 20 whooping cranes. The growth factor is 1.0374, and the growth rate is 0.0374, or about 3.74% per year.

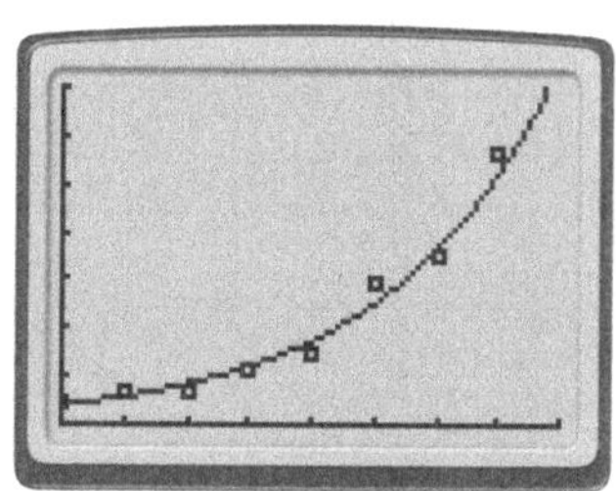

Step 3: The model is very good fit for the data. Some data points are a little above the curve and some a little below, but the fit is close and reflects the trend very well.

(B) Use the data about automobiles in Example 1 Part B.

Step 1:

Step 2:

Step 3:

Reflect

10. **Discussion** What does the model for the whooping crane population predict for the population in 2040? Do you think it is possible that the whooping crane will be removed from the endangered species list any time in the next few decades? Explain.

Your Turn

11. Using the data in YourTurn Exercise 6, complete the three steps described at the beginning of Example 3. Also, tell whether you would expect the trend indicated by your model to continue for a time after the data shown, or whether you expect that it would soon change, and explain your answer.

Elaborate

12. **Discussion** How does making a prediction from a model help make a decision or judgment based on a given set of data?

13. Describe the process for obtaining a regression model using a graphing calculator.

14. **Essential Question Check-In** How can a scatter plot of a data set help you determine the best type of model to choose for the data?

Evaluate: Homework and Practice

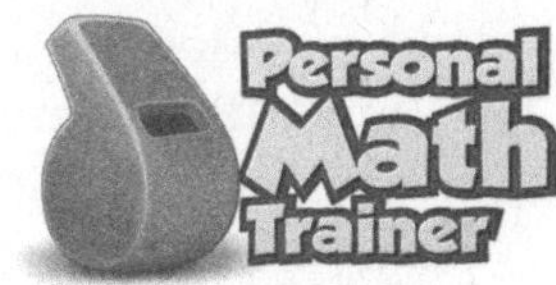

- Online Homework
- Hints and Help
- Extra Practice

1. Match each scatter plot with the most appropriate model from the following. Do not use any choice more than once.

I. quadratic, $a > 0$

II. quadratic, $a < 0$

III. linear, $a < 0$

IV. exponential, $b > 1$

V. exponential, $b < 1$

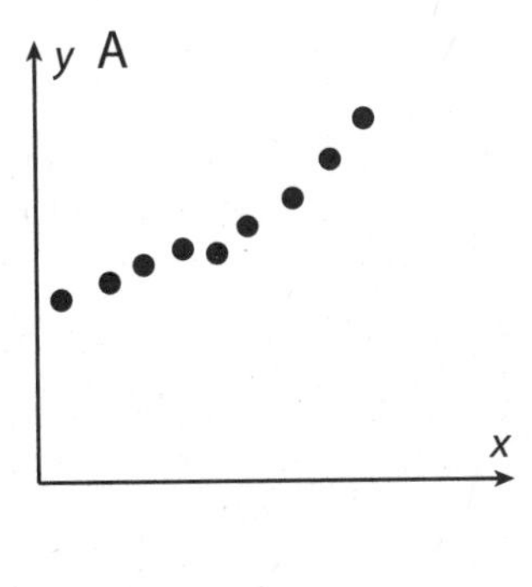

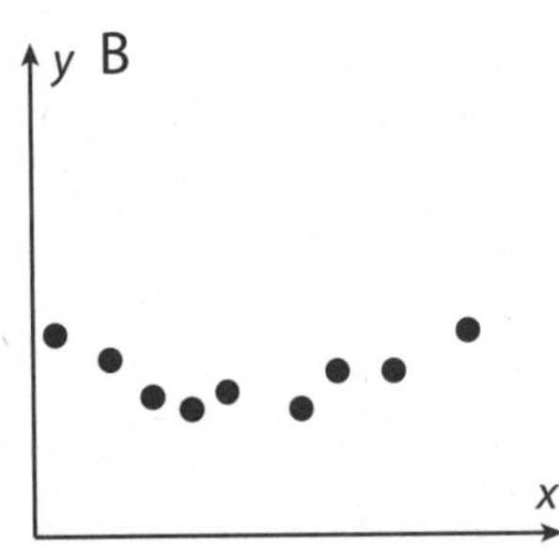

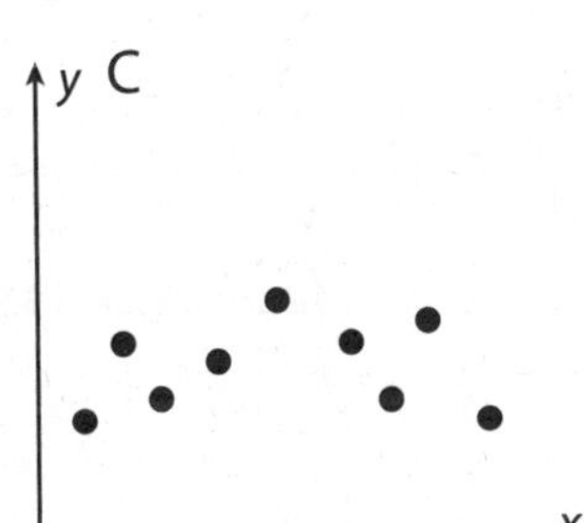

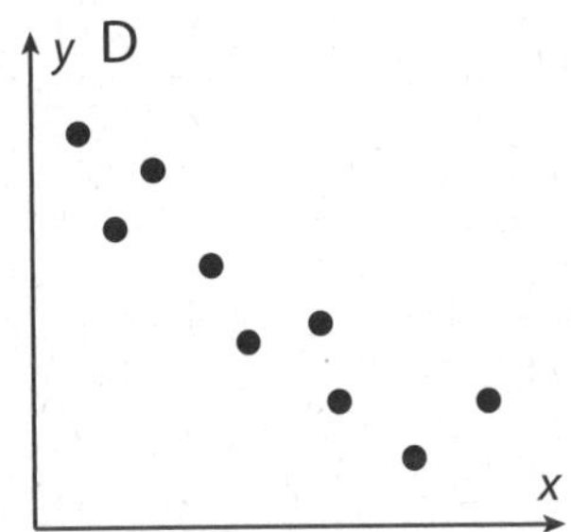

For the data set given:

a. Create a scatter plot of the data. What kind of model do you think is most appropriate? Why? What characteristics do you expect that this model will have?

b. Use a graphing calculator to find an equation of the regression model. Then interpret the model, including the meaning of important parameters.

c. Graph the regression model with its scatter plot using a graphing calculator. How well does the model fit?

d. Answer the question following the data.

2. **Population Demographics** The data set shows the number of Americans living in multigenerational households.

Year	Number (in Millions)
1950	32
1960	27
1970	26
1980	28
1990	35
2000	42
2010	52

What does the model predict for the number in 2020? in 2040? Are these numbers reasonable? Explain.

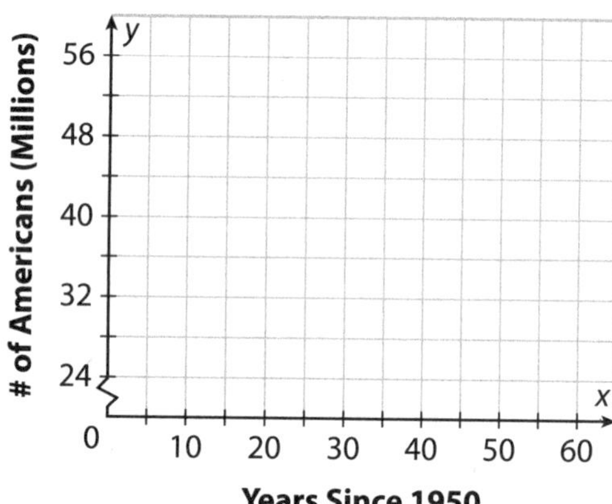

3. **Cycling** The data set shows the inseam length for different frame sizes for road bicycles.

Frame Size (cm)	Inseam Length (cm)
46	69
48	71
49	74
51	76
53	79
54	81
58	86
60	89
61	91

Jarrell has an inseam of 84 cm, but the table does not give a frame size for him. He graphs the model on a graphing calculator and finds that a y-value of 84 is closest to an x-value of 56. He decides he needs a 56 cm frame. Do you think this is a reasonable conclusion. Explain.

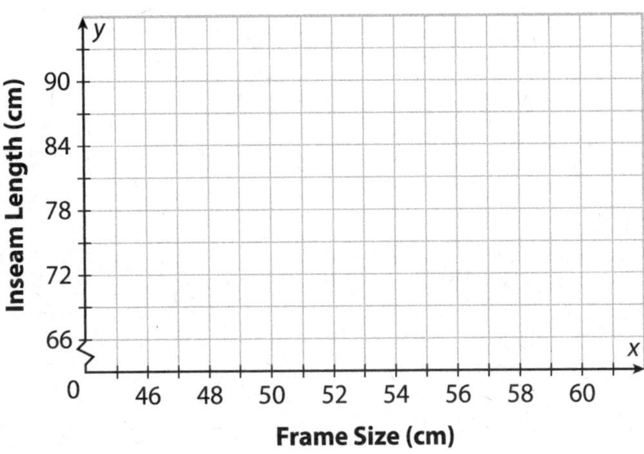

4. Population Geography The data set shows the percent of the U.S. population living in central cities.

What does your model predict for the percent of the population living in central cities in 2010? How much confidence would you have in this prediction? Explain. Given that the actual number for 2010 was about 36.9%, does this support your judgment?

Year	% of Population
1910	21.2
1920	24.2
1930	30.8
1940	32.5
1950	32.8
1960	32.3
1970	31.4
1980	30.0

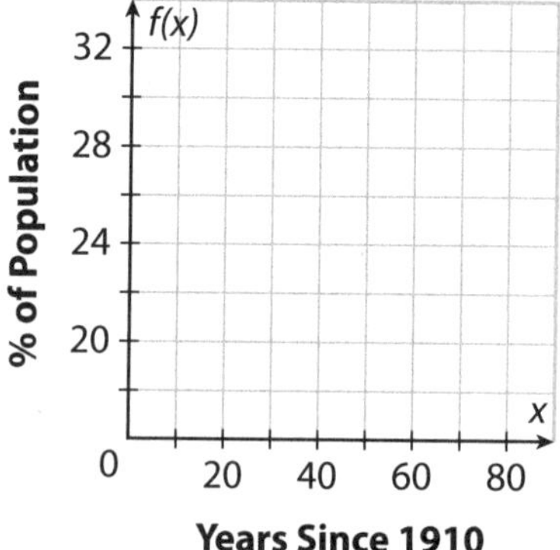

5. Animal Migration The data set shows the number of bald eagles counted passing a particular location on a migration route. Predict the number of bald eagles in 2033. How much confidence do you have in this prediction?

Year	Number of Eagles
1973	41
1978	79
1983	384
1988	261
1993	1725
1998	3289
2003	3356

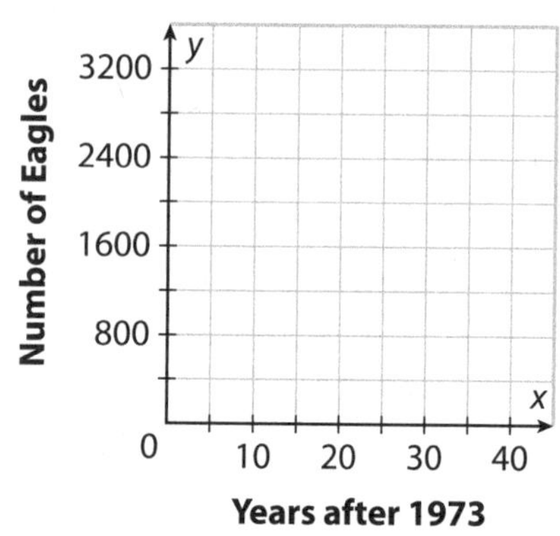

6. **Smart Phones** The data set shows the percent of the world's population owning a smart phone.

a. Create a scatter plot of the data.

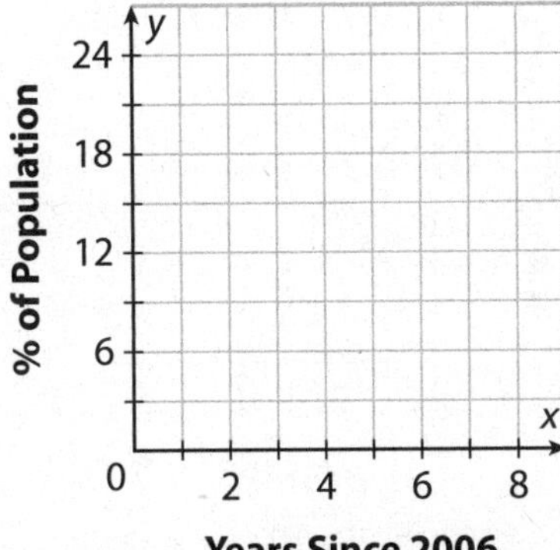

Year	% of Population
2006	1
2007	3
2008	4
2009	5
2010	7
2011	11
2012	16
2013	22

b. Use a graphing calculator to find equations for both exponential and quadratic regression models. Then interpret the models, including the meaning of important parameters.

c. Graph both regression models with the scatter plot using a graphing calculator. Do both models fit the data well? Does one seem significantly better than the other?

d. For how long after the data set do you think either model will be a good predictor? Explain your reasoning.

7. **Stock Market** The data set gives the U.S. stock market's average annual return rates for each of the last 11 decades.

Decade Ending	Annual Return Rate (%)
1910	9.96
1920	4.20
1930	14.95
1940	−0.63
1950	8.72
1960	19.28
1970	7.78
1980	5.82
1990	17.57
2000	18.17
2010	3.1

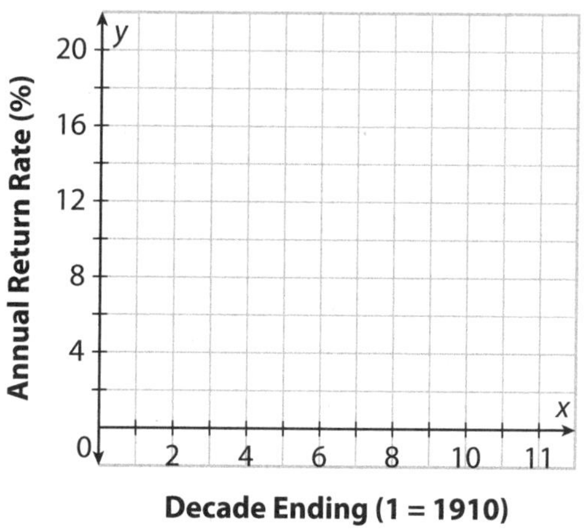

a. Make a scatter plot of the data.

b. What kind of model do you think is most appropriate for the data? Explain your reasoning.

c. Do the data give any kind of prediction about investing in the stock market for the future? Explain.

H.O.T. Focus on Higher Order Thinking

8. Explain the Error Out of curiosity, Julia enters the stock market data from Exercise 7 into her calculator and performs linear and quadratic regression. As she expected, there is almost no fit to the data at all. She then tries exponential regression, and get the message "ERR: DOMAIN." Why does she get this message?

9. Critical Thinking A student enters the road bicycle data from Exercise 3, accidentally performs quadratic regression instead of linear, and has the calculator graph the regression model with the scatter plot. The student sees this graph:

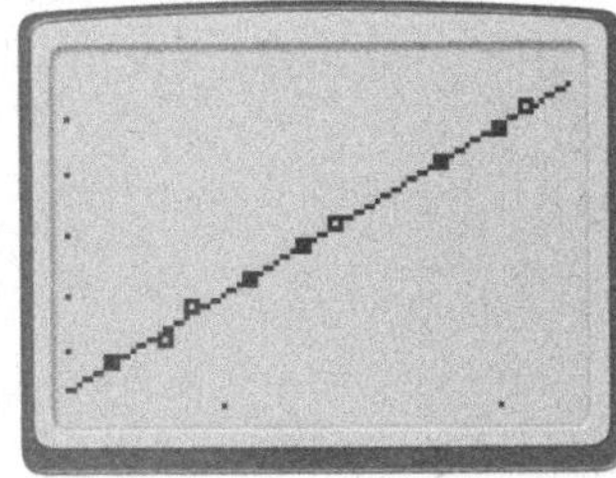

The model graphed by the calculator is $y = -0.001241x^2 + 1.587x - 1.549$. It is obviously a very close fit to the data, and looks almost identical to the linear model. Explain how this can be true. (*Hint*: What happens when you zoom out?)

10. Critical Thinking A graphing calculator returns a linear regression equation $y = ax + b$, a quadratic regression equation $y = ax^2 + bx + c$, and an exponential regression equation $y = a \cdot b^x$. For exponential regression, a is always positive, which is not true of the other models. How does an exponential model differ from the other two models regarding translations of a parent function?

11. Extension In past work, you have used the correlation coefficient r, which indicates how well a line fits the data. The closer $|r|$ is to 1, the better the fit, and the closer to 0, the worse the fit. When you perform quadratic or exponential regression with a calculator, the *coefficient of determination* r^2 or R^2 serves a similar purpose. Return to the smart phone data in Exercise 6, and perform the quadratic and exponential regression again. (Make sure that "Diagnostics" is turned on in the Catalog menu of your calculator first.) Which model is the closer fit to the data, that is, for which model is R^2 closest to 1?

Lesson Performance Task

A student is given $50 to invest. The student chooses the investment very well. The following data shows the amount that the investment is worth over several years.

x (Years)	y (Dollars)
0	50
1	147
2	462
3	1411
4	4220
5	4431
6	4642

a. Determine an appropriate model for the data. If it is reasonable to break up the data so that you can use different models over different parts of the time, then do so. Explain the reasoning for choosing your model(s).

b. Write a situation that may reflect the given data.

STUDY GUIDE REVIEW

Modeling with Exponential and Other Functions

MODULE 14

Essential Question: How can modeling with exponential and other functions help you to solve real-world problems?

Key Vocabulary
exponential regression
(regresión exponencial)

KEY EXAMPLE *(Lesson 14.1)*

What type of function is illustrated in the table?

x	−1	0	1	2	3
$f(x)$	9	3	1	$\frac{1}{3}$	$\frac{1}{9}$

Whenever x increases by 1, $f(x)$ is multiplied by the common ratio of $\frac{1}{3}$. Since this ratio is less than 1, the table represents an exponential decay function.

KEY EXAMPLE *(Lesson 14.2)*

Create a scatter plot for the data in the table. Treat age as the independent variable x and median BMI as the dependent variable y. Then, use a graphing calculator to find an appropriate regression model of the data. Explain why you chose that particular type of function.

Age	2	3	4	5	6	7	8	9	10
Median BMI	16.5	16.0	15.2	15.2	15.6	15.6	15.8	16.0	16.3

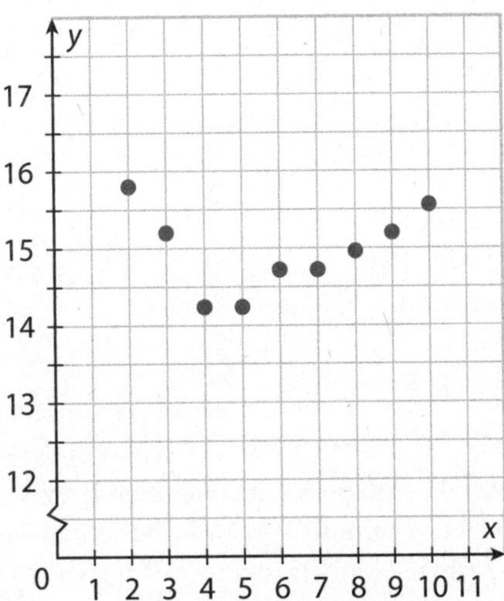

A quadratic regression of the data on a graphing calculator produces the equation $y = 0.0636x^2 - 0.7503x + 17.5867$. A quadratic function was chosen because the data points generally lie on a curve that approximates a parabola.

EXERCISES

Choose the type of function (linear, quadratic, or exponential) you would use to model the data. *(Lesson 14.2)*

1.

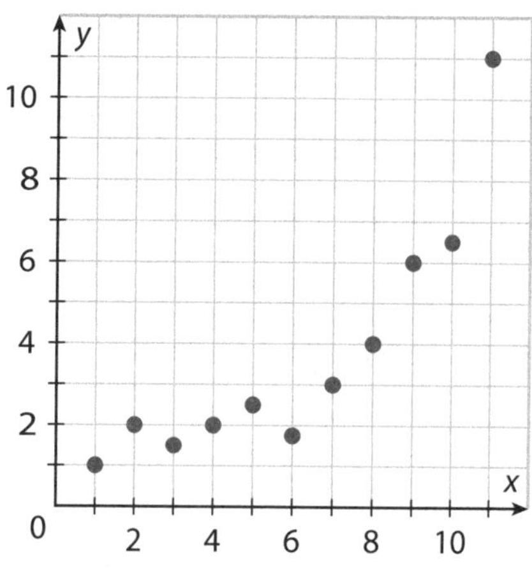

2.

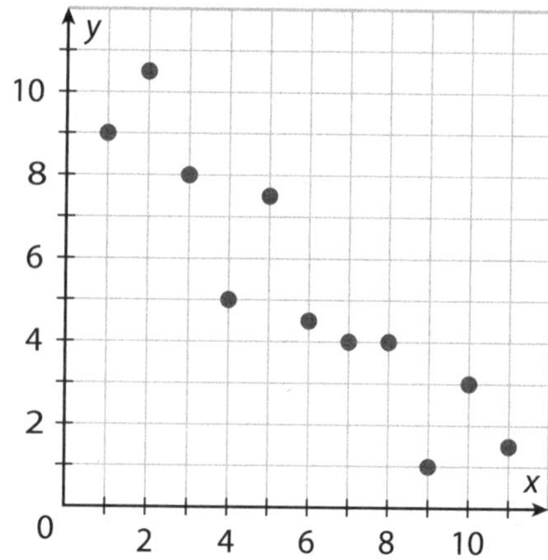

3. **How does the Product of Powers Property of Exponents help you in deciding whether a given set of data is exponential?** *(Lesson 14.1)*

MODULE PERFORMANCE TASK

Double Your Money

Jenna was a game show participant and won $10,000! She eventually wants to use the money as a down payment on a home but is not quite ready for such a big commitment. She decides to invest the money and plans to use it once the amount reaches $20,000. She researches various investment opportunities and would like to choose the one that will let her money double fastest.

Plan	Interest Rate	Compounding
Plan A	5.2%	Quarterly
Plan B	4.8%	Monthly
Plan C	4.25%	Continuously

To the nearest tenth of a year, how long will each plan take to double Jenna's money? Which should Jenna choose? Use your own paper to complete the task. Be sure to write down all your data and assumptions. Then use graphs, numbers, words, or algebra to explain how you reached your conclusion.

Ready to Go On?

14.1–14.2 Modeling with Exponential an Other Functions

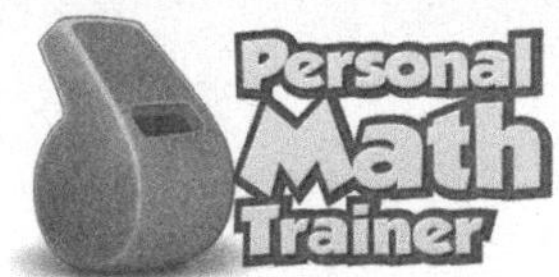

- Online Homework
- Hints and Help
- Extra Practice

The isotope X has a half-life of 10 days. Complete the table showing the decay of a sample of X. *(Lesson 14.1, 14.2)*

1.

Number of Half-Lives	Number of Days (t)	Percent of Isotope Remaining (p)
0	0	100
1	10	50
2	20	
3		
4		

2. Write the decay rate per half-life, r, as a fraction.

3. Write an expression for the number of half-lives in t days.

4. Write a function that models this situation. The function $p(t)$ should give the percent of the isotope remaining after t days.

ESSENTIAL QUESTION

5. What are two ways you can find an exponential function that models a given set of data?

Assessment Readiness

1. Which exponential function has a graph that passes through $(1, 2)$ and $(3, 50)$?

A. $y = \frac{3}{5} \cdot 5^x$

B. $y = \frac{2}{5} \cdot 5^x$

C. $y = \frac{1}{2} \cdot 5^x$

D. $y = \frac{1}{3} \cdot 5^x$

2. What is the asymptote of the graph of the function $y = -2\left(\frac{1}{4}\right)^{x+1} + 7$?

A. $y = 7$

B. $y = \frac{1}{4}$

C. $y = -2$

D. $y = 2$

3. What are the real-number solutions of the equation $y^3 - 5y^2 = 0$?

A. 0 and 5

B. 5 and 10

C. −5 and 0

D. There are no real-number solutions.

4. What type of function is illustrated in the table?

x	4	5	6	7	8
$f(y)$	−4	−3	−4	−7	−12

A. Exponential

B. Quadratic

C. Logarithmic

D. Linear

5. How can you use the pattern formed by the points on a scatterplot to determine whether a linear, quadratic, or exponential function is most likely to be the best fit for the data?

Logarithmic Functions

MODULE 15

Essential Question: How can you use logarithmic functions to solve real-world problems?

TEKS

LESSON 15.1
Defining and Evaluating a Logarithmic Function

TEKS A2.2.A, A2.2.B, A2.2.C, A2.5.B, A2.5.C, A2.7.I

LESSON 15.2
Graphing Logarithmic Functions

TEKS A2.2.A, A2.5.A, A2.5.B

REAL WORLD VIDEO
Check out some of the considerations that go into to determining the proper dosage of a medication and learn about the role of logarithmic functions in this process.

MODULE PERFORMANCE TASK PREVIEW

What's the Dosage?

Scientists working in the pharmaceutical industry discover, develop, and test drugs for everything from relieving a headache to controlling high blood pressure. One important question regarding a specific drug is how long the drug stays in a person's system. How can logarithmic functions be used to answer this question? Let's find out!

Are YOU Ready?

Complete these exercises to review skills you will need for this module.

- Online Homework
- Hints and Help
- Extra Practice

Exponents

Example 1 Rewrite $x^{-3}y^2$ using only positive exponents.

$$x^{-3}y^2 = \frac{y^2}{x^3}$$

Rewrite each expression using only positive exponents.

1. $x^{-4}y^{-2}$
2. x^8y^{-5}
3. $\frac{x^{-1}}{y^{-2}}$

Rational and Radical Exponents

Example 2 Write $\sqrt[4]{a^6b^4}$ using rational exponents.

$$\sqrt[4]{a^6b^4} = \left(a^6b^4\right)^{\frac{1}{4}}$$ $\sqrt[n]{x} = x^{\frac{1}{n}}$

$$= a^{\frac{3}{2}}b$$ Power of a Product Property

Write the expression using rational exponents.

4. $\sqrt[6]{a^3b^{12}}$
5. $\sqrt[3]{ab^2}$
6. $\sqrt[4]{81a^{12}b^8}$

Graphing Linear Nonproportional Relationships

Example 3 Graph $y = -3x + 1$.

Graph the y–intercept of $(0, 1)$.

Use the slope $\frac{-3}{1}$ to plot a second point.

Draw a line through the two points.

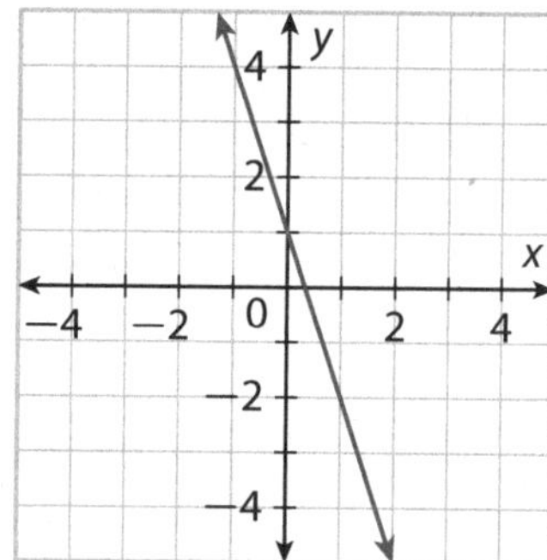

Graph each equation.

7. $y = \frac{2}{3}x - 4$

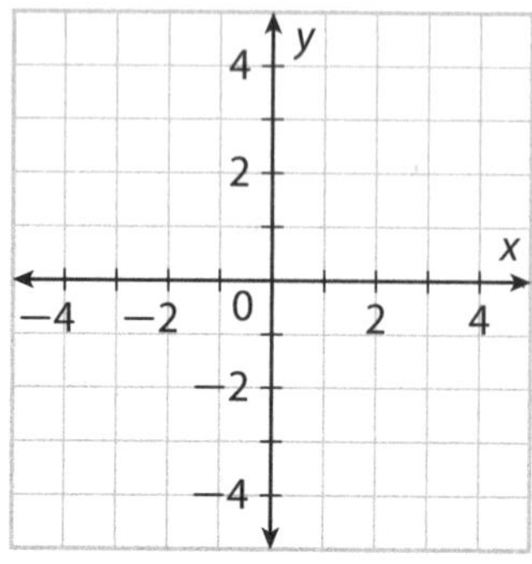

8. $y = 2x + 3$

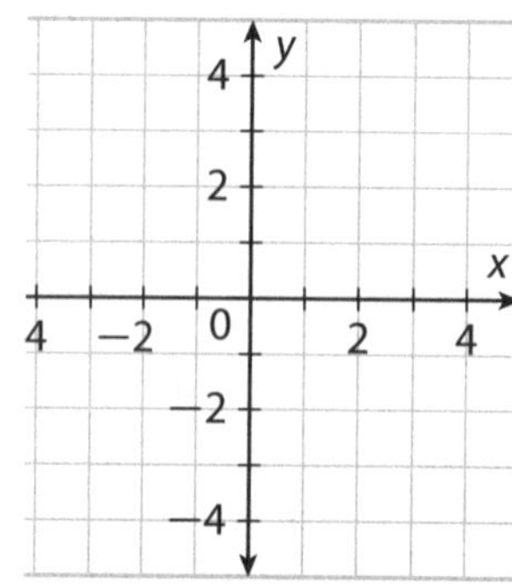

Name________________________ Class____________ Date________

15.1 Defining and Evaluating a Logarithmic Function

Resource Locker

Essential Question: What is the inverse of the exponential function $f(x) = b^x$ where $b > 0$ and $b \neq 1$, and what is the value of $f^{-1}(b^m)$ for any real number m?

A2.5.C Rewrite exponential equations as their corresponding logarithmic equations and logarithmic equations as their corresponding exponential equations. Also A2.2.A, A2.2.B, A2.2.C, A2.5.B, A2.7.I

Explore Understanding Logarithmic Functions as Inverses of Exponential Functions

An exponential function such as $f(x) = 2^x$ accepts values of the exponent as inputs and delivers the corresponding power of 2 as the outputs. The inverse of an exponential function is called a **logarithmic function**. For $f(x) = 2^x$, the inverse function is written $f^{-1}(x) = \log_2 x$, which is read either as "the logarithm with base 2 of x" or simply as "log base 2 of x." It accepts powers of 2 as inputs and delivers the corresponding exponents as outputs.

(A) Graph $f^{-1}(x) = \log_2 x$ using the graph of $f(x) = 2^x$ shown. Begin by reflecting the labeled points on the graph of $f(x) = 2^x$ across the line $y = x$ and labeling the reflected points with their coordinates. Then draw a smooth curve through the reflected points.

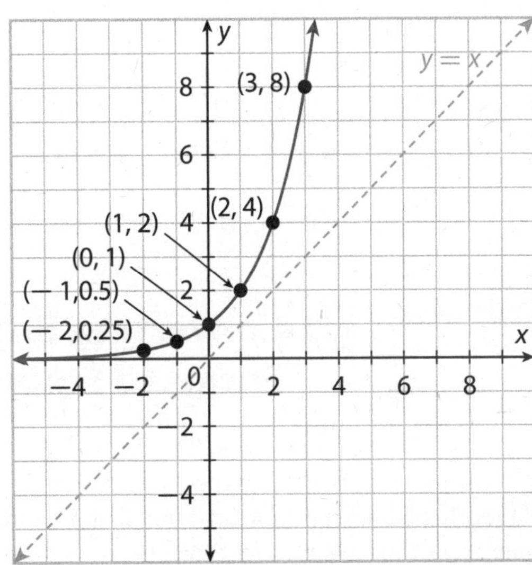

(B) Using the labeled points on the graph of $f^{-1}(x)$, complete the following statements.

$f^{-1}(0.25) = \log_2$ ☐ = ☐

$f^{-1}(0.5) = \log_2$ ☐ = ☐

$f^{-1}(1) = \log_2$ ☐ = ☐

$f^{-1}(2) = \log_2$ ☐ = ☐

$f^{-1}(4) = \log_2$ ☐ = ☐

$f^{-1}(8) = \log_2$ ☐ = ☐

Reflect

1. Explain why the domain of $f(x) = 2^x$ doesn't need to be restricted in order for its inverse to be a function.

2. State the domain and range of $f^{-1}(x) = \log_2 x$ using set notation.

3. Identify any intercepts and asymptotes for the graph of $f^{-1}(x) = \log_2 x$.

4. Is $f^{-1}(x) = \log_2 x$ an increasing function or a decreasing function?

5. How does $f^{-1}(x) = \log_2 x$ behave as x increases without bound? As x decreases toward 0?

6. Based on the inverse relationship between $f(x) = 2^x$ and $f^{-1}(x) = \log_2 x$, complete this statement:

$f^{-1}(16) = \log_2 \square = \square$ because $f(\square) = \square$.

Explain 1 Converting Between Exponential and Logarithmic Forms of Equations

In general, the exponential function $f(x) = b^x$, where $b > 0$ and $b \neq 1$, has the logarithmic function $f^{-1}(x) = \log_b x$ as its inverse. For instance, if $f(x) = 3^x$, then $f^{-1}(x) = \log_3 x$, and if $f(x) = \left(\frac{1}{4}\right)^x$, then $f^{-1}(x) = \log_{\frac{1}{4}} x$. The inverse relationship between exponential functions and logarithmic functions also means that you can write any exponential equation as a logarithmic equation and any logarithmic equation as an exponential equation.

Exponential Equation	Logarithmic Equation
$b^x = a$	$\log_b a = x$

$$b > 0, b \neq 1$$

Example 1 **Complete the table by writing each given equation in its alternate form.**

Exponential Equation	Logarithmic Equation
$4^3 = 64$	?
?	$\log_5 \frac{1}{25} = -2$
$\left(\frac{2}{3}\right)^p = q$	?
?	$\log_{\frac{1}{2}} m = n$

Think of each equation as involving an exponential function or a logarithmic function. Identify the function's base, input, and output. For the inverse function, use the same base but switch the input and output.

Think of the equation $4^3 = 64$ as involving an exponential function with base 4. The input is 3, and the output is 64. So, the inverse function (a logarithmic function) also has base 4, but its input is 64, and its output is 3.

Think of the equation $\log_5 \frac{1}{25} = -2$ as involving a logarithmic function with base 5. The input is $\frac{1}{25}$, and the output is −2. So, the inverse function (an exponential function) also has base 5, but its input is −2, and its output is $\frac{1}{25}$.

Think of the equation $\left(\frac{2}{3}\right)^p = q$ as involving an exponential function with base $\frac{2}{3}$. The input is p, and the output is q. So, the inverse function (a logarithmic function) also has base $\frac{2}{3}$, but its input is q, and its output is p.

Think of the equation $\log_{\frac{1}{2}} m = n$ as involving a logarithmic function with base $\frac{1}{2}$. The input is m, and the output is n. So, the inverse function (an exponential function) also has base $\frac{1}{2}$, but its input is n, and its output is m.

Exponential Equation	Logarithmic Equation
$4^3 = 64$	$\log_4 64 = 3$
$5^{-2} = \frac{1}{25}$	$\log_5 \frac{1}{25} = -2$
$\left(\frac{2}{3}\right)^p = q$	$\log_{\frac{2}{3}} q = p$
$\left(\frac{1}{2}\right)^n = m$	$\log_{\frac{1}{2}} m = n$

Ⓑ

Exponential Equation	Logarithmic Equation
$3^5 = 243$	
	$\log_4 \frac{1}{64} = -3$
$\left(\frac{3}{4}\right)^r = s$	
	$\log_{\frac{1}{5}} v = w$

Think of the equation $3^5 = 243$ as involving an exponential function with base 3. The input is ____, and the output is ____. So, the inverse function (a logarithmic function) also has base 3, but its input is ____, and its output is ____.

Think of the equation $\log_4 \frac{1}{64} = -3$ as involving a logarithmic function with base ____. The input is $\frac{1}{64}$, and the output is ____. So, the inverse function (an exponential function) also has base ____, but its input is ____, and its output is $\frac{1}{64}$.

Think of the equation $\left(\frac{3}{4}\right)^r = s$ as involving an exponential function with base ____. The input is ____, and the output is s. So, the inverse function (a logarithmic function) also has base ____, but its input is s, and its output is ____.

Think of the equation $\log_{\frac{1}{5}} v = w$ as involving a logarithmic function with base ____. The input is ____, and the output is ____. So, the inverse function (an exponential function) also has base ____, but its input is ____, and its output is ____.

Reflect

7. A student wrote the logarithmic form of the exponential equation $5^0 = 1$ as $\log_5 0 = 1$. What did the student do wrong? What is the correct logarithmic equation?

____

____

Your Turn

8. Complete the table by writing each given equation in its alternate form.

Exponential Equation	Logarithmic Equation
$10^4 = 10{,}000$	
	$\log_2 \frac{1}{16} = -4$
$\left(\frac{2}{5}\right)^c = d$	
	$\log_{\frac{1}{3}} x = y$

Explain 2 Evaluating Logarithmic Functions by Thinking in Terms of Exponents

The logarithmic function $f(x) = \log_b x$ accepts a power of b as an input and delivers an exponent as an output. In cases where the input of a logarithmic function is a recognizable power of b, you should be able to determine the function's output. You may find it helpful first to write a logarithmic equation by letting the output equal y and then to rewrite the equation in exponential form. Once the bases on each side of the exponential equation are equal, you can equate their exponents to find y.

Example 2

(A) If $f(x) = \log_{10} x$, find $f(1000)$, $f(0.01)$, and $f\left(\sqrt{10}\right)$.

$f(1000) = y$

$\log_{10} 1000 = y$

$10^y = 1000$

$10^y = 10^3$

$y = 3$

So, $f(1000) = 3$.

$f(0.01) = y$

$\log_{10} 0.01 = y$

$10^y = 0.01$

$10^y = 10^{-2}$

$y = -2$

So, $f(0.01) = -2$.

$f\left(\sqrt{10}\right) = y$

$\log_{10} \sqrt{10} = y$

$10^y = \sqrt{10}$

$10^y = 10^{\frac{1}{2}}$

$y = \frac{1}{2}$

So, $f\left(\sqrt{10}\right) = \frac{1}{2}$.

(B) If $f(x) = \log_{\frac{1}{2}} x$, find $f(4)$, $f\left(\frac{1}{32}\right)$ and $f\left(2\sqrt{2}\right)$.

$f(4) = y$

$\log_{\frac{1}{2}} 4 = y$

$\left(\frac{1}{2}\right)^y = 4$

$\left(\frac{1}{2}\right)^y = 2^{\square}$

$\left(\frac{1}{2}\right)^y = \left(\frac{1}{2}\right)^{\square}$

$y = \square$

So, $f(4) = \square$.

$f\left(\frac{1}{32}\right) = y$

$\log_{\frac{1}{2}} \frac{1}{32} = y$

$\left(\frac{1}{2}\right)^y = \frac{1}{32}$

$\left(\frac{1}{2}\right)^y = \frac{1}{2^{\square}}$

$\left(\frac{1}{2}\right)^y = \left(\frac{1}{2}\right)^{\square}$

$y = \square$

So, $f\left(\frac{1}{32}\right) = \square$.

$f\left(2\sqrt{2}\right) = y$

$\log_{\frac{1}{2}} 2\sqrt{2} = y$

$\left(\frac{1}{2}\right)^y = 2\sqrt{2}$

$\left(\frac{1}{2}\right)^y = \sqrt{2^2 \cdot 2}$

$\left(\frac{1}{2}\right)^y = \sqrt{2^{\square}}$

$\left(\frac{1}{2}\right)^y = 2^{\square}$

$\left(\frac{1}{2}\right)^y = \left(\frac{1}{2}\right)^{\square}$

$y = \square$

So $f\left(2\sqrt{2}\right) = \square$.

Your Turn

9. If $f(x) = \log_7 x$, find $f(343)$, $f\left(\frac{1}{49}\right)$, and $f\left(\sqrt{7}\right)$.

10. If $f(x) = \log_{\frac{1}{3}} x$, find $f(27)$, $f\left(\frac{1}{81}\right)$, and $f\left(9\sqrt{3}\right)$.

Explain 3 Evaluating Logarithmic Functions Using a Scientific Calculator

You can use a scientific calculator to find the logarithm of any positive number x when the logarithm's base is either 10 or e. When the base is 10, you are finding what is called the *common logarithm* of x, and you use the calculator's LOG key because $\log_{10} x$ is also written as $\log x$ (where the base is understood to be 10). When the base is e, you are finding what is called the *natural logarithm* of x, and you use the calculator's LN key because $\log_e x$ is also written as $\ln x$.

Example 3 **Use a scientific calculator to find the common logarithm and the natural logarithm of the given number. Verify each result by evaluating the appropriate exponential expression.**

 13

First, find the common logarithm of 13. Round the result to the thousandths place and raise 10 to that number to confirm that the power is close to 13.

So, $\log 13 \approx 1.114$.

Next, find the natural logarithm of 13. Round the result to the thousandths place and raise e to that number to confirm that the power is close to 13.

ln(13)
2.564949357
e^2.565
13.00065837

So, $\ln 13 \approx 2.565$.

 0.42

First, find the common logarithm of 0.42. Round the result to the thousandths place and raise 10 to that number to confirm that the power is close to 0.42.

$\log 0.42 \approx$ ____

$10^{\text{\_\_\_\_}} \approx 0.42$

Next, find the natural logarithm of 0.42. Round the result to the thousandths place and raise e to that number to confirm that the power is close to 0.42.

$\ln 0.42 \approx$ ____

$e^{\text{\_\_\_\_}} \approx 0.42$

Reflect

11. For any $x > 1$, why is $\log x < \ln x$?

Your Turn

Use a scientific calculator to find the common logarithm and the natural logarithm of the given number. Verify each result by evaluating the appropriate exponential expression.

12. 0.25

13. 4

Explain 4 Evaluating a Logarithmic Model

There are standard scientific formulas that involve logarithms, such as the formulas for the acidity level (pH) of a liquid and the intensity level of a sound. It's also possible to develop your own models involving logarithms by finding the inverses of exponential growth and decay models.

Example 4

(A) The acidity level, or pH, of a liquid is given by the formula $\text{pH} = \log \frac{1}{[\text{H}^+]}$ where $[\text{H}^+]$ is the concentration (in moles per liter) of hydrogen ions in the liquid. In a typical chlorinated swimming pool, the concentration of hydrogen ions ranges from 1.58×10^{-8} moles per liter to 6.31×10^{-8} moles per liter. What is the range of the pH for a typical swimming pool?

Using the pH formula, substitute the given values of $[\text{H}^+]$.

$$\text{pH} = \log\left(\frac{1}{6.31 \times 10^{-8}}\right)$$

$$\approx \log 15{,}800{,}000$$

$$\approx 7.2$$

$$\text{pH} = \log\left(\frac{1}{1.58 \times 10^{-8}}\right)$$

$$\approx \log 63{,}300{,}000$$

$$\approx 7.8$$

So, the pH of a swimming pool ranges from 7.2 to 7.8.

 *Lactobacillus acidophilus* is one of the bacteria used to turn milk into yogurt. The population P of a colony of 3500 bacteria at time t (in minutes) can be modeled by the function $P(t) = 3500(2)^{\frac{t}{73}}$. How long does it take the population to reach 1,792,000?

Step 1 Solve $P = 3500(2)^{\frac{t}{73}}$ for t.

Write the model. $P = 3500(2)^{\frac{t}{73}}$

Divide both sides by 3500. $\frac{P}{\square} = (2)^{\frac{t}{73}}$

Rewrite in logarithmic form. $\log_2 \frac{P}{\square} = \frac{t}{73}$

Multiply both sides by 73. $73 \log_2 \frac{P}{\square} = t$

Step 2 Use the logarithmic model to find t when $P = 1{,}792{,}000$.

$t = 73 \log_2 \frac{P}{\square}$

$= 73 \log_2 \frac{1{,}792{,}000}{\square}$

$= 73 \log_2 \square$

$= 73(\square)$

$= \square$

So, the bacteria population will reach 1,792,000 in ______ minutes, or about ______ hours.

Reflect

14. **Discussion** Describe two ways you can use the solution in Part B to find the time it takes for the bacteria population to reach 3,584,000.

____

____

____

____

____

Your Turn

15. The intensity level L (in decibels, dB) of a sound is given by the formula $L = 10 \log \frac{I}{I_0}$ where I is the intensity (in watts per square meter, W/m^2) of the sound and I_0 is the intensity of the softest audible sound, about 10^{-12} W/m^2. What is the intensity level of a rock concert if the sound has an intensity of 3.2 W/m^2?

16. The mass (in milligrams) of beryllium-11, a radioactive isotope, in a 500-milligram sample at time t (in seconds) is given by the function $m(t) = 500e^{-0.05t}$. When will there be 90 milligrams of beryllium-11 remaining?

Elaborate

17. What is a logarithmic function? Give an example.

18. How can you turn an exponential model that gives y as a function of x into a logarithmic model that gives x as a function of y?

19. **Essential Question Check-In** Write the inverse of the exponential function $f(x) = b^x$ where $b > 0$ and $b \neq 1$.

Evaluate: Homework and Practice

- Online Homework
- Hints and Help
- Extra Practice

1. Complete the input-output table for $f(x) = \log_2 x$. Plot and label the ordered pairs from the table. Then draw the complete graph of $f(x)$.

x	$f(x)$
0.25	
0.5	
1	
2	
4	
8	

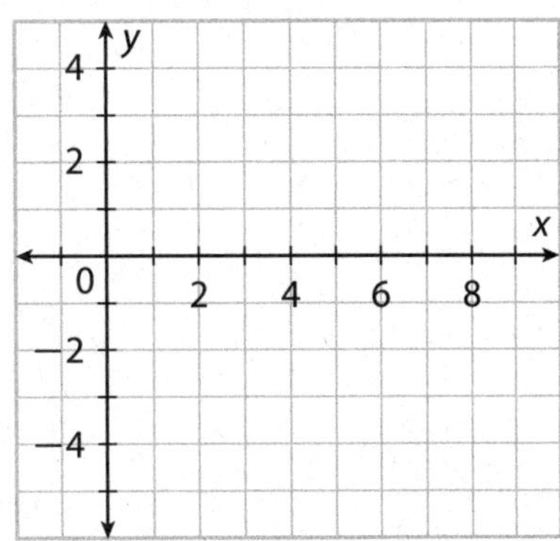

2. Use the graph of $f(x) = \log_2 x$ to do the following.

   a. State the function's domain and range using set notation.

   b. Identify the function's end behavior.

   c. Identify the graph's x- and y-intercepts.

   d. Identify the graph's asymptotes.

   e. Identify the intervals where the function has positive values and where it has negative values.

   f. Identify the intervals where the function is increasing and where it is decreasing

3. Consider the exponential function $f(x) = 3^x$.

a. State the function's domain and range using set notation.

b. Describe any restriction you must place on the domain of the function so that its inverse is also a function.

c. Write the rule for the inverse function.

d. State the inverse function's domain and range using set notation.

4. Consider the logarithmic function $f(x) = \log_4 x$.

a. State the function's domain and range using set notation.

b. Describe any restriction you must place on the domain of the function so that its inverse is also a function.

c. Write the rule for the inverse function.

d. State the inverse function's domain and range using set notation.

Write the given exponential equation in logarithmic form.

5. $5^3 = 125$

6. $\left(\frac{1}{10}\right)^{-2} = 100$

7. $3^m = n$

8. $\left(\frac{1}{2}\right)^p = q$

Write the given logarithmic equation in exponential form.

9. $\log_6 1296 = 4$

10. $\log_{\frac{1}{4}} \frac{1}{64} = 3$

11. $\log_8 x = y$

12. $\log_{\frac{2}{3}} c = d$

13. If $f(x) = \log_3 x$, find $f(243)$, $f\left(\frac{1}{27}\right)$, and $f\left(\sqrt{27}\right)$.

14. If $f(x) = \log_6 x$, find $f(36)$, $f\left(\frac{1}{6}\right)$ and $f\left(6\sqrt[3]{6}\right)$

15. If $f(x) = \log_{\frac{1}{4}} x$, find $f\left(\frac{1}{64}\right)$, $f(256)$, and $f\left(\sqrt[3]{16}\right)$

Use a scientific calculator to find the common logarithm and the natural logarithm of the given number. Verify each result by evaluating the appropriate exponential expression.

16. 19

17. 9

18. 0.6

19. 0.31

20. The acidity level, or pH, of a liquid is given by the formula $\text{pH} = \log \frac{1}{[\text{H}^+]}$ where $[\text{H}^+]$ is the concentration (in moles per liter) of hydrogen ions in the liquid. What is the pH of iced tea with a hydrogen ion concentration of 0.000158 mole per liter?

21. The intensity level L (in decibels, dB) of a sound is given by the formula $L = 10 \log \frac{I}{I_0}$ where I is the intensity (in watts per square meter, W/m^2) of the sound and I_0 is the intensity of the softest audible sound, about 10^{-12} W/m^2. What is the intensity level of a lawn mower if the sound has an intensity of 0.00063 W/m^2?

22. *Bacillus megaterium* is one the largest bacteria known. The population P of a colony of 1000 bacteria at time t (in minutes) can be modeled by the function $P(t) = 1000e^{0.0277t}$. How long does it take the population to reach 1,000,000?

23. Thorium-230 is a radioactive isotope used in the dating of coral and some cave formations. The percentage of thorium-230 remaining in a sample after time t (in years) is modeled by the function $p(t) = 100(2)^{-\frac{t}{75{,}000}}$. How long has a coral reef been growing if the oldest portion has 1.5625% of its thorium-230 remaining?

24. Match each liquid with its pH given the concentration of hydrogen ions in the liquid.

Liquid	Hydrogen Ion Concentration	pH
A. Cocoa	5.2×10^{-7}	______ 3.5
B. Cider	7.9×10^{-4}	______ 3.3
C. Ginger Ale	4.9×10^{-4}	______ 2.4
D. Honey	1.3×10^{-4}	______ 4.5
E. Buttermilk	3.2×10^{-5}	______ 6.3
F. Cranberry juice	4.0×10^{-3}	______ 6.4
G. Pinneapple juice	3.1×10^{-4}	______ 3.1
H. Tomato juice	6.3×10^{-2}	______ 1.2
I. Carrot juice	4.0×10^{-7}	______ 3.9

H.O.T. Focus on Higher Order Thinking

25. Explain the Error Jade is taking a chemistry test and has to find the pH of a liquid given that its hydrogen ion concentration is 7.53×10^{-9} moles per liter. She writes the following.

$$\begin{aligned} pH &= \ln \frac{1}{[H^+]} \\ &= \ln \frac{1}{7.53 \times 10^{-9}} \\ &\approx 18.7 \end{aligned}$$

She knows that the pH scale ranges from 1 to 14, so her answer of 18.7 must be incorrect, but she runs out of time on the test. Explain her error and find the correct pH.

26. **Multi-step** Exponential functions have the general form $f(x) = ab^{x-h} + k$ where a, b, h, and k are constants, $a \neq 0$, $b > 0$, and $b \neq 1$.

a. State the domain and range of $f(x)$ using set notation.

b. Show how to find $f^{-1}(x)$. Give a description of each step you take.

c. State the domain and range of $f^{-1}(x)$ using set notation.

27. **Justify Reasoning** Evaluate each expression without using a calculator. Explain your reasoning.

a. $\ln e^2$

b. $10^{\log 7}$

c. $4^{\log_2 5}$

Lesson Performance Task

Skydivers use an instrument called an altimeter to determine their height above Earth's surface. An altimeter measures atmospheric pressure and converts it to altitude based on the relationship between pressure and altitude. One model for atmospheric pressure P (in kilopascals, kPa) as a function of altitude a (in kilometers) is $P = 100e^{-a/8}$.

a. Since an altimeter measures pressure directly, pressure is the independent variable for an altimeter. Rewrite the model $P = 100e^{-a/8}$ so that it gives altitude as a function of pressure.

b. To check the function in part a, use the fact that atmospheric pressure at Earth's surface is about 100 kPa.

c. Suppose a skydiver deploys the parachute when the altimeter measures 87 kPa. Use the function in part a to determine the skydiver's altitude. Give your answer in both kilometers and feet. (1 kilometer $\approx$ 3281 feet)

Name________________________ Class______________ Date________

15.2 Graphing Logarithmic Functions

Essential Question: How is the graph of $g(x) = a\log_b(x - h) + k$ where $b > 0$ and $b \neq 1$ related to the graph of $f(x) = \log_b x$?

Resource Locker

A2.5.A Determine the effects on the key attributes on the graphs of $f(x) = log_b(x)$ where b is 2, 10, and e when $f(x)$ is replaced by $af(x)$, $f(x) + d$, and $f(x - c)$ for specific positive and negative real values of a, c, and d. Also A2.2.A, A2.5.B, A2.7.I

Explore 1 Graphing and Analyzing Parent Logarithmic Functions

The graph of the logarithmic function $f(x) = \log_2 x$, which you analyzed in the previous lesson, is shown. In this Explore, you'll graph and analyze other basic logarithmic functions.

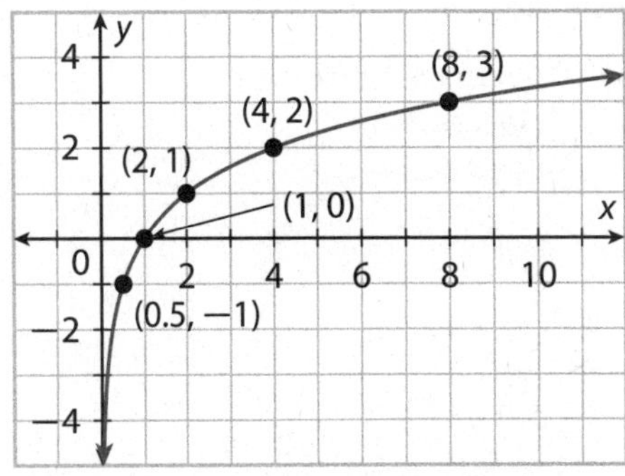

Ⓐ Complete the table for the function $f(x) = \log x$. (Remember that when the base of a logarithmic function is not specified, it is understood to be 10.) Then plot and label the ordered pairs from the table and draw a smooth curve through the points to obtain the graph of the function.

x	$f(x) = \log x$
0.1	
1	
10	

Ⓑ Complete the table for the function $f(x) = \ln x$. (Remember that the base of this function is e). Then plot and label the ordered pairs from the table and draw a smooth curve through the points to obtain the graph of the function.

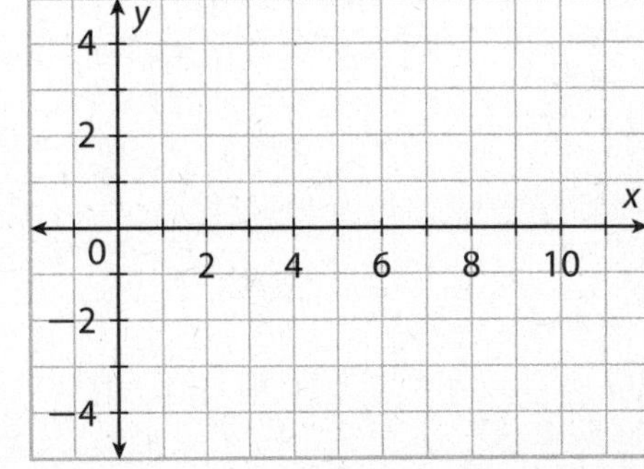

x	$f(x) = \ln x$
$\frac{1}{e} \approx 0.368$	
1	
$e \approx 2.72$	
$e^2 \approx 7.39$	

(C) Analyze the two graphs from Steps A and B, and then complete the table.

Function	$f(x) = \log_2(x)$	$f(x) = \log x$	$f(x) = \ln x$
Domain	$\{x \mid x > 0\}$		
Range	$\{y \mid 0 < y < \infty\}$		
End behavior	As $x \to +\infty$, $f(x) \to +\infty$. As $x \to 0^+$, $f(x) \to -\infty$.		
Vertical and horizontal asymptotes	Vertical asymptote at $x = 0$; no horizontal asymptote		
Intervals where increasing or decreasing	Increasing throughout its domain		
Intercepts	x-intercept at $(1, 0)$; no y-intercepts		
Intervals where positive or negative	Positive on $(1, +\infty)$; negative on $(0, 1)$		

Reflect

1. What similarities do you notice about all logarithmic functions of the form $f(x) = \log_b x$ where $b > 1$? What differences do you notice?

Explore 2 Predicting Transformations of the Graphs of Parent Logarithmic Functions

You can graph the logarithmic function $f(x) = \log_b x$ where $b > 0$ and $b \neq 1$ on a graphing calculator by specifying the base when you enter the function's rule using the LOG key after pressing the Y= key. For instance, the first calculator screen shows how to enter the function $f(x) = \log_2 x$, and the second screen shows the function's graph. Notice that the graph passes through the point $(2, 1)$ as you would expect.

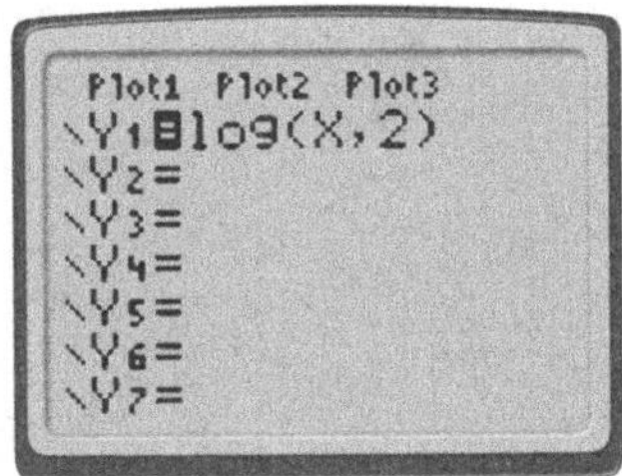

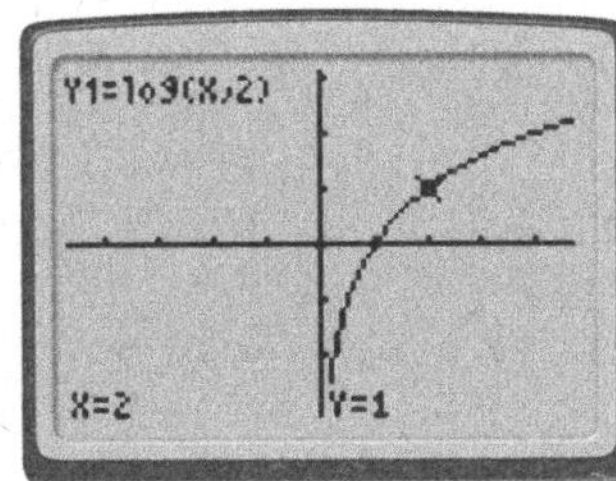

In this Explore, you will predict the effects of parameters on the graphs of logarithmic functions with bases 2, 10, and *e*. You will then confirm your predictions by graphing the transformed functions on a graphing calculator.

(A) Predict the effect of the parameter h on the graph of $g(x) = \log_b (x - h)$ for each function.

a. The graph of $g(x) = \log_2 (x - 2)$ is a ______________ of the graph of

$f(x) = \log_2 x$ [right/left/up/down] 2 units.

b. The graph of $g(x) = \log_2 (x + 2)$ is a ______________ of the graph of

$f(x) = \log_2 x$ [right/left/up/down] 2 units.

c. The graph of $g(x) = \log (x - 1)$ is a ______________ of the graph of

$f(x) = \log x$ [right/left/up/down] 1 unit.

d. The graph of $g(x) = \log (x + 1)$ is a ______________ of the graph of

$f(x) = \log x$ [right/left/up/down] 1 unit.

e. The graph of $g(x) = \ln (x - 3)$ is a ______________ of the graph of

$f(x) = \ln x$ [right/left/up/down] 3 units.

f. The graph of $g(x) = \ln (x + 3)$ is a ______________ of the graph of

$f(x) = \ln x$ [right/left/up/down] 3 units.

Check your predictions using a graphing calculator.

(B) Predict the effect of the parameter k on the graph of $g(x) = \log_b x + k$ for each function.

a. The graph of $g(x) = \log_2 x + 3$ is a ______________________ of the graph of $f(x) = \log_2 x$ [right/up/left/down] 3 units.

b. The graph of $g(x) = \log_2 x - 3$ is a ______________________ of the graph of $f(x) = \log_2 x$ [right/up/left/down] 3 units.

c. The graph of $g(x) = \log x + 2$ is a ______________________ of the graph of $f(x) = \log x$ [right/up/left/down] 2 units.

d. The graph of $g(x) = \log x - 2$ is a ______________________ of the graph of $f(x) = \log x$ [right/up/left/down] 2 units.

e. The graph of $g(x) = \ln x + 1$ is a ______________________ of the graph of $f(x) = \ln x$ [right/left/up/down] 1 unit.

f. The graph of $g(x) = \ln x - 1$ is a ______________________ of the graph of $f(x) = \ln x$ [right/left/up/down] 1 unit.

Check your predictions using a graphing calculator.

(C) Predict the effect of the parameter a on the graph of $g(x) = a\log_b x$ for each function.

a. The graph of $g(x) = 2\log_2 x$ is a ________________ of the graph of $f(x) = \log_2 x$ by a factor of _____.

b. The graph of $g(x) = -\frac{1}{2}\log_2 x$ is a ______________________ of the graph of $f(x) = \log_2 x$ by a factor of _____ as well as a ____________ across the ________.

c. The graph of $g(x) = -3\log x$ is a ________________ of the graph of $f(x) = \log x$ by a factor of _____ as well as a ____________ across the ________.

d. The graph of $g(x) = \frac{1}{3}\log x$ is a ______________________ of the graph of $f(x) = \log x$ by a factor of _____.

e. The graph of $g(x) = 4\ln x$ is a ________________ of the graph of $f(x) = \ln x$ by a factor of _____.

f. The graph of $g(x) = -\frac{1}{4}\ln x$ is a ______________________ of the graph of $f(x) = \ln x$ by a factor of _____ as well as a ____________ across the ________.

Check your predictions using a graphing calculator.

(D) Complete the table by identifying which parameters $(a, h, \text{and/or } k)$ affect the attributes of each logarithmic function listed in the table. When necessary, indicate specifically whether a positive or negative value of a parameter affects an attribute. (For end behavior, consider both x increasing without bound and x decreasing toward 0 from the right.)

Function	$g(x) = a \log_2 (x - h) + k$	$g(x) = a \log (x - h) + k$	$g(x) = a \ln (x - h) + k$
Parameters that affect the domain			
Parameters that affect the range			
Parameters that affect the end behavior			
Parameters that affect the vertical asymptote			
Parameters that affect whether the function is increasing or decreasing			
Parameters that affect the x-intercept			
Parameters that affect the intervals where the function is positive or negative			

Reflect

2. In Step D, do the effects of the parameters on the attributes of the logarithmic functions depend on the base of the function? Explain.

3. In Step D, would your answers change if the logarithmic functions had bases between 0 and 1 instead of bases greater than 1? Explain.

Explain 1 Graphing Combined Transformations of $f(x) = \log_b x$ Where $b > 1$

When graphing transformations of $f(x) = \log_b x$ where $b > 1$, it helps to consider the effect of the transformations on the following features of the graph of $f(x)$: the vertical asymptote, $x = 0$, and two reference points, $(1, 0)$ and $(b, 1)$. The table lists these features as well as the corresponding features of the graph of $g(x) = a \log_b (x - h) + k$.

Function	$f(x) = \log_b x$	$g(x) = a \log_b (x - h) + k$
Asymptote	$x = 0$	$x = h$
Reference point	$(1, 0)$	$(1 + h, k)$
Reference point	$(b, 1)$	$(b + h, a + k)$

Example 1 **Identify the transformations of the graph of $f(x) = \log_b x$ that produce the graph of the given function $g(x)$. Then graph $g(x)$ on the same coordinate plane as the graph of $f(x)$ by applying the transformations to the asymptote $x = 0$ and to the reference points $(1, 0)$ and $(b, 1)$. Also state the domain and range of $g(x)$ using set notation.**

(A) $g(x) = -2 \log_2 (x - 1) - 2$

The transformations of the graph of $f(x) = \log_2 x$ that produce the graph of $g(x)$ are as follows:

- a vertical stretch by a factor of 2
- a reflection across the x-axis
- a translation of 1 unit to the right and 2 units down

Note that the translation of 1 unit to the right affects only the x-coordinates of points on the graph of $f(x)$, while the vertical stretch by a factor of 2, the reflection across the x-axis, and the translation of 2 units down affect only the y-coordinates.

Function	$f(x) = \log_2 x$	$g(x) = -2 \log_2 (x - 1) - 2$
Asymptote	$x = 0$	$x = 1$
Reference point	$(1, 0)$	$(1 + 1, -2(0) - 2) = (2, -2)$
Reference point	$(2, 1)$	$(2 + 1, -2(1) - 2) = (3, -4)$

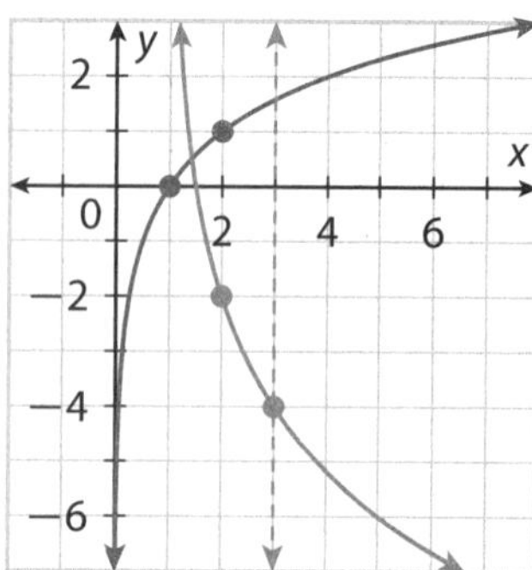

Domain: $\{x \mid x > 1\}$

Range: $\{y \mid \infty < y < +\infty\}$

(B) $g(x) = 2 \log (x + 2) + 4$

The transformations of the graph of $f(x) = \log x$ that produce the graph of $g(x)$ are as follows:

- a vertical stretch by a factor of 2
- a translation of 2 units to the left and 4 units up

Note that the translation of 2 units to the left affects only the x-coordinates of points on the graph of $f(x)$, while the vertical stretch by a factor of 2 and the translation of 4 units up affect only the y-coordinates.

Function	$f(x) = \log x$	$g(x) = 2 \log (x + 2) + 4$
Asymptote	$x = 0$	$x = \square$
Reference point	$(1, 0)$	$(\square - 2, 2(\square) + 4) = (\square, \square)$
Reference point	$(10, 1)$	$(\square - 2, 2(\square) + 4) = (\square, \square)$

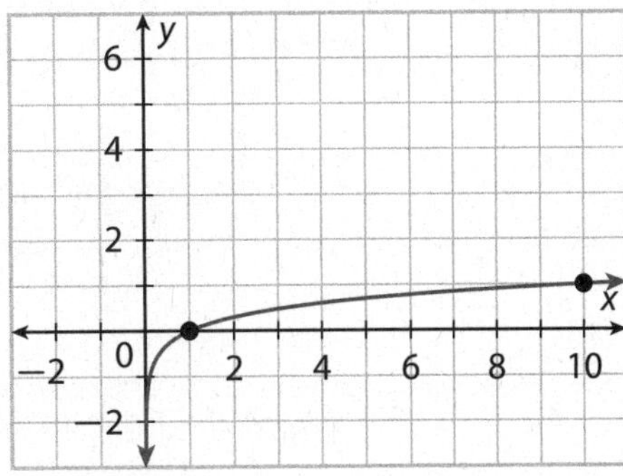

Domain: $\{x \mid x > \square\}$

Range: $\{y \mid -\infty < y < \square\}$

Your Turn

Identify the transformations of the graph of $f(x) = \log_b x$ that produce the graph of the given function $g(x)$. Then graph $g(x)$ on the same coordinate plane as the graph of $f(x)$ by applying the transformations to the asymptote $x = 0$ and to the reference points $(1, 0)$ and $(b, 1)$. Also state the domain and range of $g(x)$ using set notation.

4. $g(x) = 3 \ln (x + 4) + 2$

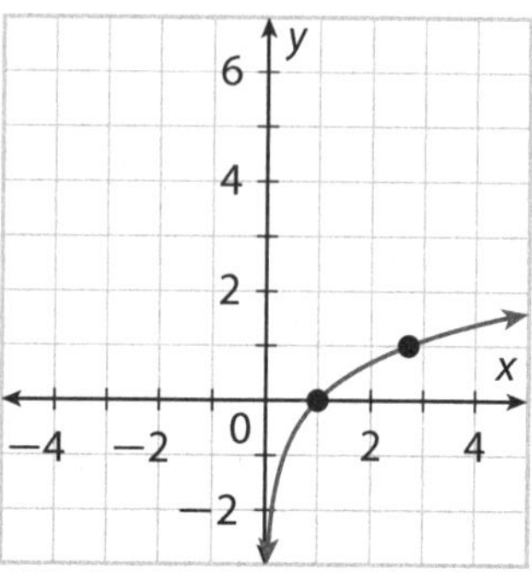

5. $g(x) = \frac{1}{2} \log_2 (x + 1) + 2$

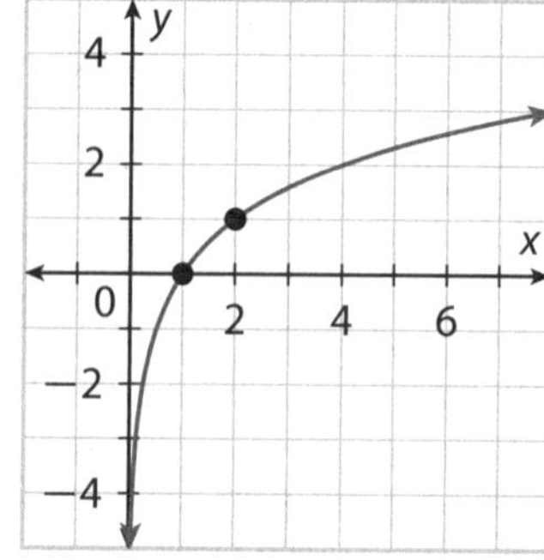

6. $g(x) = -3 \log (x - 1) - 4$

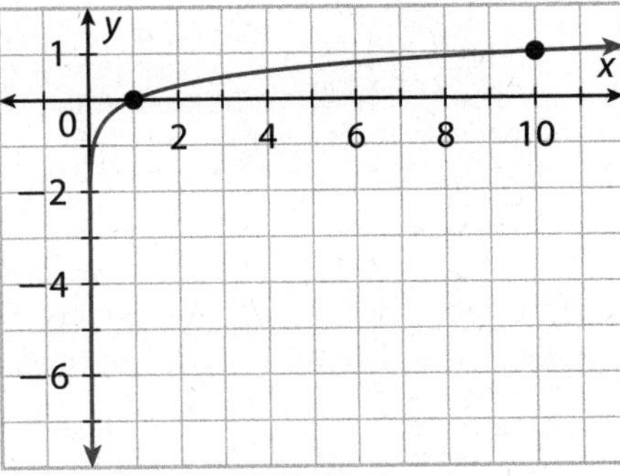

Explain 2 Writing, Graphing, and Analyzing a Logarithmic Model

You can obtain a logarithmic model for real-world data either by performing logarithmic regression on the data or by finding the inverse of an exponential model if one is available.

Example 2 **A biologist studied a population of foxes in a forest preserve over a period of time. The table gives the data that the biologist collected.**

Years Since Study Began	Fox Population
0	55
2	72
3	99
5	123
8	151
12	234
15	336
18	475

From the data, the biologist obtained the exponential model $P = 62(1.12)^t$ where P is the fox population at time t (in years since the study began). The biologist is interested in having a model that gives the time it takes the fox population to reach a certain level.

(A) One way to obtain the model that the biologist wants is to perform logarithmic regression on a graphing calculator using the data set but with the variables switched (that is, the fox population is the independent variable and time is the dependent variable). After obtaining the logarithmic regression model, graph it on a scatter plot of the data. Analyze the model in terms of whether it is increasing or decreasing as well as its average rate of change from $P = 100$ to $P = 200$, from $P = 200$ to $P = 300$, and from $P = 300$ to $P = 400$. Do the model's average rates of change increase, decrease, or stay the same? What does this mean for the fox population?

Using a graphing calculator, enter the population data into one list (L1) and the time data into another list (L2).

L1	L2	L3 2
55	0	------
72	2	
99	3	
123	5	
151	8	
234	12	
336	15	

L2(1)=0

Perform logarithmic regression by pressing the STAT key, choosing the **CALC** menu, and selecting **9:LnReg**. Note that the calculator's regression model is a natural logarithmic function.

```
LnReg
 y=a+blnx
 a=-35.57466967
 b=8.658691119
 r²=.9867920221
 r=.9933740595
```

So, the model is $t = -35.6 + 8.66 \ln P$. Graphing this model on a scatter plot of the data visually confirms that the model is a good fit for the data.

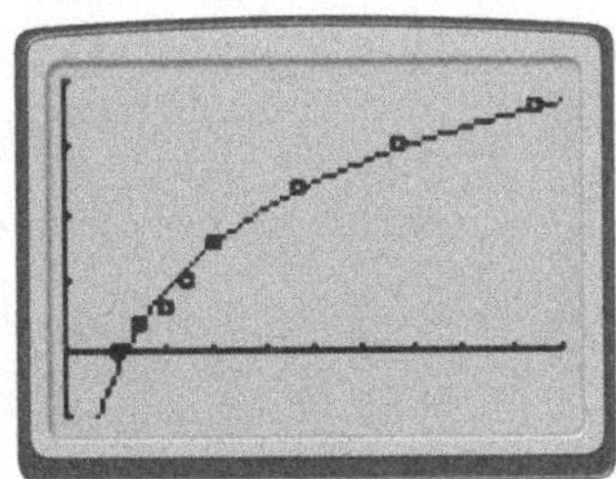

From the graph, you can see that the function is increasing. To find the model's average rates of change, divide the change in t (the dependent variable) by the change in P (the independent variable):

Average rate of change $= \dfrac{t_2 - t_1}{P_2 - P_1}$

Population	Number of Years to Reach That Population	Average Rate of Change
100	$t = -35.6 + 8.66 \ln 100 \approx 4.3$	
200	$t = -35.6 + 8.66 \ln 200 \approx 10.3$	$\dfrac{10.3 - 4.3}{200 - 100} = \dfrac{6.0}{100} = 0.060$
300	$t = -35.6 + 8.66 \ln 300 \approx 13.8$	$\dfrac{13.8 - 10.3}{300 - 200} = \dfrac{3.5}{100} = 0.035$
400	$t = -35.6 + 8.66 \ln 400 \approx 16.3$	$\dfrac{16.3 - 13.8}{400 - 300} = \dfrac{2.5}{100} = 0.025$

The model's average rates of change are decreasing. This means that as the fox population grows, it takes less time for the population to increase by another 100 foxes.

(B) Another way to obtain the model that the biologist wants is to find the inverse of the exponential model. Find the inverse model and compare it with the logarithmic regression model.

In order to compare the inverse of the biologist's model, $P = 62(1.12)^t$, with the logarithmic regression model, you must rewrite the biologist's model with base e so that the inverse will involve a natural logarithm. This means that you want to find a constant c such that $e^c = 1.12$. Writing the exponential equation $e^c = 1.12$ in logarithmic form gives $c = \ln 1.12$, so $c =$ ☐ to the nearest thousandth.

Replacing 1.12 with $e^{☐}$ in the biologist's model gives $P = 62\left(e^{☐}\right)^t$, or $P = 62\,e^{☐\,t}$. Now find the inverse of this function.

Write the equation. $P = 62e^{☐\,t}$

Divide both sides by 62. $\frac{P}{62} = e^{☐\,t}$

Write in logarithmic form. $\ln \frac{P}{62} = ☐\,t$

Divide both sides by ☐. ☐ $\ln \frac{P}{62} = t$

So, the inverse of the exponential model is $t =$ ☐ $\ln \frac{P}{62}$. To compare this model with the logarithmic regression model, use a graphing calculator to graph both $y =$ ☐ $\ln \frac{x}{62}$ and $y = -35.6 + 8.66 \ln x$. You observe that the graphs [roughly coincide/significantly diverge], so the models are [basically equivalent/very different].

Reflect

7. Discussion In a later lesson, you will learn the quotient property of logarithms, which states that $\log_b \frac{m}{n} = \log_b m - \log_b n$ for any positive numbers m and n. Explain how you can use this property to compare the two models in Example 3.

Your Turn

8. Maria made a deposit in a bank account and left the money untouched for several years. The table lists her account balance at the end of each year.

Years Since the Deposit Was Made	Account Balance
0	$1000.00
1	$1020.00
2	$1040.40
3	$1061.21

a. Write an exponential model for the account balance as a function of time (in years since the deposit was made).

b. Find the inverse of the exponential model after rewriting it with a base of e. Describe what information the inverse gives.

c. Perform logarithmic regression on the data (using the account balance as the independent variable and time as the dependent variable). Compare this model with the inverse model from part b.

Elaborate

9. Which transformations of $f(x) = \log_b(x)$ change the function's end behavior (both as x increases without bound and as x decreases toward 0 from the right)? Which transformations change the location of the graph's x-intercept?

10. How are reference points helpful when graphing transformations of $f(x) = \log_b(x)$?

11. What are two ways to obtain a logarithmic model for a set of data?

12. **Essential Question Check-In** Describe the transformations you must perform on the graph of $f(x) = \log_b(x)$ to obtain the graph of $g(x) = a\log_b(x - h) + k$.

Evaluate: Homework and Practice

- Online Homework
- Hints and Help
- Extra Practice

1. Graph the logarithmic functions $f(x) = \log_2 x$, $f(x) = \log x$, and $f(x) = \ln x$ on the same coordinate plane. To distinguish the curves, label the point on each curve where the y-coordinate is 1.

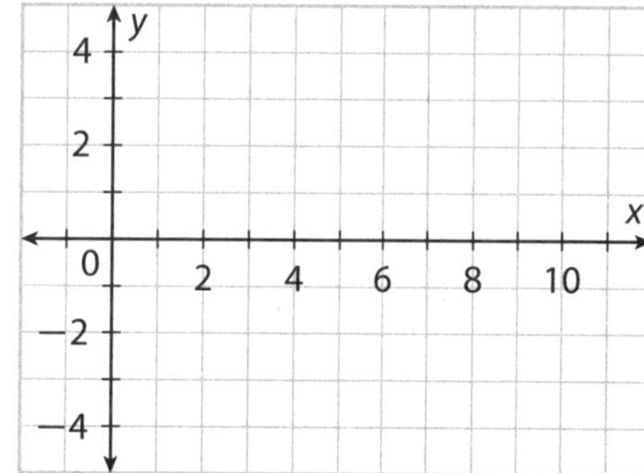

2. Describe the attributes that the logarithmic functions $f(x) = \log_2 x$, $f(x) = \log x$, and $f(x) = \ln x$ have in common and the attributes that make them different. Attributes should include domain, range, end behavior, asymptotes, intercepts, intervals where the functions are positive and where they are negative, intervals where the functions are increasing and where they are decreasing, and the average rate of change on an interval.

3. For each of the six functions, describe how its graph is a transformation of the graph of $f(x) = \log_2 x$. Also identify what attributes of $f(x) = \log_2 x$ change as a result of the transformation. Attributes to consider are the domain, the range, the end behavior, the vertical asymptote, the x-intercept, the intervals where the function is positive and where it is negative, and whether the function increases or decreases throughout its domain.

a. $g(x) = \log_2 x - 5$

b. $g(x) = 4 \log_2 x$

c. $g(x) = \log_2 (x + 6)$

d. $g(x) = -\frac{3}{4} \log_2 x$

e. $g(x) = \log_2 x + 7$

f. $g(x) = \log_2 (x - 8)$

4. For each of the six functions, describe how its graph is a transformation of the graph of $f(x) = \log x$. Also identify what attributes of $f(x) = \log x$ change as a result of the transformation. Attributes to consider are the domain, the range, the end behavior, the vertical asymptote, the x-intercept, the intervals where the function is positive and where it is negative, and whether the function increases or decreases throughout its domain.

a. $g(x) = -3 \log x$

b. $g(x) = \log (x - 5)$

c. $g(x) = \log x + 1$

d. $g(x) = \log(x + 4)$

e. $g(x) = 0.5 \log x$

f. $g(x) = \log x - 2$

5. For each of the six functions, describe how the graph is a transformation of the graph of $f(x) = \ln x$. Also identify what attributes of $f(x) = \ln x$ change as a result of the transformation. Attributes to consider are the domain, the range, the end behavior, the vertical asymptote, the x-intercept, the intervals where the function is positive and where it is negative, and whether the function increases or decreases throughout its domain.

a. $g(x) = \ln(x + 6)$

b. $g(x) = \ln x - 1$

c. $g(x) = \frac{3}{2} \ln x$

d. $g(x) = \ln x + 8$

e. $g(x) = -\frac{2}{3} \ln x$

f. $g(x) = \ln(x - 4)$

Identify the transformations of the graph of $f(x) = \log_b x$ that produce the graph of the given function $g(x)$. Then graph $g(x)$ on the same coordinate plane as the graph of $f(x)$ by applying the transformations to the asymptote $x = 0$ and to the reference points $(1, 0)$ and $(b, 1)$. Also state the domain and range of $g(x)$ using set notation.

6. $g(x) = -4 \log_2 (x + 2) + 1$

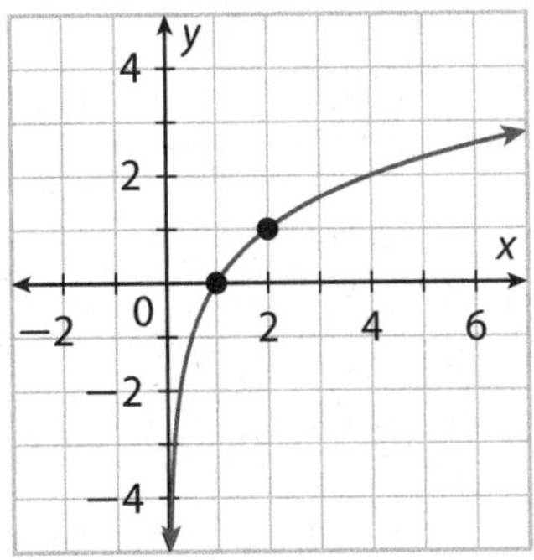

7. $g(x) = \frac{1}{2} \ln (x + 2) - 3$

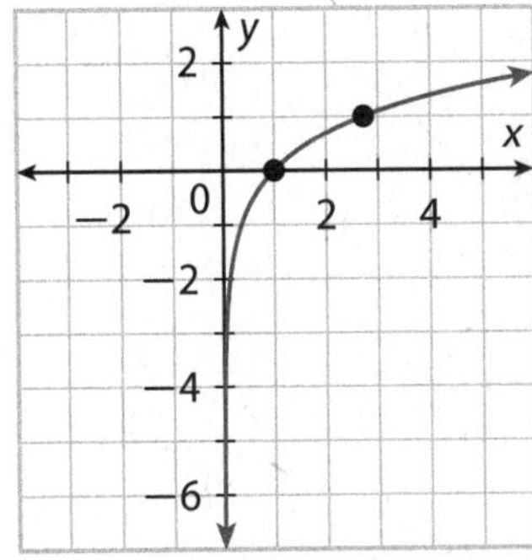

8. $g(x) = 3 \log (x - 1) - 1$

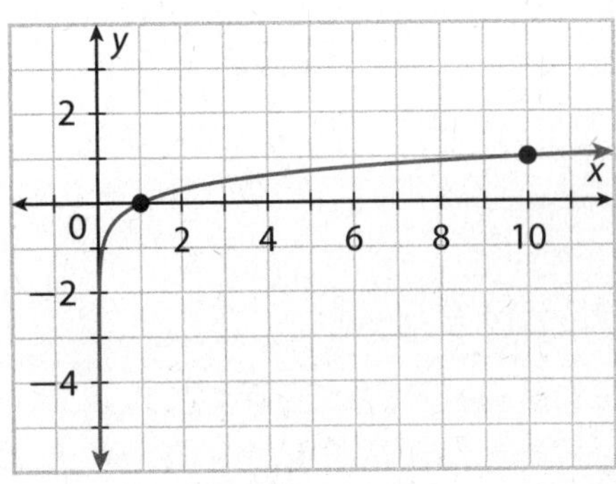

9. $f(x) = \frac{1}{2}\log_2(x - 1) - 2$

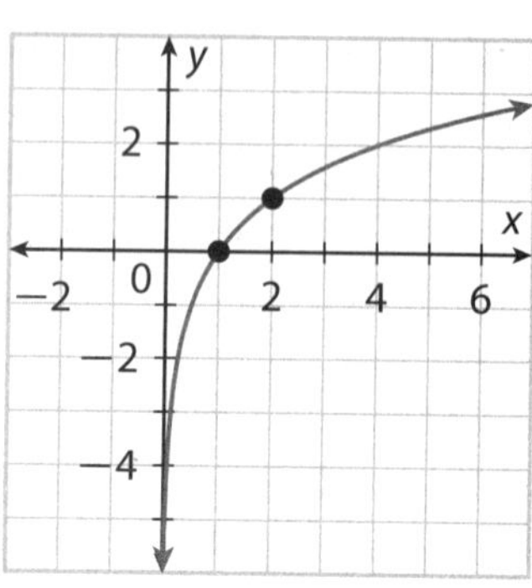

10. $g(x) = -4\ln(x - 4) + 3$

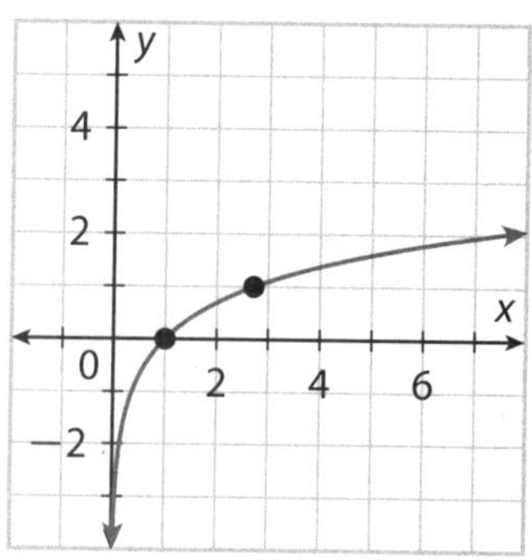

11. $g(x) = -2\log(x + 2) + 5$

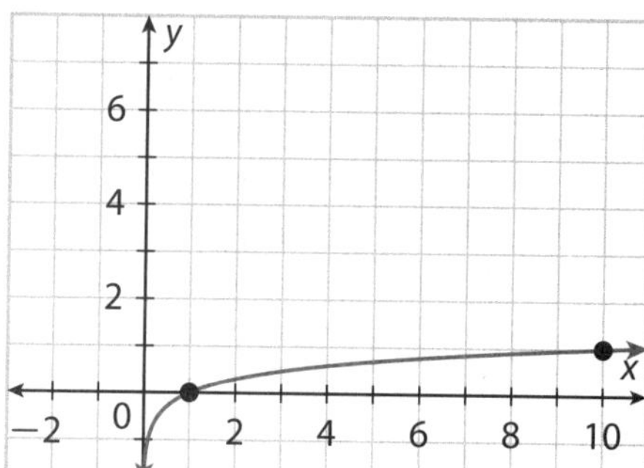

12. The radioactive isotope fluorine-18 is used in medicine to produce images of internal organs and detect cancer. It decays to the stable element oxygen-18. The table gives the percent of fluorine-18 that remains in a sample over a period of time.

Time (hours)	Percent of Fluorine-18 Remaining
0	100
1	68.5
2	46.9
3	32.1

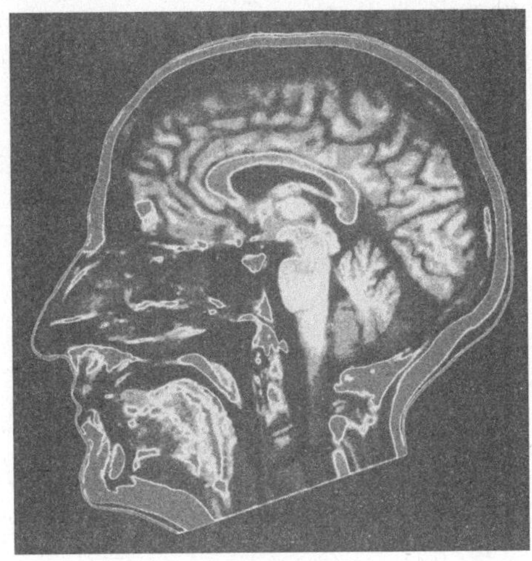

a. Write an exponential model for the percent of fluorine-18 remaining as a function of time (in hours).

b. Find the inverse of the exponential model after rewriting it with a base of *e*. Describe what information the inverse gives.

c. Perform logarithmic regression on the data (using the percent of fluorine-18 remaining as the independent variable and time as the dependent variable). Compare this model with the inverse model from part b.

13. During the period between 2001–2011, the average price of an ounce of gold doubled every 4 years. In 2001, the average price of gold was about $270 per ounce.

Year	Average Price of an Ounce of Gold
2001	$271.04
2002	$309.73
2003	$363.38
2004	$409.72
2005	$444.74
2006	$603.46
2007	$695.39
2008	$871.96
2009	$972.35
2010	$1224.53
2011	$1571.52

a. Write an exponential model for the average price of an ounce of gold as a function of time (in years since 2001).

b. Find the inverse of the exponential model after rewriting it with a base of *e*. Describe what information the inverse gives.

c. Perform logarithmic regression on the data in the table (using the average price of gold as the independent variable and time as the dependent variable). Compare this model with the inverse model from part b.

H.O.T. Focus on Higher Order Thinking

14. Multiple Representations For the function $g(x) = \log(x - h)$, what value of the parameter h will cause the function to pass through the point $(7, 1)$? Answer the question in two different ways: once by using the function's rule, and once by thinking in terms of the function's graph.

15. Explain the Error A student drew the graph of $g(x) = 2\log_{\frac{1}{2}}(x - 2)$ as shown. Explain the error that the student made, and draw the correct graph.

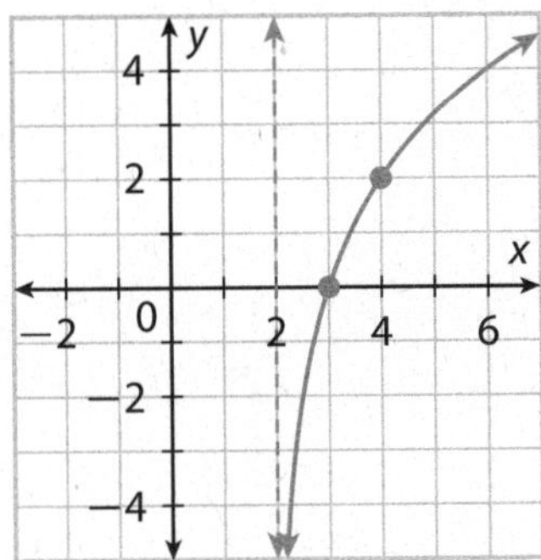

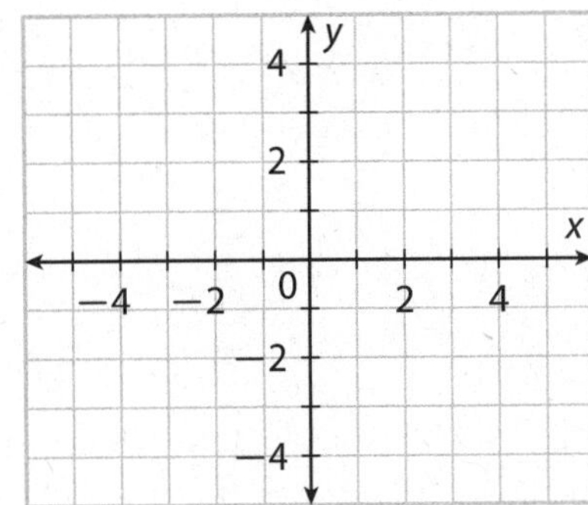

16. Construct Arguments Prove that $\log_{\frac{1}{b}} x = -\log_b x$ for any positive value of b not equal to 1. Begin the proof by setting $\log_{\frac{1}{b}} x$ equal to m and rewriting the equation in exponential form.

Lesson Performance Task

Given the following data about the heights of chair seats and table tops for children, make separate scatterplots of the ordered pairs (age of child, chair seat height) and the ordered pairs (age of child, table top height). Explain why a logarithmic model would be appropriate for each data set. Perform a logarithmic regression on each data set, and describe the transformations needed to obtain the graph of the model from the graph of the parent function $f(x) = \ln x$.

Age of Child (years)	Chair Seat Height (inches)	Table Top Height (inches)
1	5	12
1.5	6.5	14
2	8	16
3	10	18
5	12	20
7.5	14	22
11	16	25

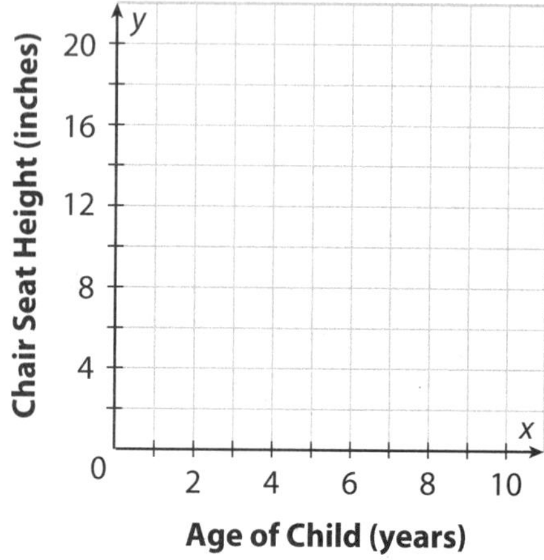

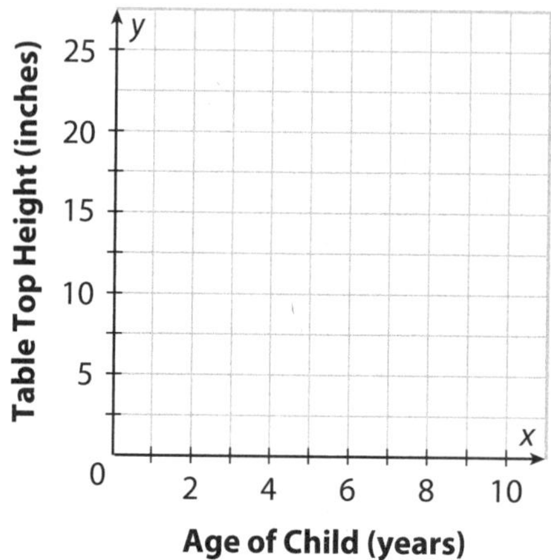

Logarithmic Functions

Essential Question: How can you use logarithmic functions to solve real-world problems?

Key Vocabulary

asymptote
(asíntota)
common logarithm
(logaritmo común)
logarithm
(logaritmo)
logarithmic function
(función logarítmica)
natural logarithm
(logaritmo natural)

KEY EXAMPLE *(Lesson 15.1)*

Evaluate $f(x) = \log_4 x$ when $x = 1024$.

$f(1024) = \log_4 1024$

$4^{f(1024)} = 1024$ Definition of logarithm

$4^{f(1024)} = 4^5$ $1024 = 4^5$

$f(1024) = 5$

KEY EXAMPLE *(Lesson 15.2)*

Graph $f(x) = 3\log_2(x + 1) - 4$.

The parameters for $f(x) = a \log_b (x - h) + k$ are:

$a = 3$

$b = 2$

$h = -1$

$k = -4$

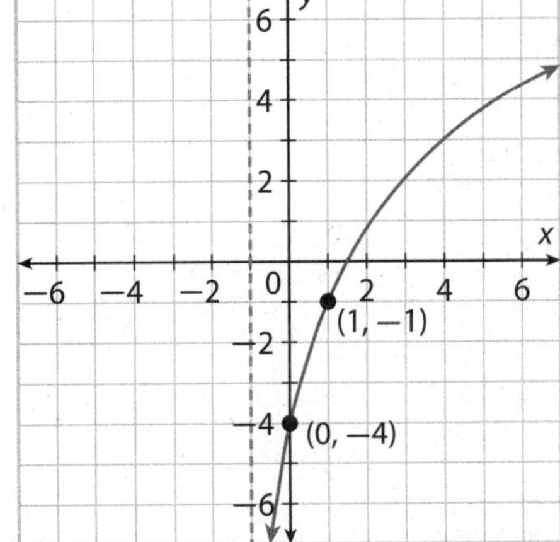

Find reference points:

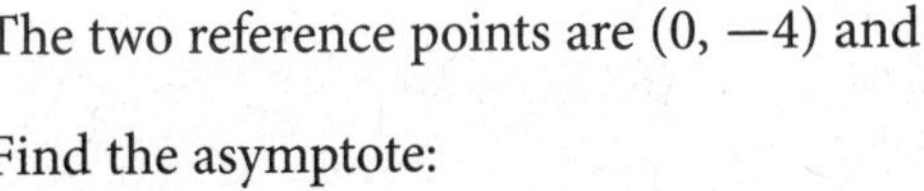

$(1 + h, k) = (0, -4)$

$(b + h, a + k) = (2 - 1, 3 - 4) = (1, -1)$

The two reference points are (0, −4) and (1, −1).

Find the asymptote:

$x = h = -1$

Plot the points and draw the asymptote. Connect the points with a curve that passes through the reference points and continually draws nearer the asymptote.

EXERCISES

Evaluate each logarithmic function for the given value. *(Lesson 15.1)*

1. $f(x) = \log_2 x$, for $f(256)$

2. $f(x) = \log_9 x$, for $f(6561)$

Graph each function. *(Lesson 15.2)*

3. $f(x) = 2 \log_3 (x - 2) + 1$

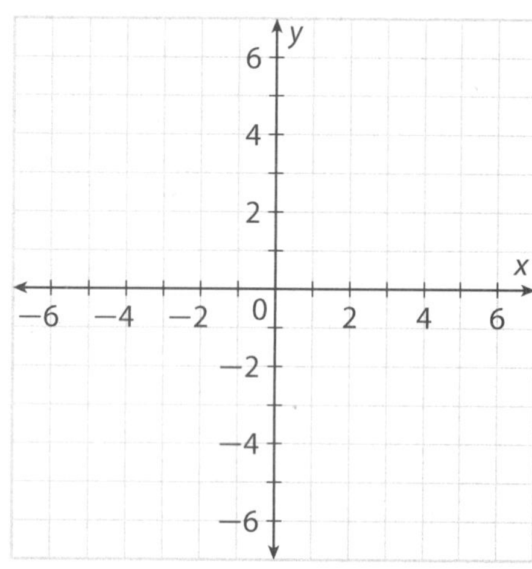

4. $f(x) = \log_5 (x + 1) - 1$

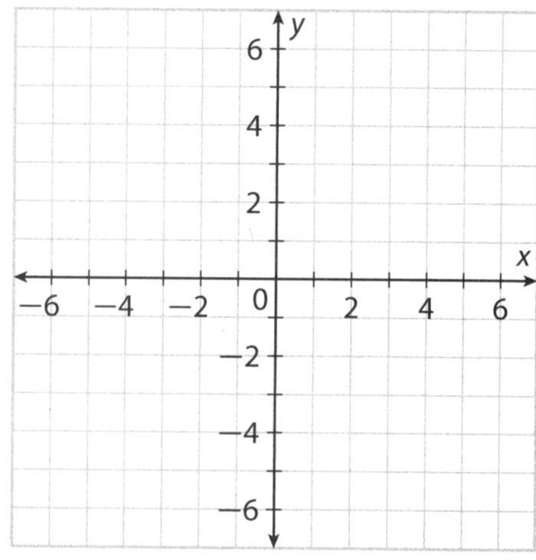

MODULE PERFORMANCE TASK

What's the Dosage?

Kira is a scientist working for a pharmaceutical lab and has developed a new drug. According to her research, 30% of the drug is eliminated from the bloodstream every 6 hours. Her initial dosage plan is to have the patient take a 1200 mg pill of the drug every 12 hours.

The patient needs to have at least 500 mg of the drug in the bloodstream at all times, but the total amount should never exceed 2500 mg. The drug should be taken for no more than 4 days. Does Kira's proposed dosage plan meet the medical requirements for the drug? Find a function that describes the amount of drug in the patient's bloodstream as a function of the number of doses.

Start by listing in the space below the information you will need to solve the problem. Then use your own paper to complete the task. Be sure to write down all your data and assumptions. Then use graphs, numbers, words, or algebra to explain how you reached your conclusion.

Ready to Go On?

15.1–15.2 Logarithmic Functions

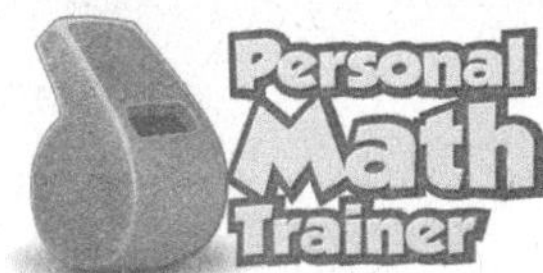

- Online Homework
- Hints and Help
- Extra Practice

Rewrite the given equation in exponential format. *(Lesson 15.1)*

1. $\log_6 x = r$

2. $\log_{\frac{3}{4}} 12x = 35y$

Evaluate each logarithmic function for the given value. *(Lesson 15.1)*

3. $f(x) = \log_5 x$ for $f(125)$

4. $f(x) = \log_3 x$ for $f(729)$

Graph each function. *(Lesson 15.2)*

5. $f(x) = -3 \log_e (x + 1) + 2$

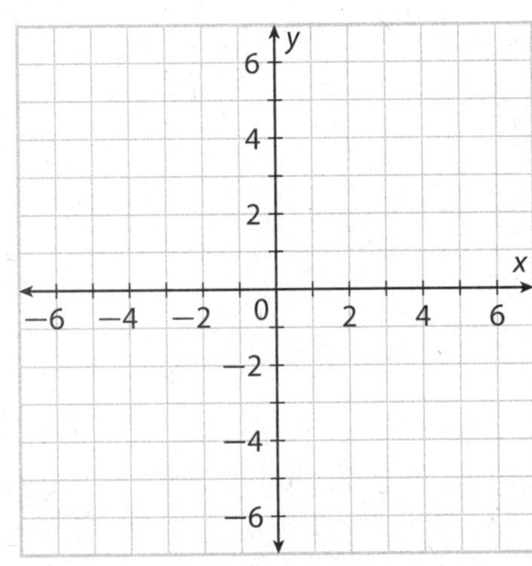

6. $f(x) = 4 \log_{10} (x + 4) - 3$

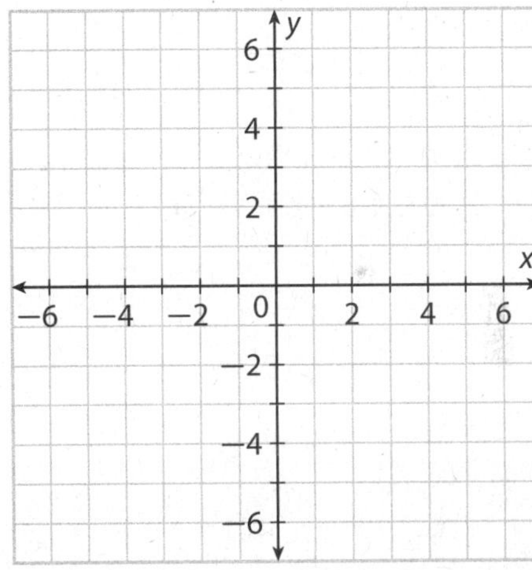

ESSENTIAL QUESTION

7. How is the graph of the logarithmic function $f(x) = \log_4 x$ related to the graph of the exponential function $g(x) = 4^x$?

MODULE 15
MIXED REVIEW

Assessment Readiness

1. What is the domain of the function $f(x) = \frac{1}{4}\log_5(x + 3) - 2$?
   A. $\{x \mid x > -3\}$
   B. $\{x \mid x > -2\}$
   C. $\{x \mid x > 2\}$
   D. $\{x \mid x > 3\}$

2. Let $f(x) = \log_4 x$. What is $f(8)$?
   A. $\frac{1}{2}$
   B. $\frac{2}{3}$
   C. $\frac{3}{2}$
   D. -2

3. Shelby is canoeing in a river. She travels 4 miles upstream and 4 miles downstream in a total of 5 hours. In still water, Shelby can travel at an average speed of 2 miles per hour. To the nearest tenth, what is the average speed of the river's current?
   A. 0.4 mph
   B. 0.6 mph
   C. 0.9 mph
   D. 1.6 mph

4. Which function shows exponential decay?
   A. $f(x) = \left(\frac{5}{4}\right)^x$
   B. $f(x) = 1.6\left(\frac{3}{4}\right)^x$
   C. $f(x) = \frac{3}{5}(1.1)^x$
   D. $f(x) = 0.2(1 + 0.03)^x$

5. Researchers have found that after 25 years of age, the average size of the pupil in a person's eye decreases. The relationship between pupil diameter d (in millimeters) and age a (in years) can be modeled by $d = -2.1158 \log_e a + 13.669$. What is the average diameter of a pupil for a person 25 years old? 50 years old? Explain how you got your answer.

Logarithmic Properties and Exponential and Logarithmic Equations

MODULE 16

TEKS

Essential Question: How do the properties of logarithms allow you to solve real-world problems?

REAL WORLD VIDEO
Scientists use radiocarbon dating and other techniques to study the fossils of mastodons and other extinct species found at the La Brea Tar Pits.

MODULE PERFORMANCE TASK PREVIEW

How Old Is That Bone?

All living organisms contain carbon. Carbon has two main isotopes, carbon-12 and carbon-14. C-14 is radioactive and decays at a steady rate. Living organisms continually replenish their stores of carbon, and the ratio between C-12 and C-14 stays relatively constant. When the organism dies, this ratio changes at a known rate as C-14 decays. How can we use a logarithmic equation and carbon dating to determine the age of a mastodon bone? Let's find out!

Are YOU Ready?

Complete these exercises to review skills you will need for this module.

Personal Math Trainer

- Online Homework
- Hints and Help
- Extra Practice

Exponents

Example 1 Simplify $\frac{40 \cdot x^6y}{5x^2y^5}$.

$\frac{40 \cdot x^6y}{5x^2y^5} = \frac{40}{5} \cdot x^{6-2}y^{1-5}$ Subtract exponents.

$= \frac{8x^4}{y^4}$ Simplify.

Simplify each expression.

1. $\frac{xy^2}{x^3y^2}$ ______
2. $\frac{18x^3y^7}{2y^5}$ ______
3. $\frac{12x^4}{8x^9y}$ ______

Multi-Step Equations

Example 2 Solve $3(5 - 2x) = -x$ for x.

$15 - 6x = -x$ Distribute the 3.

$15 = 5x$ Add $6x$ to both sides.

$3 = x$ Divide both sides by 5.

The solution is $x = 3$.

Solve.

4. $5(4x + 9) = 2x$ ______
5. $3(x + 12) = 2(4 - 2x)$ ______
6. $(x - 2)^2 = 4(x + 1)$ ______

Equations Involving Exponents

Example 3 Solve $2x^{\frac{1}{3}} - 1 = 3$ for x.

$2x^{\frac{1}{3}} = 4$ Add 1 to both sides.

$x^{\frac{1}{3}} = 2$ Divide both sides by 2.

$\left(x^{\frac{1}{3}}\right)^3 = (2)^3$ Raise both sides to the power of 3.

$x = 8$ Simplify.

Solve.

7. $3x^{\frac{1}{4}} + 2 = 11$ ______
8. $8x^{\frac{1}{2}} + 20 = 100$ ______
9. $4x^{\frac{1}{3}} + 15 = 35$ ______

Name ______________________ Class ______________ Date ________

16.1 Properties of Logarithms

Resource Locker

Essential Question: What are the properties of logarithms?

A2.5.C Rewrite exponential equations as their corresponding logarithmic equations and logarithmic equations as their corresponding exponential equations. Also A2.5.B

Explore 1 Investigating the Properties of Logarithms

You can use a scientific calculator to evaluate a logarithmic expression.

(A) Evaluate the expressions in each set using a scientific calculator.

Set A	Set B
$\log\frac{10}{e} \approx$ ______	$\frac{1}{\log e} \approx$ ______
$\ln 10 \approx$ ______	$1 + \log e \approx$ ______
$\log e^{10} \approx$ ______	$1 - \log e \approx$ ______
$\log 10e \approx$ ______	$10 \log e \approx$ ______

(B) Match the expressions in Set A to the equivalent expressions in Set B.

$\log\frac{10}{e} =$ ______

$\ln 10 =$ ______

$\log e^{10} =$ ______

$\log 10e =$ ______

Reflect

1. How can you check the results of evaluating the logarithmic expressions in Set A? Use this method to check each.

2. **Discussion** How do you know that $\log e$ and $\ln 10$ are reciprocals? Given that the expressions are reciprocals, show another way to represent each expression.

Explore 2 Proving the Properties of Logarithms

A logarithm is the exponent to which a base must be raised in order to obtain a given number. So $\log_b b^m = m$. It follows that $\log_b b^0 = 0$, so $\log_b 1 = 0$. Also, $\log_b b^1 = 1$, so $\log_b b = 1$. Additional properties of logarithms are the Product Property of Logarithms, the Quotient Property of Logarithms, the Power Property of Logarithms, and the Change of Base Property of Logarithms.

Properties of Logarithms			
For any positive numbers $a, m, n, b\ (b \neq 1)$, and $c\ (c \neq 1)$, the following properties hold.			
Definition-Based Properties	$\log_b b^m = m$	$\log_b 1 = 0$	$\log_b b = 1$
Product Property of Logarithms	$\log_b mn = \log_b m + \log_b n$		
Quotient Property of Logarithms	$\log_b \frac{m}{n} = \log_b m - \log_b n$		
Power Property of Logarithms	$\log_b m^n = n\log_b m$		
Change of Base Property of Logarithms	$\log_c a = \frac{\log_b a}{\log_b c}$		

Given positive numbers m, n, and $b\ (b \neq 1)$, prove the Product Property of Logarithms.

Ⓐ Let $x = \log_b m$ and $y = \log_b n$. Rewrite the expressions in exponential form.

$m =$ ☐

$n =$ ☐

Ⓑ Substitute for m and n.

$\log_b mn = \log_b \left(\square \right)$

Ⓒ Use the Product of Powers Property of Exponents to simplify.

$\log_b \left(b^x \cdot b^y\right) = \log_b b^{\square}$

Ⓓ Use the definition of a logarithm $\log_b b^m = m$ to simplify further.

$\log_b b^{x+y} =$ ☐

Ⓔ Substitute for x and y.

$x + y =$ ☐

Reflect

3. Prove the Power Property of Logarithms. Justify each step of your proof.

Explain 1 Using the Properties of Logarithms

Logarithmic expressions can be rewritten using one or more of the properties of logarithms.

Example 1 **Express each expression as a single logarithm. Simplify if possible. Then check your results by converting to exponential form and evaluating.**

(A) $\log_3 27 - \log_3 81$

$\log_3 27 - \log_3 81 = \log_3\left(\frac{27}{81}\right)$	Quotient Property of Logarithms
$= \log_3\left(\frac{1}{3}\right)$	Simplify.
$= \dfrac{\log\left(\frac{1}{3}\right)}{\log 3}$	Change of Base Property of Logarithms
$\approx \dfrac{-0.477}{0.477}$	Evaluate the logarithms.
$= -1$	Simplify.

Check:

$$\log_3\left(\frac{1}{3}\right) = -1$$
$$\frac{1}{3} = 3^{-1}$$
$$\frac{1}{3} = \frac{1}{3}$$

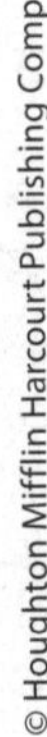

(B) $\log_5\left(\frac{1}{25}\right) + \log_5 625$

$\log_5\left(\frac{1}{25}\right) + \log_5 625 = \log_5\left(\frac{1}{25} \square 625\right)$ ______ Property of Logarithms

$= \log_5 \square$ Simplify.

$= \frac{\log \square}{\log \square}$ Change of Base Property of Logarithms

$\approx \frac{\square}{\square}$ Evaluate the logarithms.

$= \square$ Simplify

Check:

$\log_5 25 = \square$

$25 = 5^{\square}$

$25 = \square$

Your Turn

Express each expression as a single logarithm. Simplify if possible.

4. $\log_4 64^3$

5. $\log_8 18 - \log_8 2$

Explain 2 Rewriting a Logarithmic Model

There are standard formulas that involve logarithms, such as the formula for measuring the loudness of sounds. The loudness of a sound $L(I)$, in decibels, is given by the function $L(I) = 10\log\left(\frac{I}{I_0}\right)$, where I is the sound's intensity in watts per square meter and I_0 is the intensity of a barely audible sound. It's also possible to develop logarithmic models from exponential growth or decay models of the form $f(t) = a(1 + r)^t$ or $f(t) = a(1 - r)^t$ by finding the inverse.

Example 2 **Solve the problems using logarithmic models.**

(A) During a concert, an orchestra plays a piece of music in which its volume increases from one measure to the next, tripling the sound's intensity. Find how many decibels the loudness of the sound increases between the two measures.

Let I be the intensity in the first measure. So $3I$ is the intensity in the second measure.

Increase in loudness $= L(3I) - L(I)$	Write the expression.
$= 10\log\left(\frac{3I}{I_0}\right) - 10\log\left(\frac{I}{I_0}\right)$	Substitute.
$= 10\left(\log\left(\frac{3I}{I_0}\right) - \log\left(\frac{I}{I_0}\right)\right)$	Distributive Property
$= 10\left(\log 3 + \log\left(\frac{I}{I_0}\right) - \log\left(\frac{I}{I_0}\right)\right)$	Product Property of Logarithms
$= 10\log 3$	Simplify.
≈ 4.77	Evaluate the logarithm.

So the loudness of sound increases by about 4.77 decibels.

(B) The population of the United States in 2012 was 313.9 million. If the population increases exponentially at an average rate of 1% each year, how long will it take for the population to double?

The exponential growth model is $P = P_0(1+ r)^t$, where P is the population in millions after t years, P_0 is the population in 2012, and r is the average growth rate.

$P_0 = 313.9$

$P = 2P_0 = \square$

$r = 0.01$

Find the inverse model of $P = P_0(1 + r)^t$.

$P = P_0(1 + r)^t$	Exponential model
$\frac{P}{P_0} = (1 + r)^t$	Divide both sides by P_0.
$\log_{1+r}\left(\frac{P}{P_0}\right) = \log_{\square}(1 + r)^t$	Take the log of both sides.

$$\log_{1+r}\left(\frac{\square}{\square}\right) = t$$ Definition of a logarithm

$$\frac{\log\left(\frac{\square}{\square}\right)}{\log\left(\square\right)} = t$$ Change of Base Property of Logarithms

Substitute and solve for t.

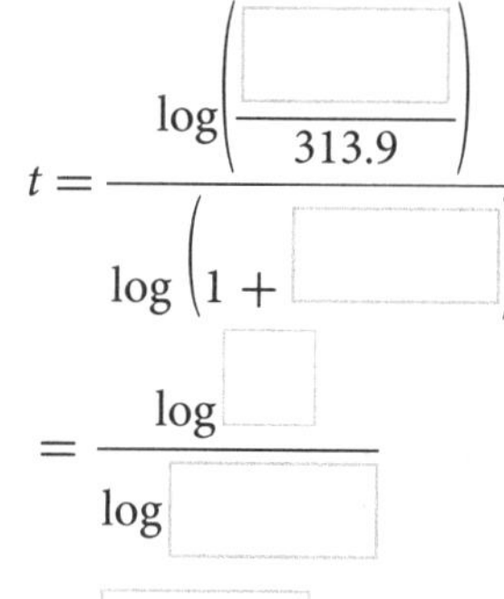

$$t = \frac{\log\left(\frac{\square}{313.9}\right)}{\log\left(1 + \square\right)}$$ Substitute.

$$= \frac{\log \square}{\log \square}$$ Simplify.

$$= \frac{\square}{\square}$$ Evaluate the logarithms.

$$= \square$$ Simplify.

The population of the United States will double in $\square$ years from 2012, or in the year $\square$.

Your Turn

6. A bank account earns 0.06% annual interest compounded monthly. The balance B of the account after t months is given by the equation $B = B_0(1.06)^t$, where B_0 is the starting balance. If the account starts with a balance of \$250, how long will it take to triple the balance of the account?

Elaborate

7. On what other properties do the proofs of the properties of logarithms rely?

8. What properties of logarithms would you use to rewrite the expression $\log_7 x + \log_7 4x$ as a single logarithm?

9. Explain how the properties of logarithms are useful in finding the inverse of an exponential growth or decay model.

10. **Essential Question Check-In** State each of the Product, Quotient, and Power Properties of Logarithms in a simple sentence.

Evaluate: Homework and Practice

- Online Homework
- Hints and Help
- Extra Practice

Express each expression as a single logarithm. Simplify if possible.

1. $\log_9 12 + \log_9 546.75$

2. $\log_2 76.8 - \log_2 1.2$

3. $\log_{\frac{2}{5}} 0.0256^3$

4. $\log_{11} 11^{23}$

5. $\log_5 5^{x+1} + \log_4 256^2$

6. $\log\left(\log_7 98 - \log_7 2\right)^x$

7. $\log_{x+1}\left(x^2 + 2x + 1\right)^3$

8. $\log_4 5 + \log_4 12 - \log_4 3.75$

Solve the problems using logarithmic models.

9. **Geology** Seismologists use the Richter scale to express the energy, or magnitude, of an earthquake. The Richter magnitude of an earthquake M is related to the energy released in ergs E shown by the formula $M = \frac{2}{3}\log\left(\frac{E}{10^{11.8}}\right)$. In 1964, an earthquake centered at Prince William Sound, Alaska, registered a magnitude of 9.2 on the Richter scale. Find the energy released by the earthquake.

10. **Astronomy** The difference between the apparent magnitude (brightness) m of a star and its absolute magnitude M is given by the formula $m - M = 5\log\frac{d}{10}$, where d is the distance of the star from the Earth, measured in parsecs. Find the distance d of the star Rho Oph from Earth, where Rho Oph has an apparent magnitude of 5.0 and an absolute magnitude −0.4.

11. The intensity of the sound of a conversation ranges from 10^{-10} watts per square meter to 10^{-6} watts per square meter. What is the range in the loudness of the conversation? Use $I_0 = 10^{-12}$ watts per square meter.

12. The intensity of sound from the stands of a football game is 25 times as great when the home team scores a touchdown as it is when the away team scores. Find the difference in the loudness of the sound when the two teams score.

13. **Finance** A stock priced at $40 increases at a rate of 8% per year. Write and evaluate a logarithmic expression for the number of years that it will take for the value of the stock to reach $50.

14. Suppose that the population of one endangered species decreases at a rate of 4% per year. In one habitat, the current population of the species is 143. After how long will the population drop below 30?

15. The population P of bacteria in a culture after t minutes is given by the equation $P = P_0\left(1.12\right)^t$, where P_0 is the initial population. If the number of bacteria starts at 200, how long will it take for the population to increase to 1000?

16. Chemistry Most swimming pool experts recommend a pH of between 7.0 and 7.6 for water in a swimming pool. Use $\text{pH} = -\log\left[\text{H}^+\right]$ and write an expression for the difference in hydrogen ion concentration over this pH range.

17. Match the logarithmic expressions to equivalent expressions.

a. $\log_2 4x$	______	$2x$
b. $\log_2 \frac{x}{4}$	______	$2 + \log_2 x$
c. $\log_2 4^x$	______	$\frac{\log x}{\log 2}$
d. $\log_2 x^4$	______	$4\log_2 x$
e. $\log_2 x$	______	$\log_2 x - 2$

18. Prove the Quotient Property of Logarithms. Justify each step of your proof.

19. Prove the Change of Base Property of Logarithms. Justify each step of your proof.

H.O.T. Focus on Higher Order Thinking

20. Multi-Step The radioactive isotope Carbon-14 decays exponentially at a rate of 0.0121% each year.

a. How long will it take 250 g of Carbon-14 to decay to 100 g?

b. The half-life for a radioactive isotope is the amount of time it takes for the isotope to reach half its initial value. What is the half-life of Carbon-14?

21. Explain the Error A student simplified the expression $\log_2 8 + \log_3 27$ as shown. Explain and correct the student's error.

$$\log_2 8 + \log_3 27 = \log(8 \cdot 27)$$
$$= \log(216)$$
$$\approx 2.33$$

22. Communicate Mathematical Ideas Explain why it is not necessary for a scientific calculator to have both a key for common logs and a key for natural logs.

23. Analyze Relationships Explain how to find the relationship between $\log_b a$ and $\log_{\frac{1}{b}} a$.

Lesson Performance Task

Given the population data for the state of Texas from 1920–2010, perform exponential regression to obtain an exponential growth model for population as a function of time (represent 1920 as Year 0).

Obtain a logarithmic model for time as a function of population two ways: (1) by finding the inverse of the exponential model, and (2) by performing logarithmic regression on the same set of data but using population as the independent variable and time as the dependent variable. Then confirm that the two expressions are equivalent by applying the properties of logarithms.

Year	U.S. Census Count
1920	4,663,228
1930	5,824,715
1940	6,414,824
1950	7,711,194
1960	9,579,677
1970	11,196,730
1980	14,229,191
1990	16,986,335
2000	20,851,820
2010	25,145,561

Name____________________ Class__________ Date________

16.2 Solving Exponential Equations

Resource Locker

Essential Question: What are some ways you can solve an equation of the form $ab^x = c$, where a and c are nonzero real numbers and b is greater than 0 and not equal to 1?

A2.5.D Solve exponential equations of the form $y = ab^x$ where a is a nonzero real number and b is greater than zero and not equal to one...

Explore Solving Exponential Equations Graphically

One way to solve exponential equations is graphically. First, graph each side of the equation separately. The point(s) at which the two graphs intersect are the solutions of the equation.

(A) First, look at the equation $275e^{0.06x} = 1000$. To solve the equation graphically, split it into two separate equations.

$y_1 =$ ☐

$y_2 =$ ☐

(B) What will the graphs of y_1 and y_2 look like?

____

____

(C) Graph y_1 and y_2 using a graphing calculator.

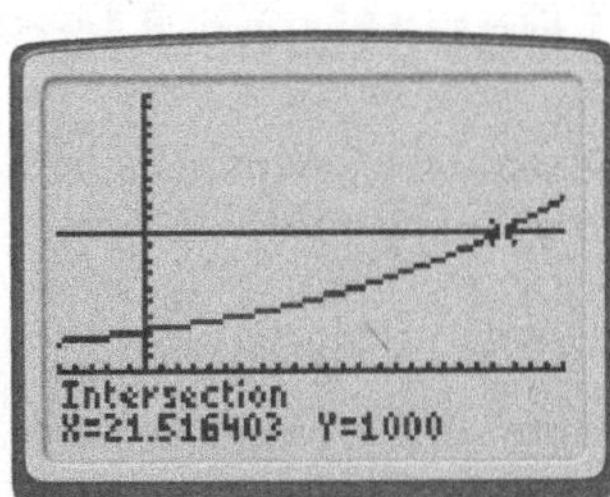

(D) The x-coordinate of the point of intersection is approximately ☐.

(E) So, the solution of the equation is $x \approx$ ☐.

(F) Now, look at the equation $10^{2x} = 10^4$. Split the equation into two separate equations.

$y_1 =$ ☐

$y_2 =$ ☐

(G) What will the graphs of y_1 and y_2 look like?

(H) Graph y_1 and y_2 using a graphing calculator.

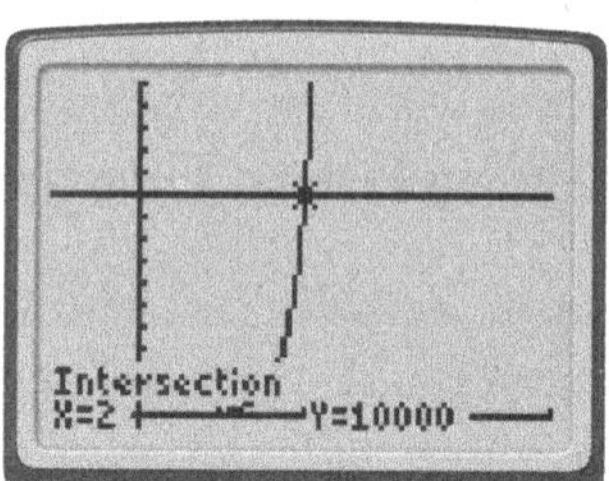

(I) The x-coordinate of the point of intersection is ☐.

(J) So, the solution of the equation is $x \approx$ ☐.

Reflect

1. How can you check the solution of an exponential equation after it is found graphically?

Explain 1 Solving Exponential Equations Algebraically

In addition to solving exponential equations graphically, exponential equations can be solved algebraically. The Property of Equality for Logarithmic Equations states that for any positive numbers x, y, and b, $(b \neq 1)$, $\log_b x = \log_b y$ if and only if $w = y$.

Example 1 **Solve the equations. Give the exact solution and an approximate solution to three decimal places.**

(A) $10 = 5e^{4x}$

$10 = 5e^{4x}$	Original equation
$2 = e^{4x}$	Divide both sides by 5.
$\ln 2 = \ln e^{4x}$	Take the natural logarithm of both sides.
$\ln 2 = 4x \ln e$	Power Property of Logarithms
$\ln 2 = 4x$	Simplify $\ln e$.
$\frac{\ln 2}{4} = \frac{4x}{4}$	Divide both sides by 4.
$\frac{\ln 2}{4} = x$	Simplify.
$0.173 \approx x$	Evaluate. Round to three decimal palces.

(B) $5^x - 4 = 7$

$5^x - 4 = 7$		Original equation
$5^x - 4 + \square = 7 + \square$		Add $\square$ to both sides.
$5^x = \square$		Simplify.
$\log 5^x = \log \square$		Take the common logarithm of both sides.
$\square = \log 11$		Power Property of Logarithms
$x = \dfrac{\log \square}{\log \square}$		Divide both sides by $\log 5$.
$x \approx \square$		Evaluate. Round to three decimal palces.

Reflect

2. Consider the equation $2^{x-3} = 85$. How can you solve this equation using logarithm base 2?

3. **Discussion** When solving an exponential equation with base *e*, what is the benefit of taking the natural logarithm of both sides of the equation?

Your Turn

Solve the equations. Give the exact solution and an approximate solution to three decimal places.

4. $2e^{x-1} + 5 = 80$

5. $6^{3x} = 12$

Explain 2 Solve a Real-World Problem by Solving an Exponential Equation

Suppose that \$250 is deposited into an account that pays 4.5% compounded quarterly. The equation $A = P\left(1 + \frac{r}{4}\right)^n$ gives the amount A in the account after n quarters for an initial investment P that earns interest at a rate r. Solve for n to find how long it will take for the account to contain at least \$500.

Analyze Information

Identify the important information.

- The initial investment P is \$ ____.
- The interest rate is ____ %, so r is ____.
- The amount A in the account after n quarters is \$ ____.

Formulate a Plan

Solve the equation for $A = P\left(1 + \frac{r}{4}\right)^n$ for ____ by substituting in the known information and using logarithms.

Solve

$\Box = \Box\left(1 + \frac{\Box}{4}\right)^n$ Substitute.

$\Box = \left(1 + \frac{0.045}{4}\right)^n$ Divide both sides by 250.

$2 = \Box^n$ Evaluate the expression in parentheses.

$\log 2 = \log 1.01125^n$ Take the common logarithm of both sides.

$\log 2 = \Box \log \Box$ Power Property of Logarithms

$\frac{\log \Box}{\log \Box} = n$ Divide both sides by log 1.01125.

$\Box \approx n$ Evaluate.

Justify and Evaluate

It will take about $\Box$ quarters, or about $\Box$ years, for the account to contain at least $500.

Check by substituting this value for n in the equation and solving for A.

$A = 250\left(1 + \frac{0.045}{4}\right)^{\Box}$ Substitute.

$= 250(\Box)^{61.96}$ Evaluate the expression in parentheses.

$\approx 250(\Box)$ Evaluate the exponent.

$\approx \Box$ Multiply.

So, the answer is reasonable.

Your Turn

6. How long will it take to triple a $250 initial investment in an account that pays 4.5% compounded quarterly?

Elaborate

7. Describe how to solve an exponential equation graphically.

8. **Essential Question Check-In** Describe how to solve an exponential equation algebraically.

Evaluate: Homework and Practice

- Online Homework
- Hints and Help
- Extra Practice

Solve the equations graphically.

1. $4e^{0.1x} = 60$

2. $120e^{2x} = 75e^{3x}$

3. $5 = 625e^{0.02x}$

Solve the equations graphically. Then check your solutions algebraically.

4. $10e^{6x} = 5e^{-3x}$

5. $450e^{0.4x} = 2000$

6. $500e^{\frac{1}{3}x} = 225e^{\frac{2}{3}x}$

Solve the equations. Give the exact solution and an approximate solution to three decimal places.

7. $6^{3x-9} - 10 = -3$

8. $7e^{3x} = 42$

9. $11^{6x+2} = 12$

10. $e^{\frac{2x-1}{3}} = 250$

11. $(10^x)^2 + 90 = 105$

12. $5^{\frac{x}{4}} = 30$

Solve.

13. The price P of a gallon of gas after t years is given by the equation $P = P_0(1 + r)^t$, where P_0 is the initial price of gas and r is the rate of inflation. If the price of a gallon of gas is currently \$3.25, how long will it take for the price to rise to \$4.00 if the rate of inflation is 10.5%?

14. **Finance** The amount A in a bank account after t years is given by the equation $A = A_0\left(1 + \frac{r}{6}\right)^{6t}$, where A_0 is the initial amount and r is the interest rate. Suppose there is \$600 in the account. If the interest rate is 4%, after how many years will the amount triple?

15. A baseball player has a 25% chance of hitting a home run during a game. For how many games will the probability of hitting a home run in every game drop to 5%?

16. **Meteorology** In one part of the atmosphere where the temperature is a constant −70 °F, pressure can be expressed as a function of altitude by the equation $P(h) = 128(10)^{-0.682h}$, where P is the atmospheric pressure in kilopascals (kPa) and h is the altitude in kilometers above sea level. The pressure ranges from 2.55 kPa to 22.9 kPa in this region. What is the range of altitudes?

17. You can choose a prize of either a \$20,000 car or one penny on the first day, double that (2 cents) on the second day, and so on for a month. On what day would you receive at least the value of the car?

18. Population The population of a small coastal resort town, currently 3400, grows at a rate of 3% per year. This growth can be expressed by the exponential equation $P = 3400(1 + 0.03)^t$, where P is the population after t years. Find the number of years it will take for the population to reach 10,000.

19. A veterinarian has instructed Harrison to give his 75-lb dog one 325-mg aspirin tablet for arthritis. The amount of aspirin A remaining in the dog's body after t minutes can be expressed by $A = 325\left(\frac{1}{2}\right)^{\frac{t}{15}}$. How long will it take for the amount of aspirin to drop to 50 mg?

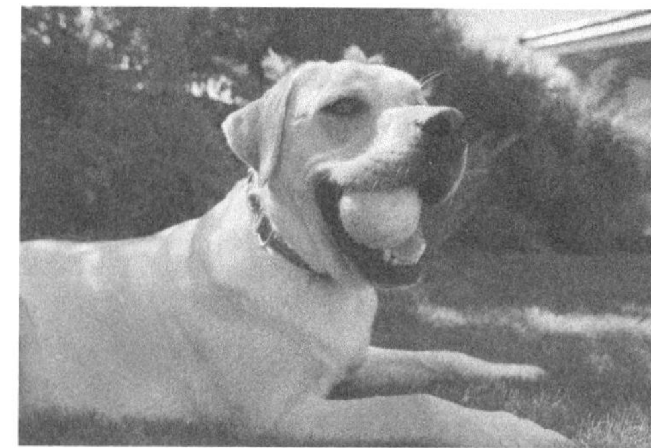

20. **Agriculture** The number of farms in Iowa (in thousands) can be modeled by $N(t) = 119(0.987)^t$, where t is the number of years since 1980. According to the model, when will the number of farms in Iowa be about 80,000?

21. Match the equations with the solutions.

a. $9e^{3x} = 27$ ________ $x \approx 1.099$

b. $9e^{x} = 27$ ________ $x \approx 1.022$

c. $9e^{3x-4} = 27$ ________ $x \approx 0.366$

d. $9e^{3x} + 2 = 27$ ________ $x \approx 1.700$

H.O.T. Focus on Higher Order Thinking

22. **Explain the Error** A student solved the equation $e^{4x} - 6 = 10$ as shown. Find and correct the student's mistake. Is there an easier way to solve the problem? Verify that both methods result in the same answer.

$$\begin{aligned} e^{4x} - 6 &= 10 \\ e^{4x} &= 16 \\ \log e^{4x} &= \log 16 \\ 4x \log e &= \log 16 \\ 4x(1) &= \log 16 \\ x &= \frac{\log 16}{4} \\ x &\approx 0.301 \end{aligned}$$

23. **Multi-Step** The amount A in an account after t years is given by the equation $A = Pe^{rt}$, where P is the initial amount and r is the interest rate.

a. Find an equation that models approximately how long it will take for the initial amount P in the account to double with the interest rate r. Write the equation in terms of the interest rate expressed as a percent.

b. The Rule of 72 states that you can find the approximate time it will take to double your money by dividing 72 by the interest rate. The rule uses 72 instead of 69 because 72 has more divisors, making it easier to calculate mentally. Use the Rule of 72 to find the approximate time it takes to double an initial investment of $300 with an interest rate of 3.75%. Determine that this result is reasonable by solving the equation $A = P_0(1.0375)^t$, where A is the amount after t years and P_0 is the initial investment.

24. Represent Real-World Problems Suppose you have an initial mass M_0 of a radioactive substance with a half-life of h. Then the mass of the parent isotopes at time t is $P(t) = M_0\left(\frac{1}{2}\right)^{\frac{t}{h}}$. Since the substance is decaying from the original parent isotopes into the new daughter isotopes while the mass of all the isotopes remains constant, the mass of the daughter isotopes at time t is $D(t) = M_0 - P(t)$. Find when the masses of the parent isotopes and daughter isotopes are equal. Explain the meaning of your answer and why it makes sense.

Lesson Performance Task

The frequency of a note on the piano, in Hz, is related to its position on the keyboard by the function $f(n) = 440 \cdot 2^{\frac{n}{12}}$, where n is the number of keys above or below the note concert A, concert A being the A key above middle C on the piano. Using this function, find the position n of the key that has a frequency of 110 Hz. Why is this number a negative value?

Name________________________ Class______________ Date________

16.3 Solving Logarithmic Equations

Essential Question: What are some ways you can solve logarithmic equations?

TEKS **A2.5.D** Solve … single logarithmic equations having real solutions. Also A2.5.E

Resource Locker

Explore Solving Logarithmic Equations Graphically

One way to solve logarithmic equations is graphically. First, graph each side of the equation separately. The point(s) at which the two graphs intersect are the solutions of the equation.

(A) Look at the equation $12.2 + 5.45\ln x = 12.5 + 5.2\ln x$. To solve the equation graphically, split it into two separate equations.

$y_1 =$ ______

$y_2 =$ ______

(B) What will the graphs of y_1 and y_2 look like?

____

(C) Graph y_1 and y_2 using a graphing calculator.

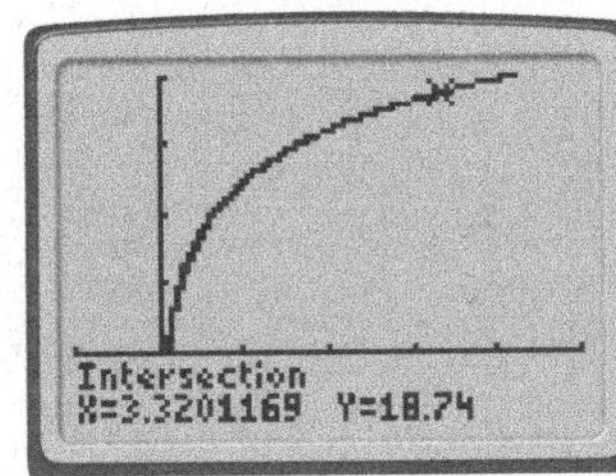

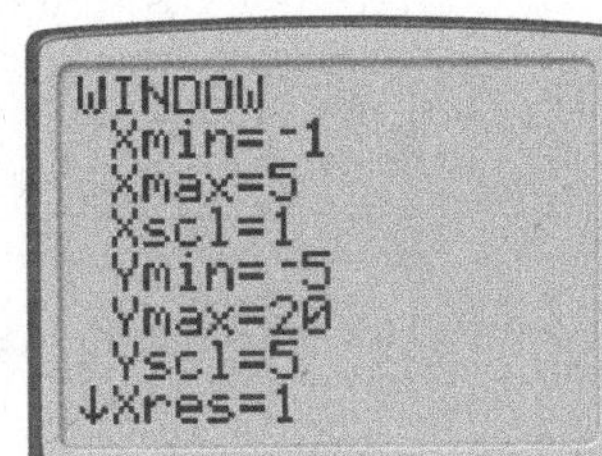

(D) The x-coordinate of the point of intersection is approximately ______.

(E) So, the solution of the equation is $x \approx$ ______.

Reflect

1. How can you check the solution of a logarithmic equation after it is found graphically?

____

____

2. How would you graph and solve a logarithmic equation where the base of the logarithmic function is not 10 or e if your calculator only graphed those bases?

____

____

Explain 1 Solving Logarithmic Equations Algebraically

In addition to solving logarithmic equations graphically, logarithmic equations can be solved algebraically. The inverse relationship between logarithmic and exponential functions allows you to rewrite $\log_b x = a$ as $b^a = x$.

Example 1 **Solve the equations. Check for extraneous solutions.**

(A) $7 + \log_3(5x - 4) = 10$

$7 + \log_3(5x - 4) = 10$	Original equation
$7 - 7 + \log_3(5x - 4) = 10 - 7$	Subtract 7 from both sides.
$\log_3(5x - 4) = 3$	Simplify.
$3^3 = 5x - 4$	Definition of a logarithm
$27 = 5x - 4$	Evaluate the exponent.
$31 = 5x$	Add 4 to both sides.
$6.2 = x$	Divide both sides by 5.

Check:

$$7 + \log_3\left(5(6.2) - 4\right) = 10$$
$$7 + \log_3(31 - 4) = 10$$
$$7 + \log_3(27) = 10$$
$$7 + \frac{\log 27}{\log 3} = 10$$
$$7 + 3 = 10$$
$$10 = 10$$

The solution is $x = 6.2$.

(B) $\log x + \log(x + 9) = 1$

$\log x + \log(x + 9) = 1$	Original equation
$\log\left(x \boxed{} (x + 9)\right) = 1$	Product Property of Logarithms
$\log\left(\boxed{}\right) = 1$	Multiply.
$\boxed{}^{1} = x^2 + 9x$	Definition of a logarithm
$\boxed{} = x^2 + 9x$	Evaluate the exponent.
$\boxed{} = x^2 + 9x - \boxed{}$	Subtract 10 from both sides.
$0 = \left(x + \boxed{}\right)\left(x - \boxed{}\right)$	Factor.
$x = \boxed{}$ or $x = \boxed{}$	Solve.

Check:

$\log(-10) + \log(-10 + 9) = 1$

$\log(\square) + \log(\square) = 1$

$\square = 1$

$x = \square$ is an extraneous solution.

$\log 1 + \log(1 + 9) = 1$

$\log 1 + \log \square = 1$

$\square + \square = 1$

$\square = 1$

The solution is $x = \square$.

Reflect

3. Explain how you would solve the equation $\log_5 45x = 1 + \log_5 3$. Then solve.

Your Turn

Solve the equations. Check for extraneous solutions.

4. $\log_4(2x + 12) + 5 = 8$

Your Turn

5. $\log_6(x - 5) = 2 - \log_6 x$

Explain 2 Solving a Real-World Problem Using a Logarithmic Equation

Given a logarithmic function $f(x)$ that models a real-world situation, find the value of x for which $f(x) = c$, where c is a constant. Check the reasonableness of the solution by graphing.

Example 2 **Solve using properties of logarithms. Then check the reasonableness of the solution by graphing.**

(A) The energy released by an earthquake can be a very large number, so a much smaller number is reported as the magnitude of the earthquake. The magnitude M of an earthquake with energy E (in ergs) is calculated using the formula $M = \frac{2}{3}\log E - 2.9$. The largest earthquake known to have occurred in Texas happened on August 16, 1931 and had a magnitude of 5.8. How much energy did the earthquake have?

Let $M = 5.8$. Substitute the value for M into the formula for the magnitude and solve for E.

$5.8 = \frac{2}{3}\log E - 2.9$ Substitute.

$5.8 + 2.9 = \frac{2}{3}\log E - 2.9 + 2.9$ Add 2.9 to both sides.

$8.7 = \frac{2}{3}\log E$	Simplify.
$13.05 = \log E$	Multiply both sides by $\frac{3}{2}$.
$E \approx 10^{13.05}$	Definition of a logarithm
$E \approx 11{,}200{,}000{,}000{,}000$	Evaluate the exponent

The earthquake had about 11,200,000,000,000 ergs of energy.

Check the reasonableness of the solution by graphing both sides of the equation on a graphing calculator.

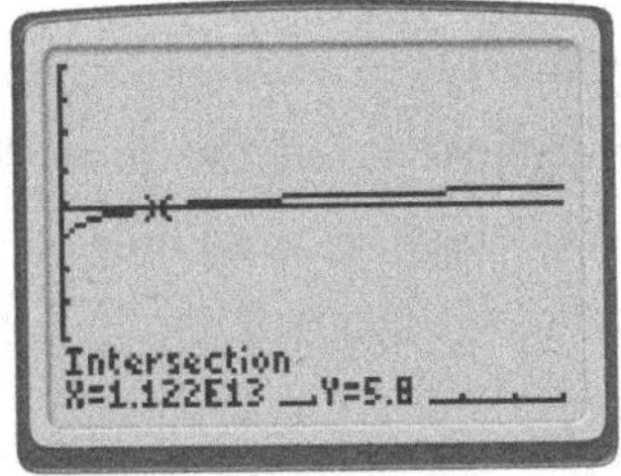

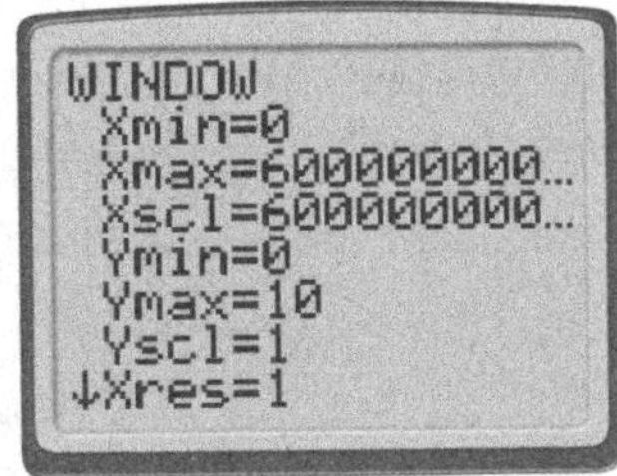

The two graphs intersect at approximately $x = 11{,}200{,}000{,}000{,}000$, so the answer is reasonable.

B An earthquake that occurred on December 7, 2013, near Volcano, Hawaii, had a magnitude of 2.9. How much energy did the earthquake have?

Let $M = 2.9$. Substitute the value for M into the formula for the magnitude and solve for E.

$\square = \frac{2}{3}\log E - 2.9$	Substitute.
$2.9 + \square = \frac{2}{3}\log E - 2.9 + \square$	Add 2.9 to both sides.
$\square = \frac{2}{3}\log E$	Simplify.
$\square = \log E$	Multiply both sides by $\frac{3}{2}$.
$E = \square^{\square}$	Definition of a logarithm
$E \approx \square$	Evaluate the exponent.

The earthquake had about $\square$ ergs of energy.

Check the reasonableness of the solution by graphing both sides of the equation on a graphing calculator.

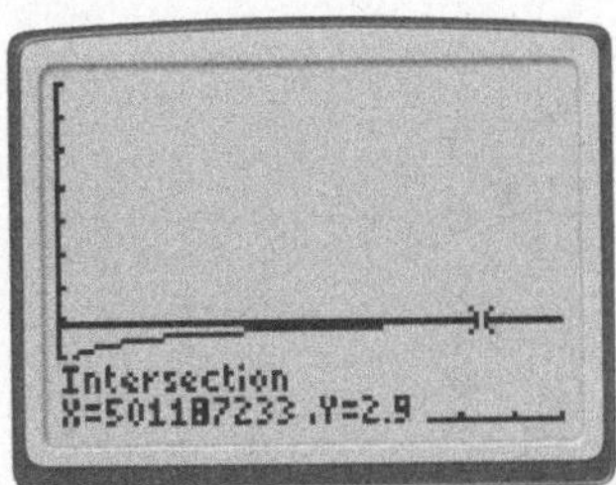

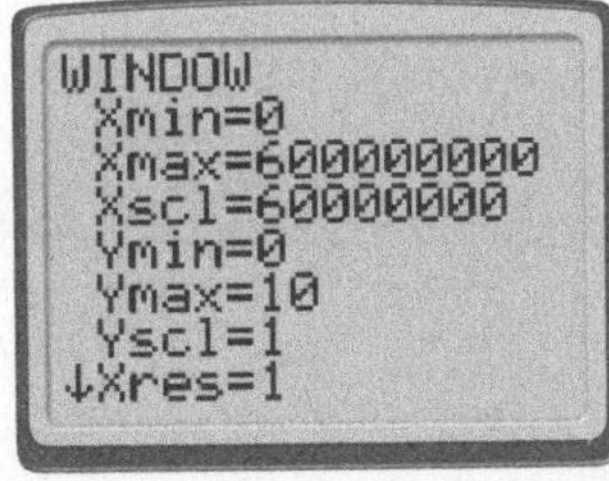

The two graphs intersect at approximately $x = \square$, so the answer is reasonable.

Reflect

6. Compare the energies of the two earthquakes.

Your Turn

7. The difference between the apparent magnitude (brightness) m of a star and its absolute magnitude M is given by the formula $m - M = 5\log\frac{d}{10}$, where d is the distance of the star from Earth, measured in parsecs. The star Antares has an apparent magnitude of 1.0 and an absolute magnitude of -5.3. Find the distance of Antares from Earth. Check the reasonableness of your answer by graphing.

Elaborate

8. Explain why it's important to check the solutions of a logarithmic equation.

9. Describe how to solve an exponential equation graphically.

10. **Essential Question Check-In** Describe how to solve a logarithmic equation algebraically.

Evaluate: Homework and Practice

- Online Homework
- Hints and Help
- Extra Practice

Solve the equations graphically.

1. $8 + \log x = 2\log x - 12$

2. $\log_6 2x = 2\log_6 x + 1$

3. $10 = 3\log_2 x$

4. $9.4 - \log_5 x = 4\log_5 x + 0.5$

5. $0.2\ln x + 5 = \ln x - 6$

6. $8\log_3 x - \frac{1}{4} = \frac{3}{4}\log_3 x + 2$

Solve the equations. Check for extraneous solutions.

7. $\log_9(4x + 5) = 2$

8. $\ln 8x = \ln 2 + 5$

9. $\log_2 x + \log_2(x - 4) = 5$

10. $\log_6 x = -\left(\log_6\left(x - \frac{1}{4}\right) + 2\right)$

11. $2.4\log_4 x = \log_4 3x + 1$

12. $\log 2x + \log(x - 5) = 2$

13. $\log_5(4x + 6) = \log_5(8x - 2)$

14. $2\ln x - 0.4 = 2.5 - 5\ln x$

Solve using properties of logarithms. Then check the reasonableness of the solution by graphing.

15. Photography On many cameras, the amount of light admitted through the lens can be controlled by changing the size of the opening, or aperture. The size of the aperture is measured as an f-stop setting. The relationship between the f-stop and the amount of light admitted can be represented by the equation $n = \log_2 \frac{1}{\ell}$, where n is the change in f-stop setting from the starting value, $\frac{f}{5.6}$. Solve the equation for ℓ when the f-stop setting is increased to $\frac{f}{16}$.

F-stop Setting	$\frac{f}{2}$	$\frac{f}{2.8}$	$\frac{f}{4}$	$\frac{f}{5.6}$	$\frac{f}{8}$	$\frac{f}{11}$	$\frac{f}{16}$
Change in F-stop Setting	−3	−2	−1	0	1	2	3

16. Astronomy A telescope's limiting magnitude m is the brightness of the faintest star that can be seen using the telescope. The limiting magnitude depends on the diameter d (in millimeters) of the telescope's objective lens. What diameter lens would be needed to view a faintest star with a brightness of 21?

Formulas for Determining Limiting Magnitude from Lens Diameter	
Standard formula	$m = 2.7 + 5\log d$

17. The equation for finding pH levels is $\text{pH} = -\log[\text{H}^+]$, where H^+ is the hydrogen ion concentration. Cow's milk has a pH level of 6.7 and goat's milk has a pH of 6.48. Find the difference in the hydrogen ion concentration between the two types of milk.

18. The sound level at a rock concert is 115 decibels. The loudness L of sound in decibels is given by the equation $L = 10\log\left(\frac{\ell}{\ell_0}\right)$, where I is the intensity of sound and ℓ_0 is the audible sound. If $\ell_0 = 10^{-12}$ decibels, what is the intensity of the sound at the rock concert?

19. The magnitude M of an earthquake with energy E (in ergs) is calculated using the formula $M = \frac{2}{3}\log E - 2.9$. How much energy does an earthquake with a magnitude of 3.8 have?

20. The brightness m of a star is given by the formula $m = 5\log\frac{d}{10} - 1.6$, where d is the distance of the star from Earth, measured in parsecs. Find the distance the star is from Earth if the brightness of the star is 3.2.

21. Match the equations with the solutions.

A. $2\log_5 x = 10$ ________ $x = 390{,}625$

B. $\log_5 2x = 10$ ________ $x = 3125$

C. $\log_5 x + 2 = 10$ ________ $x = 1562.5$

D. $2\log_5 2x = 10$ ________ $x = 4{,}882{,}812.5$

H.O.T. Focus on Higher Order Thinking

22. **Multi-Step** Charles collected data on the atmospheric pressure (ranging from 4 to 15 pounds per square inch [psi]) and the corresponding altitude above the surface of Earth (ranging from 1 to 30,000 feet). He used exponential and linear regression to write two functions that give the altitude above the surface of the Earth given the atmospheric pressure, x.

$f(x) = 66{,}990 - 24{,}747 \ln x$

$g(x) = -2870x + 40{,}393$

a. At what atmospheric pressure(s) do the equations give the same altitude?

b. At what altitude(s) above Earth do these atmospheric pressure(s) occur?

23. Draw Conclusions Solve the equation $\log_2(7x + 1) = \log_2(2 - x)$ by subtracting the logarithms. Is there an easier way to solve this equation? Explain.

24. Explain the Error A student found the solutions of the equation $\log_6 x + \log_6(x + 5) = 2$ to be $x = 4$ and $x = -9$. Explain the error in the student's reasoning.

Lesson Performance Task

A telescope's limiting magnitude m is the brightness of the faintest star that can be seen using the telescope. The limiting magnitude depends on the diameter d (in millimeters) of the telescope's objective lens. Limiting magnitude can be calculated in different ways. The table gives two formulas relating m to d. One is a standard formula used in astronomy. The other is a proposed new formula based on data gathered from users of telescopes of various lens diameters. For what lens diameter do the two formulas give the same limiting magnitude? Solve by graphing.

Formulas for Determining Limiting Magnitude from Lens Diameter	
Standard formula	$m = 2.7 + 5\log d$
Proposed formula	$m = 4.5 + 4.4\log d$

STUDY GUIDE REVIEW
Logarithmic Properties and Exponential and Logarithmic Equations

MODULE 16

Essential Question: How do the properties of logarithms allow you to solve real-world problems?

Key Vocabulary
exponential equation
(ecuación exponencial)
logarithmic equation
(ecuación logarítmica)

KEY EXAMPLE *(Lesson 16.1)*

Simplify: $\log_5 5^{x+2} + \log_2 16^3$.

Apply properties of logarithms.

$$\log_5 5^{x+2} + \log_2 16^3 = (x+2)\log_5 5 + 3\log_2 16$$

$$= (x+2) + 3\left(\frac{\log 16}{\log 2}\right)$$ $\log_5 5 = 1$; Change of Base Property

$$\approx x + 2 + 3\left(\frac{1.204}{0.301}\right)$$ Evaluate the logarithms.

$$= x + 2 + 3(4)$$ Simplify.

$$= x + 14$$

KEY EXAMPLE *(Lesson 16.2)*

Solve the equation: $4^{3x+1} = 6$.

$$4^{3x+1} = 6$$

$$\log 4^{3x+1} = \log 6$$ Take the log of both sides.

$$(3x+1)\log 4 = \log 6$$ Power Property of Logarithms.

$$3x + 1 = \frac{\log 6}{\log 4}$$ Divide both sides by log 4.

$$3x = \frac{\log 6}{\log 4} - 1$$ Isolate the variable term.

$$x = \frac{1}{3}\left(\frac{\log 6}{\log 4} - 1\right) \approx 0.0975$$ Solve for x.

KEY EXAMPLE *(Lesson 16.3)*

Solve the equation: $\log_2(3x - 7) = 5$.

$$\log_2(3x - 7) = 5$$

$$3x - 7 = 2^5$$ Definition of logarithm.

$$3x - 7 = 32$$ Simplify.

$$x = 13$$ Add 7 to both sides and divide by 3.

EXERCISES

Use properties of logarithms to simplify. *(Lesson 16.1)*

1. $\log_{\frac{3}{5}} 0.216^4$

2. $\log_4 4^{x-2} + \log_3 243^2$

3. $\log_8 0.15625^x$

4. $\log 10^{2x+1} + \log_3 9$

Solve each equation. *(Lesson 16.2, 16.3)*

5. $5^x = 50$

6. $6^{x+2} = 45$

7. $\log_3(2x - 5) = 2$

8. $\log_4 x + \log_4(x + 6) = 2$

MODULE PERFORMANCE TASK

How Old Is That Bone?

The La Brea Tar Pits in Los Angeles contain one of the best preserved collections of Pleistocene vertebrates, including over 660 species of organisms. An archeologist working at the La Brea Tar Pits wants to assess the age of a mastodon bone fragment she discovered. She measures that the fragment has 22% as much carbon-14 as typical living tissue. Given that the half-life of carbon-14 is 5370 years, what is the bone fragment's age?

Start by listing in the space below the information you will need to solve the problem. Then use your own paper to complete the task. Be sure to write down all your data and assumptions. Then use graphs, numbers, words, or algebra to explain how you reached your conclusion.

Ready to Go On?

16.1–16.3 Logarithmic Properties and Exponential and Logarithmic Equations

- Online Homework
- Hints and Help
- Extra Practice

Use properties of logarithms to simplify. *(Lesson 16.1)*

1. $\log_{\frac{6}{5}} 2.0736^5$

2. $\log_2 3.2 - \log_2 0.025$

Solve each equation. Give the exact solution and an approximate solution to three decimal places. *(Lesson 16.2)*

3. $7^{2x} = 30$

4. $5^{2x-1} = 20$

Solve each equation. Check for extraneous solutions. *(Lesson 16.3)*

5. $3\log(x - 4) = 6$

6. $\log(6x^2) - \log 2x = 1$

ESSENTIAL QUESTION

7. How do you solve a logarithmic equation algebraically?

MODULE 16
MIXED REVIEW

Assessment Readiness

1. What is the approximate solution of the equation $8^{x+1} = 12$?
   A. 0.195
   B. 0.837
   C. 1.19
   D. 2.19

2. What is the solution of the equation $\log_9 x^2 = 5$?
   A. $\frac{5}{2}$
   B. 9
   C. 81
   D. 243

3. Which function has an inverse that is **not** a function?
   A. $f(x) = 4x^3 - 1$
   B. $f(x) = \sqrt{3x} + 2$
   C. $f(x) = 4x^2 + 2$
   D. $f(x) = 2^x - 1$

4. At a constant temperature, the pressure, P, of an enclosed gas is inversely proportional to the volume, V, of the gas. If $P = 50$ pounds per square inch when $V = 30$ cubic inches, what is the pressure when the volume is 125 cubic inches?
   A. 6 pounds per square inch
   B. 12 pounds per square inch
   C. 20 pounds per square inch
   D. 75 pounds per square inch

5. $A = P(1 + r)^n$ gives amount A in an account after n years after an initial investment P that earns interest at an annual rate r. How long will it take for \$250 to increase to \$500 at 4% annual interest? Explain how you got your answer.

UNIT 6
MIXED REVIEW

Assessment Readiness

- Online Homework
- Hints and Help
- Extra Practice

1. What is the formula for the geometric sequence 5, 10, 20, 40, 80, ... ?

A. $a_n = 5(2)^n$

B. $a_n = 10(2n)$

C. $a_n = 5(2)^{n-1}$

D. $a_n = 5(2)^{n+1}$

2. What type of function would you use to model a population that doubles every 5 years?

A. linear

B. quadratic

C. regression

D. exponential

3. Let $f(x) = \log_6 x$. What is $f(216)$?

A. 3

B. 18

C. 9

D. 648

4. What is the inverse function of $f(x) = 8x^3 + 2$?

A. $f^{-1}(x) = \dfrac{\sqrt[2]{x-2}}{2}$

B. $f^{-1}(x) = \dfrac{\sqrt[3]{x-2}}{2}$

C. $f^{-1}(x) = \dfrac{\sqrt[3]{x-2}}{8}$

D. $f^{-1}(x) = \dfrac{\sqrt[3]{x+2}}{2}$

5. What is the solution of the equation $\log_3(4x + 1) = 4$?

A. $x = \dfrac{11}{4}$

B. $x = \dfrac{7}{12}$

C. $x = 20$

D. $x = 0$

6. A circular plot of land has a radius $3x - 1$. What is the polynomial representing the area of the land? *(Lesson 7.2)*

7. The number of bacteria growing in a petri dish after n hours can be modeled by $b(t) = b_0 r^n$, where b_0 is the initial number of bacteria and r is the rate at which the bacteria grow. If the number of bacteria quadruples after 1 hour, how many hours will it take to produce 51,200 bacteria if there are initially 50 bacteria? *(Lesson 16.2)*

Performance Tasks

★ **8.** The amount of freight transported by rail in the United States was about 580 billon *ton-miles* in 1960 and has been increasing at a rate of 2.32% per year since then.

A. Write and graph a function representing the amount of freight, in billions of ton-miles, transported annually (1960 = year 0).

B. In what year would you predict that the number of ton-miles would have exceeded or would exceed 1 trillion (1000 billion)?

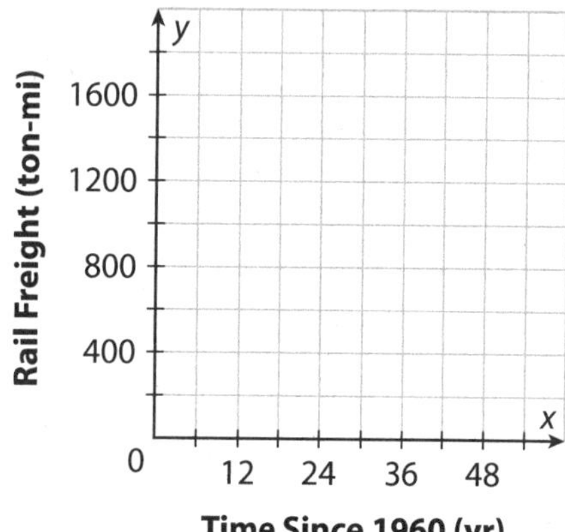

★★ **9.** In one part of the atmosphere where the temperature is a constant −70°F, pressure can be expressed as a function of altitude by the equation $P(h) = 128(10)^{-0.0682h}$, where P is the atmospheric presser in kilopascals (kPa) and h is the altitude in kilometers above sea level. The pressure ranges from 2.55 kPa to 22.9 kPa in this region.

A. What are the lowest and highest altitudes where this model is appropriate?

B. A kilopascal is 0.145 psi. Would the model predict a sea-level pressure less than or greater than the actual sea-level pressure, 14.7 psi? Explain.

★★★**10.** The loudness of sound is measured on a logarithmic scale according to the formula $L = 10\log\left(\frac{I}{I_0}\right)$, where L is the loudness of sound in decibels (dB), I is the intensity of sound, and I_0 is the intensity of the softest audible sound.

Sound	Intensity
Jet takeoff	$10^{15} \cdot I_0$
Jackhammer	$10^{12} \cdot I_0$
Hair dryer	$10^{7} \cdot I_0$
Whisper	$10^{3} \cdot I_0$
Leaves rustling	$10^{2} \cdot I_0$
Softest audible sound	I_0

A. Find the loudness in decibels of each sound listed in the table.

B. The sound at a rock concert is found to have a loudness of 110 decibels. Find the intensity of this sound. Where should this sound be placed in the table in order to preserve the order from least to greatest intensity?

C. A decibel is $\frac{1}{10}$ of a bel. Is a jet plane louder than a sound that measures 20 bels? Explain.

MATH IN CAREERS

Nuclear Medicine Technologist The radioactive properties of the isotope technetium-99m can be used in combination with a tin compound to map circulatory system disorders. Technetium-99m has a half-life of 6 hours.

a. Write an exponential decay function that models this situation. The function $p(t)$ should give the percent of the isotope remaining after t hours.

b. Describe domain, range, and the end behavior of $p(t)$ as t increases without bound for the function found in part a.

c. Write the inverse of the decay function. Use a common logarithm for your final function.

d. How long does it take until 5% of the technetium-99m remains? Round to the nearest tenth of an hour.

Glossary/Glosario

A

ENGLISH	SPANISH	EXAMPLES
absolute value of a complex number The absolute value of $a + bi$ is the distance from the origin to the point (a, b) in the complex plane and is denoted $\lvert a + bi \rvert = \sqrt{a^2 + b^2}$.	**valor absoluto de un número complejo** El valor absoluto de $a + bi$ es la distancia desde el origen hasta el punto (a, b) en el plano complejo y se expresa $\lvert a + bi \rvert = \sqrt{a^2 + b^2}$.	$\lvert 2 + 3i \rvert = \sqrt{2^2 + 3^2} = \sqrt{13}$
absolute value of a real number The absolute value of x is the distance from zero to x on a number line, denoted $\lvert x \rvert$. $\lvert x \rvert = \begin{cases} x & \text{if } x \geq 0 \\ -x & \text{if } x < 0 \end{cases}$	**valor absoluto de un número real** El valor absoluto de x es la distancia desde cero hasta x en una recta numérica y se expresa $\lvert x \rvert$. $\lvert x \rvert = \begin{cases} x & \text{si } x \geq 0 \\ -x & \text{si } x < 0 \end{cases}$	$\lvert 3 \rvert = 3$ $\lvert -3 \rvert = 3$
absolute-value function A function whose rule contains absolute-value expressions.	**función de valor absoluto** Función cuya regla contiene expresiones de valor absoluto.	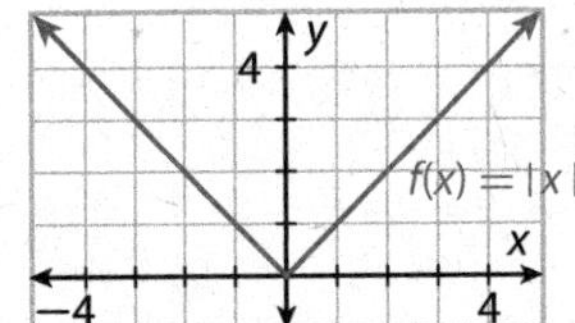
accuracy The closeness of a given measurement or value to the actual measurement or value.	**exactitud** Cercanía de una medida o un valor a la medida o el valor real.	
acute angle An angle that measures greater than 0° and less than 90°.	**ángulo agudo** Ángulo que mide más de 0° y menos de 90°.	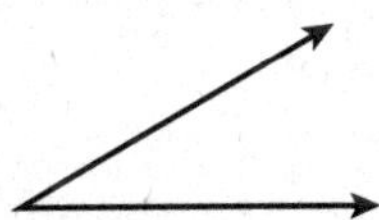
additive inverse of a matrix A matrix where each entry is the matrix. Two matrices are additive inverses if their sum is the zero matrix.	**inverso aditivo de una matriz** Matriz en la cual cada entrada es el opuesto de cada entrada en otra matriz. Dos matrices son inversos aditivos si su suma es la matriz cero.	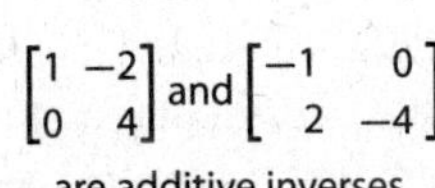are additive inverses.
address The location of an entry in a matrix, given by the row and column in which the entry appears. In matrix A, the address of the entry in row i and column j is a_{ij}.	**dirección** Ubicación de una entrada en una matriz, indicada por la fila y la columna en las que aparece la entrada. En la matriz A, la dirección de la entrada de la fila i y la columna j es a_{ij}.	In the matrix $A = \begin{bmatrix} 2 & 3 \\ 4 & 1 \end{bmatrix}$, the address of the entry 2 is a_{11}, the address of the entry 3 is a_{12}.
amplitude The amplitude of a periodic function is half the difference of the maximum and minimum values (always positive).	**amplitud** La amplitud de una función periódica es la mitad de la diferencia entre los valores máximo y mínimo (siempre positivos).	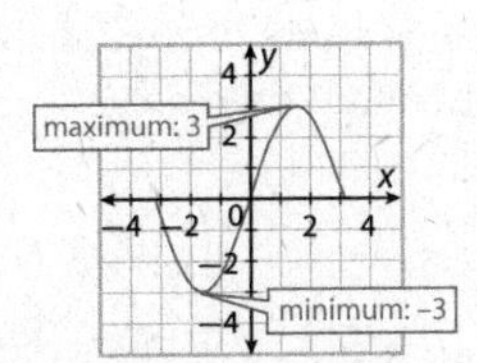amplitude $= \frac{1}{2}\left[3 - (-3)\right] = 3$

ENGLISH	SPANISH	EXAMPLES
angle of depression The angle formed by a horizontal line and a line of sight to a point below.	**ángulo de depresión** Ángulo formado por una recta horizontal y una línea visual a un punto inferior.	
angle of elevation The angle formed by a horizontal line and a line of sight to a point above.	**ángulo de elevación** Ángulo formado por una recta horizontal y una línea visual a un punto superior.	
angle of rotation An angle formed by a rotating ray, called the terminal side, and a stationary reference ray, called the initial side.	**ángulo de rotación** Ángulo formado por un rayo en rotación, denominado lado terminal, y un rayo de referencia estático, denominado lado inicial.	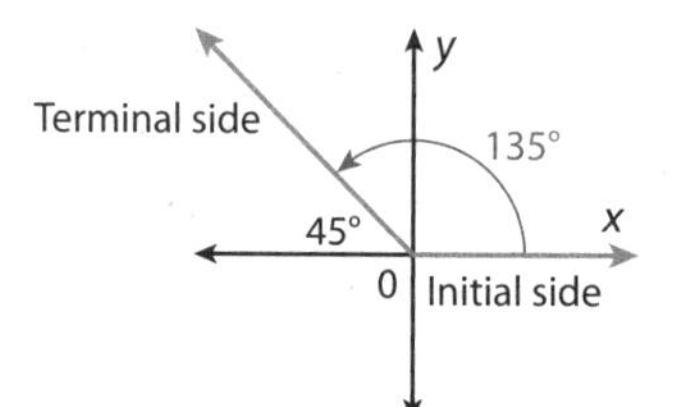
arc An unbroken part of a circle consisting of two points on the circle, called the endpoints, and all the points on the circle between them.	**arco** Parte continua de un círculo formada por dos puntos del círculo denominados extremos y todos los puntos del círculo comprendidos entre éstos.	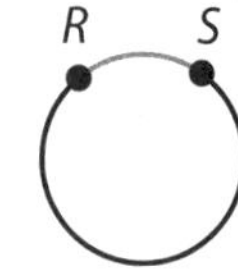
arithmetic sequence A sequence whose successive terms differ by the same nonzero number *d*, called the *common difference.*	**sucesión aritmética** Sucesión cuyos términos sucesivos difieren en el mismo número distinto de cero *d*, denominado *diferencia común*.	4, 7, 10, 13, 16, ... $+3 \; +3 \; +3 \; +3$ $d = 3$
arithmetic series The indicated sum of the terms of an arithmetic sequence.	**serie aritmética** Suma indicada de los términos de una sucesión aritmética.	$4 + 7 + 10 + 13 + 16 + ...$
asymptote A line that a graph approaches as the value of a variable becomes extremely large or small.	**asíntota** Línea recta a la cual se aproxima una gráfica a medida que el valor de una variable se hace sumamente grande o pequeño.	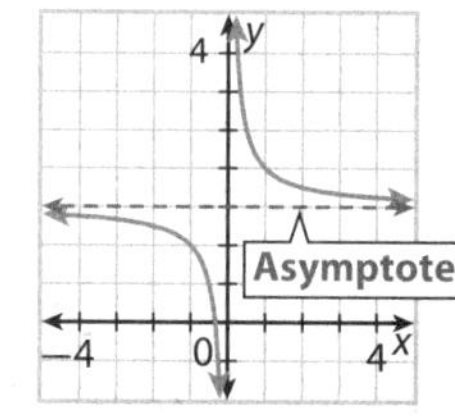
augmented matrix A matrix that consists of the coefficients and the constant terms in a system of linear equations.	**matriz aumentada** Matriz formada por los coeficientes y los términos constantes de un sistema de ecuaciones lineales.	System of equations: $3x + 2y = 5$, $2x - 3y = 1$ Augmented matrix: $\left[\begin{array}{cc\|c} 3 & 2 & 5 \\ 2 & -3 & 1 \end{array}\right]$
average rate of change The ratio of the change in the function values, $f(x_2) - f(x_1)$ to the change in the x-values, $x_2 - x_1$.	**tasa de cambio promedio** Razón entre el cambio en los valores de la función, $f(x_2) - f(x_1)$ y el cambio en los valores de x, $x_2 - x_1$.	
axis of symmetry A line that divides a plane figure or a graph into two congruent reflected halves.	**eje de simetría** Línea que divide una figura plana o una gráfica en dos mitades reflejadas congruentes.	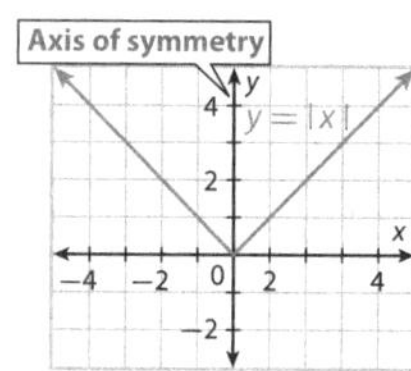

ENGLISH	SPANISH	EXAMPLES

B

ENGLISH	SPANISH	EXAMPLES
base of an exponential function The value of b in a function of the form $f(x) = ab^x$, where a and b are real numbers with $a \neq 0$, $b > 0$, and $b \neq 1$.	**base de una función exponencial** Valor de b en una función del tipo $f(x) = ab^x$, donde a y b son números reales con $a \neq 0$, $b > 0$, y $b \neq 1$.	$f(x) = 5(2)^x$ (2 = base)
biased sample A sample that does not fairly represent the population.	**muestra no representativa** Muestra que no representa adecuadamente una población.	
binomial A polynomial with two terms.	**binomio** Polinomio con dos términos.	$x + y$ $2a^2 + 3$ $4m^3n^2 + 6mn^4$
binomial experiment A probability experiment consists of n identical and independent trials whose outcomes are either successes or failures, with a constant probability of success p and a constant probability of failure q, where $q = 1 - p$ or $p + q = 1$.	**experimento binomial** Experimento de probabilidades que comprende n pruebas idénticas e independientes cuyos resultados son éxitos o fracasos, con una probabilidad constante de éxito p y una probabilidad constante de fracaso q, donde $q = 1 - p$ o $p + q = 1$.	A multiple-choice quiz has 10 questions with 4 answer choices. The number of trials is 10. If each question is answered randomly, the probability of success for each trial is $\frac{1}{4} = 0.25$ and the probability of failure is $\frac{3}{4} = 0.75$.
binomial probability In a binomial experiment, the probability of r successes $(0 \leq r \leq n)$ is $P(r) = {}_nC_r \cdot p^r q^{n-r}$.	**probabilidad binomial** En un experimento binomial, la probabilidad de r éxitos $(0 \leq r \leq n)$ es $P(r) = {}_nC_r \cdot p^r q^{n-r}$.	In the binomial experiment above, the probability of randomly guessing 6 problems correctly is $P = {}_{10}C_6\,(0.25)^6\,(0.75)^4 \approx 0.016$.
Binomial Theorem For any positive integer n, $(x + y)^n = {}_nC_0\,x^n y^0 + {}_nC_1\,x^{n-1} y^1 + {}_nC_2\,x^{n-2} y^2 + \ldots + {}_nC_{n-1}\,x^1 y^{n-1} + {}_nC_n\,x^0 y^n$.	**Teorema de los binomios** Dado un entero positivo n, $(x + y)^n = {}_nC_0\,x^n y^0 + {}_nC_1\,x^{n-1} y^1 + {}_nC_2\,x^{n-2} y^2 + \ldots + {}_nC_{n-1}\,x^1 y^{n-1} + {}_nC_n\,x^0 y^n$.	$(x + 2)^4 = {}_4C_0\,x^4 2^0 + {}_4C_1\,x^3 2^1 + {}_4C_2\,x^2 2^2 + {}_4C_3\,x^1 2^3 + {}_4C_4 x^0 2^4 = x^4 + 8x^3 + 24x^2 + 32x + 16$
box-and-whisker plot A method of showing how data is distributed by using the median, quartiles, and minimum and maximum values; also called a *box plot*.	**gráfica de mediana y rango** Método para demostrar la distribución de datos utilizando la mediana, los cuartiles y los valores mínimos y máximos; también llamado *gráfica de caja*.	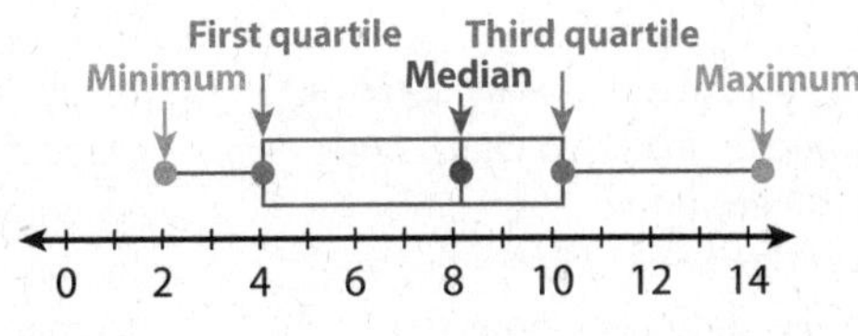
branch of a hyperbola One of the two symmetrical parts of the hyperbola.	**rama de una hipérbola** Una de las dos partes simétricas de la hipérbola.	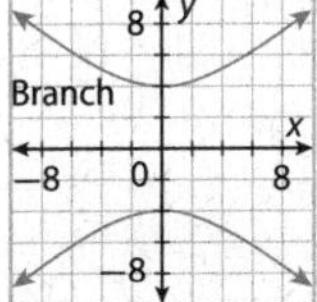

Glossary/Glosario

© Houghton Mifflin Harcourt Publishing Company

ENGLISH	SPANISH	EXAMPLES

C

ENGLISH	SPANISH	EXAMPLES
census A survey of an entire population.	**censo** Estudio de una población entera.	
circumference The distance around a circle.	**circunferencia** Distancia alrededor del círculo.	Radius Diameter Center Circumference
closure A set of numbers is said to be closed, or to have closure, under a given operation if the result of the operation on any two numbers in the set is also in the set.	**cerradura** Se dice que un conjunto de números es cerrado, o tiene cerradura, respecto de una operación determinada, si el resultado de la operación entre dos numerous cualesquiera del conjunto también está en el conjunto.	The natural numbers are closed under addition because the sum of two natural numbers is always a natural number.
cluster sample A sample in which the population is first divided into groups, a sample of the groups is randomly chosen, and all members of the chosen groups are surveyed.	**muestra por grupos** Muestra en la que la población se divide primeramente en grupos, se elige al azar una muestra de los grupos y se estudia a todos los miembros de los grupos elegidos.	
coefficient matrix The matrix of the coefficients of the variables in a linear system of equations.	**matriz de coeficientes** Matriz de los coeficientes de las variables en un sistema lineal de ecuaciones.	System of equations: $2x + 3y = 11$, $5x - 4y = 16$ Coefficient matrix: $\begin{bmatrix} 2 & 3 \\ 5 & -4 \end{bmatrix}$
coefficient of determination The number R^2, with $0 \leq R^2 \leq 1$, that shows the fraction of the data that are close to the curve of best fit and, thus, how well the curve fits the data.	**coeficiente de determinación** El número R^2, con $0 \leq R^2 \leq 1$, que muestra la fracción de los datos cercanos a la línea de mejor ajuste y, por lo tanto, cuánto se ajusta la línea de mejor ajuste a los datos.	
combination A selection of a group of objects in which order is *not* important. The number of combinations of *r* objects chosen from a group of *n* objects is denoted $_nC_r$.	**combinación** Selección de un grupo de objetos en la cual el orden *no* es importante. El número de combinaciones de *r* objetos elegidos de un grupo de *n* objetos se expresa así: $_nC_r$.	For 4 objects *A*, *B*, *C*, and *D*, there are $_4C_2 = 6$ different combinations of 2 objects: *AB*, *AC*, *AD*, *BC*, *BD*, *CD*.
combined variation A relationship containing both direct and inverse variation.	**variación combinada** Relación que contiene variaciones directas e inversas.	$y = \frac{kz}{x}$, where *k* is the constant of variation
common difference In an arithmetic sequence, the nonzero constant difference of any term and the previous term.	**diferencia común** En una sucesión aritmética, diferencia constante distinta de cero entre cualquier término y el término anterior.	In the arithmetic sequence 3, 5, 7, 9, 11, ..., the common difference is 2.

ENGLISH	SPANISH	EXAMPLES
common logarithm A logarithm whose base is 10, denoted $\log_{10}$ or just log.	**logaritmo común** Logaritmo de base 10, que se expresa $\log_{10}$ o simplemente log.	$\log 100 = \log_{10} 100 = 2$, since $10^2 = 100$.
common ratio In a geometric sequence, the constant ratio of any term and the previous term.	**razón común** En una sucesión geométrica, la razón constante *r* entre cualquier término y el término anterior.	In the geometric sequence 32, 16,18, 4, 2 ..., the common ratio is $\frac{1}{2}$.
complement of an event All outcomes in the sample space that are not in an event *E*, denoted $\bar{E}$.	**complemento de un suceso** Todos los resultados en el espacio muestral que no están en el suceso *E* y se expresan $\bar{E}$.	In the experiment of rolling a number cube, the complement of rolling a 3 is rolling a 1, 2, 4, 5, or 6.
completing the square A process used to form a perfect-square trinomial. To complete the square of $x^2 + bx$, add $\left(\frac{b}{2}\right)^2$.	**completar el cuadrado** Proceso utilizado para formar un trinomio cuadrado perfecto. Para completar el cuadrado de $x^2 + bx$, hay que sumar $\left(\frac{b}{2}\right)^2$.	$x^2 + 6x + \blacksquare$ Add $\left(\frac{6}{2}\right)^2 = 9$. $x^2 + 6x + 9$ $(x + 3)^2$ *is a perfect square.*
complex conjugate The complex conjugate of any complex number $a + bi$, denoted $\overline{a + bi}$, is $a - bi$.	**conjugado complejo** El conjugado complejo de cualquier número complejo $a + bi$, expresado como $\overline{a + bi}$, es $a - bi$.	$\overline{4 + 3i} = 4 - 3i$ $\overline{4 - 3i} = 4 + 3i$
complex fraction A fraction that contains one or more fractions in the numerator, the denominator, or both.	**fracción compleja** Fracción que contiene una o más fracciones en el numerador, en el denominador, o en ambos.	$\dfrac{\frac{1}{2}}{1 + \frac{2}{3}}$
complex number Any number that can be written as $a + bi$, where *a* and *b* are real numbers and $i = \sqrt{-1}$.	**número complejo** Todo número que se puede expresar como $a + bi$, donde *a* y *b* son números reales e $i = \sqrt{-1}$.	$4 + 2i$ $5 + 0i = 5$ $0 - 7i = -7i$
complex plane A set of coordinate axes in which the horizontal axis is the real axis and the vertical axis is the imaginary axis; used to graph complex numbers.	**plano complejo** Conjunto de ejes cartesianos en el cual el eje horizontal es el eje real y el eje vertical es el eje imaginario; se utiliza para representar gráficamente números complejos.	Imaginary axis $2i$ $0 + 0i$ Real axis -2 2 $-2i$
composite figure A plane figure made up of triangles, rectangles, trapezoids, circles, and other simple shapes, or a three-dimensional figure made up of prisms, cones, pyramids, cylinders, and other simple three-dimensional figures.	**figura compuesta** Figura plana compuesta por triángulos, rectángulos, trapecios, círculos y otras formas simples, o figura tridimensional compuesta por prismas, conos, pirámides, cilindros y otras figuras tridimensionales simples.	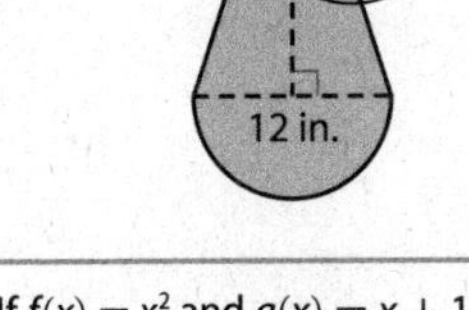
composition of functions The composition of functions *f* and *g*, written as $(f \circ g)(x)$ and defined as $f(g(x))$ uses the output of $g(x)$ as the input for $f(x)$.	**composición de funciones** La composición de las funciones *f* y *g*, expresada como $(f \circ g)(x)$ y definida como $f(g(x))$ utiliza la salida de $g(x)$ como la entrada para $f(x)$.	If $f(x) = x^2$ and $g(x) = x + 1$, the composite function $(f \circ g)(x) = (x + 1)^2$.

Glossary/Glosario

Glossary/Glosario

ENGLISH	SPANISH	EXAMPLES
compound event An event made up of two or more simple events.	**suceso compuesto** Suceso formado por dos o más sucesos simples.	In the experiment of tossing a coin and rolling a number cube, the event of the coin landing heads and the number cube landing on 3.
compression A transformation that pushes the points of a graph horizontally toward the *y*-axis or vertically toward the *x*-axis.	**compresión** Transformación que desplaza los puntos de una gráfica horizontalmente hacia el eje *y* o verticalmente hacia el eje *x*.	
conditional probability The probability of event *B*, given that event *A* has already occurred or is certain to occur, denoted $P(B \mid A)$; used to find probability of dependent events.	**probabilidad condicional** Probabilidad del suceso *B*, dado que el suceso *A* ya ha ocurrido o es seguro que ocurrirá, expresada como $P(B \mid A)$; se utiliza para calcular la probabilidad de sucesos dependientes.	
conditional relative frequency The ratio of a joint relative frequency to a related marginal relative frequency in a two-way table.	**frecuencia relativa condicional** Razón de una frecuencia relativa conjunta a una frecuencia relativa marginal en una tabla de doble entrada.	
congruent Having the same size and shape, denoted by $\cong$.	**congruente** Que tiene el mismo tamaño y forma, expresado por $\cong$.	P, Q, R, S $\overline{PQ} \cong \overline{RS}$
conic section A plane figure formed by the intersection of a double right cone and a plane. Examples include circles, ellipses, hyperbolas, and parabolas.	**sección cónica** Figura plana formada por la intersección de un cono regular doble y un plano. Algunos ejemplos son círculos, elipses, hipérbolas y parábolas.	Circle Ellipse Parabola Hyperbola
conjugate axis The axis of symmetry of a hyperbola that separates the two branches of the hyperbola.	**eje conjugado** Eje de simetría de una hipérbola que separa las dos ramas de la hipérbola.	Conjugate axis
constant function A function of the form $f(x) = c$, where *c* is a constant.	**función constante** Función del tipo $f(x) = c$, donde *c* es una constante.	y = 3
constant matrix The matrix of the constants in a linear system of equations.	**matriz de constantes** Matriz de las constantes de un sistema lineal de ecuaciones.	System of equations: $\begin{cases} 2x + 3y = 11 \\ 5x - 4y = 16 \end{cases}$ Constant matrix: $\begin{bmatrix} 11 \\ 16 \end{bmatrix}$
constant of variation The constant *k* in direct, inverse, joint, and combined variation equations.	**constante de variación** La constante *k* en ecuaciones de variación directa, inversa, conjunta y combinada.	$y = 5x$ constant of variation

ENGLISH	SPANISH	EXAMPLES
constraint One of the inequalities that define the feasible region in a linear-programming problem.	**restricción** Una de las desigualdades que definen la región factible en un problema de programación lineal.	Constraints: $x > 0$ $y > 0$ $x + y \leq 8$ $3x + 5y \leq 30$ Feasible region 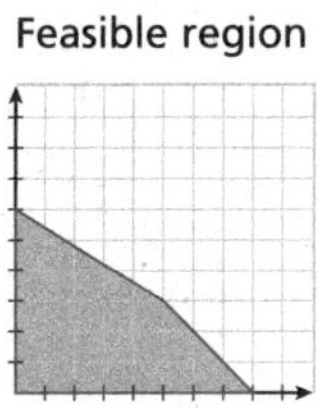
continuous data Data that can take on any real-value measurement within an interval.	**datos continuos** Datos obtenidos por medición que pueden asumir cualquier valor real dentro de un intervalo.	The quantity of water in a glass as the water evaporates is continuous data.
continuous function A function whose graph is an unbroken line or curve with no gaps or breaks.	**función continua** Función cuya gráfica es una línea recta o curva continua, sin espacios ni interrupciones.	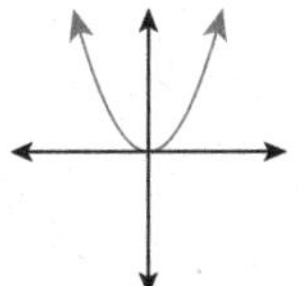
contradiction An equation that has no solutions.	**contradicción** Ecuación que no tiene soluciones.	$x + 1 = x$ $1 = 0x$
control group In a controlled experiment, the group that does not receive treatment. This group is used for comparison.	**grupo de control** En un experimento controlado, el grupo que no está expuesto a la manipulación; es el grupo que sirve como comparación.	
controlled experiment An experiment in which two groups are studied under conditions that are identical except for one variable.	**experimento controlado** Experimento en el que se estudia a dos grupos bajo condiciones que son idénticas, excepto por una variable.	
converge An infinite series converges when the partial sums approach a fixed number.	**convergir** Una sucesión o serie infinita converge cuando las sumas parciales se aproximan a un número fijo.	$\frac{1}{2} + \frac{1}{4} + \frac{1}{8} + \frac{1}{16} + \ldots$ converges to 1.
convenience sample A sample based on members of the population that are readily available.	**muestra de conveniencia** Una muestra basada en miembros de la población que están fácilmente disponibles.	A reporter surveys people he personally knows.
correlation A measure of the strength and direction of the relationship between two variables or data sets.	**correlación** Medida de la fuerza y dirección de la relación entre dos variables o conjuntos de datos.	Positive correlation 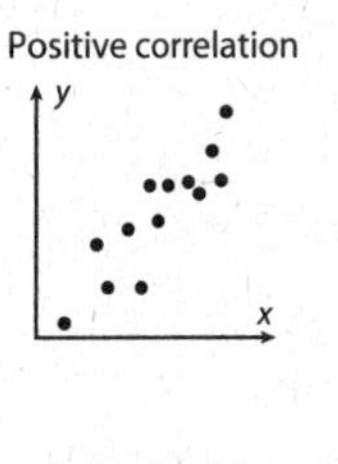 No correlation 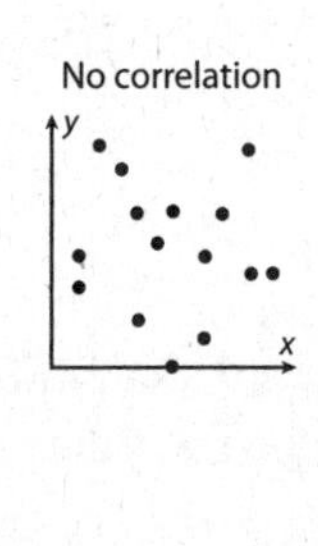 Negative correlation

Glossary/Glosario

ENGLISH	SPANISH	EXAMPLES
correlation coefficient A number r, where $-1 \leq r \leq 1$, that describes how closely the points in a scatter plot cluster around the least–squares line.	**coeficiente de correlación** Número r, donde $-1 \leq r \leq 1$, que describe a qué distancia de la recta de mínimos cuadrados se agrupan los puntos de un diagrama de dispersión.	An r–value close to 1 describes a strong positive correlation. An r–value close to 0 describes a weak correlation or no correlation. An r–value close to -1 describes a strong negative correlation.
cosecant In a right triangle, the cosecant of angle A is the ratio of the length of the hypotenuse to the length of the side opposite A. It is the reciprocal of the sine function.	**cosecante** En un triángulo rectángulo, la cosecante del ángulo A es la razón entre la longitud de la hipotenusa y la longitud del cateto opuesto a A. Es la inversa de la función seno.	A opposite hypotenuse $\csc A = \frac{\text{hypotenuse}}{\text{opposite}} = \frac{1}{\sin A}$
cosine In a right triangle, the cosine of angle A is the ratio of the length of the side adjacent to angle A to the length of the hypotenuse. It is the reciprocal of the secant function.	**coseno** En un triángulo rectángulo, el coseno del ángulo A es la razón entre la longitud del cateto adyacente al ángulo A y la longitud de la hipotenusa. Es la inversa de la función secante.	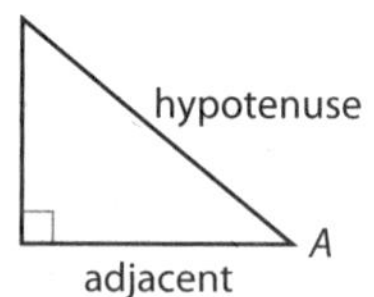 $\cos A = \frac{\text{adjacent}}{\text{hypotenuse}} = \frac{1}{\sec A}$
cotangent In a right triangle, the cotangent of angle A is the ratio of the length of the side adjacent to A to the length of the side opposite A. It is the reciprocal of the tangent function.	**cotangente** En un triángulo rectángulo, la cotangente del ángulo A es la razón entre la longitud del cateto adyacente a A y la longitud del cateto opuesto a A. Es la inversa de la función tangente.	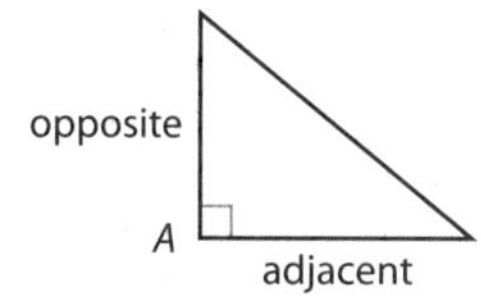 $\cot A = \frac{\text{adjacent}}{\text{opposite}} = \frac{1}{\tan A}$
coterminal angles Two angles in standard position with the same terminal side.	**ángulos coterminales** Dos ángulos en posición estándar con el mismo lado terminal.	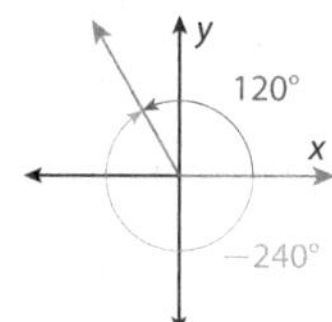
counterexample An example that proves that a conjecture or statement is false.	**contraejemplo** Ejemplo que demuestra que una conjetura o enunciado es falso.	
co-vertices of a hyperbola The endpoints of the conjugate axis.	**co-vértices de una hipérbola** Extremos de un eje conjugado.	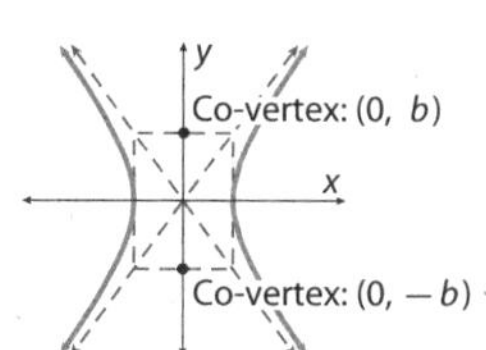
co-vertices of an ellipse The endpoints of the minor axis.	**co-vértices de una elipse** Extremos del eje menor.	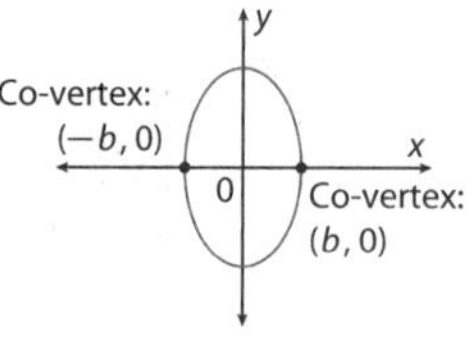

ENGLISH	SPANISH	EXAMPLES
critical values Values that separate the number line into intervals that either contain solutions or do not contain solutions.	**valores críticos** Valores que separan la recta numérica en intervalos que contienen o no contienen soluciones.	
cube-root function The function $f(x) = \sqrt[3]{x}$.	**función de raíz cúbica** La función $f(x) = \sqrt[3]{x}$.	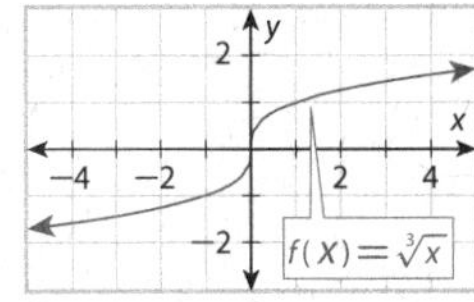
cubic function A polynomial function of degree 3.	**función cúbica** Función polinomial de grado 3.	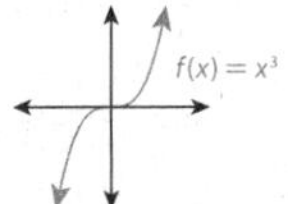
cycle of a periodic function The shortest repeating part of a periodic graph or function.	**ciclo de una función periódica** La parte repetida más corta de una gráfica o función periódica.	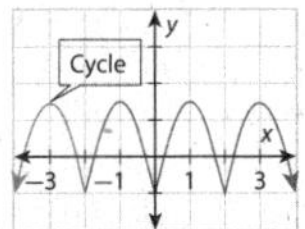

D

ENGLISH	SPANISH	EXAMPLES
decay factor The base $1 - r$ in an exponential expression.	**factor decremental** Base $1 - r$ en una expresión exponencial.	$2(0.93)^t$ ↑ decay factor (representing $1 - 0.07$)
decay rate The constant percent decrease, in decimal form, in an exponential decay function.	**tasa de disminución** Disminución porcentual constante, en forma decimal, en una función de disminución exponencial.	In the function $f(t) = a(1 - 0.2)^t$, 0.2 is the decay rate.
decreasing A function is decreasing on an interval if $f(x_1) > f(x_2)$ when $x_1 > x_2$ for any x-values x_1 and x_2 from the interval.	**decreciente** Una función es decreciente en un intervalo si $f(x_1) > f(x_2)$ cuando $x_1 > x_2$ dados los valores de x, x_1 y x_2, pertenecientes al intervalo.	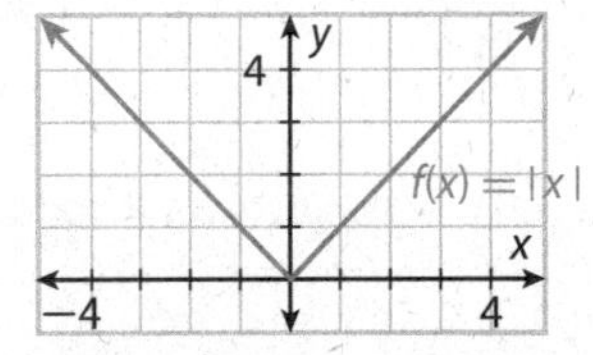 $f(x)$ is decreasing on the interval $x < 0$.
degenerate conic A degenerate conic is formed when a plane passes through the vertex of a hollow double cone. A point, a line, and a pair of intersecting lines are all degenerate conics.	**cónica degenerada** Una cónica degenerada se forma cuando un plano atraviesa el vértice de un cono doble hueco. Un punto, una línea y un par de líneas secantes son cónicas degeneradas.	A point is a circle with no radius.
degree of a monomial The sum of the exponents of the variables inthe monomial.	**grado de un monomio** Suma de los exponentes de las variables del monomio.	$4x^2y^5z^3$ Degree: $2 + 5 + 3 = 10$ 5 Degree: 0 $(5 = 5x^0)$
degree of a polynomial The degree of the term of the polynomial with the greatest degree.	**grado de un polinomio** Grado del término del polinomio con el grado máximo.	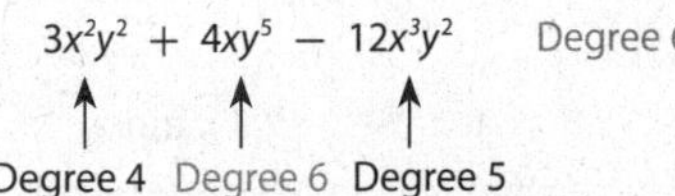

ENGLISH	SPANISH	EXAMPLES
dependent events Events for which the occurrence or nonoccurrence of one event affects the probability of the other event.	**sucesos dependientes** Dos sucesos son dependientes si el hecho de que uno de ellos se cumpla o no afecta la probabilidad del otro.	From a bag containing 3 red marbles and 2 blue marbles, drawing a red marble, and then drawing a blue marble without replacing the first marble.
dependent system A system of equations that has infinitely many solutions.	**sistema dependiente** Sistema de ecuaciones que tiene infinitamente muchas soluciones.	$\begin{cases} x + y = 3 \\ 2x + 2y = 6 \end{cases}$
dependent variable The output of a function; a variable whose value depends on the value of the input, or independent variable.	**variable dependiente** Salida de una función; variable cuyo valor depende del valor de la entrada, o variable independiente.	$y = 2x + 1$ ↑ dependent variable
determinant A real number associated with a square matrix. The determinant of $A = \begin{bmatrix} a & b \\ c & d \end{bmatrix}$ is $\lvert A \rvert = ad - bc$.	**determinante** Número real asociado con una matriz cuadrada. El determinante de $A = \begin{bmatrix} a & b \\ c & d \end{bmatrix}$ es $\lvert A \rvert = ad - bc$.	$\begin{vmatrix} 2 & -1 \\ 3 & 4 \end{vmatrix} = 2(4) - (-1)(3) = 11$
difference of two squares A polynomial of the form $a^2 - b^2$, which may be written as the product $(a + b)(a - b)$.	**diferencia de dos cuadrados** Polinomio del tipo $a^2 - b^2$, que se puede expresar como el producto $(a + b)(a - b)$.	$x^2 - 4 = (x + 2)(x - 2)$
dimensions of a matrix A matrix with *m* rows and *n* columns has dimensions $m \times n$, read "*m* by *n*."	**dimensiones de una matriz** Una matriz con *m* filas y *n* columnas tiene dimensiones $m \times n$, expresadas "*m* por *n*".	$\begin{bmatrix} -3 & 2 & 1 & -1 \\ 4 & 0 & -5 & 2 \end{bmatrix}$ Dimensions 2×4
direct variation A linear relationship between two variables, *x* and *y*, that can be written in the form $y = kx$, where *k* is a nonzero constant.	**variación directa** Relación lineal entre dos variables, *x* e *y*, que puede expresarse en la forma $y = kx$, donde *k* es una constante distinta de cero.	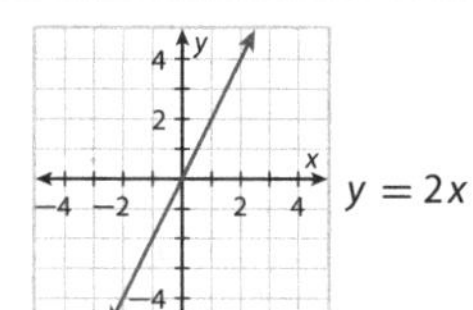
directrix A fixed line used to define a *parabola*. Every point on the parabola is equidistant from the directrix and a fixed point called the *focus*.	**directriz** Línea fija utilizada para definir una *parábola*. Cada punto de la parábola es equidistante de la directriz y de un punto fijo denominado *foco*.	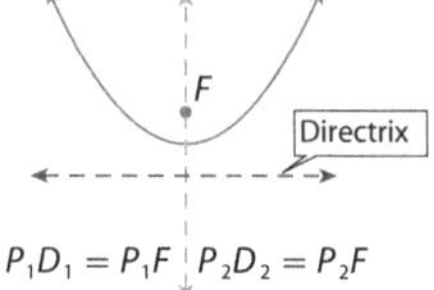
discontinuous function A function whose graph has one or more jumps, breaks, or holes.	**función discontinua** Función cuya gráfica tiene uno o más saltos, interrupciones u hoyos.	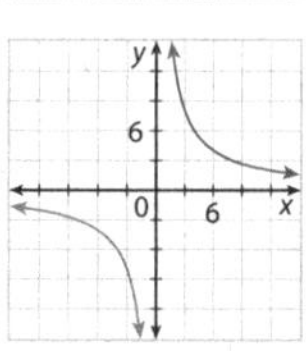
discrete data Data that cannot take on any real-value measurement within an interval.	**datos discretos** Datos que no admiten cualquier medida de valores reales dentro de un intervalo.	the number of pennies in a jar over time

ENGLISH	SPANISH	EXAMPLES
discriminant The discriminant of the quadratic equation $ax^2 + bx + c = 0$ is $b^2 - 4ac$.	**discriminante** El discriminante de la ecuación cuadrática $ax^2 + bx + c = 0$ es $b^2 - 4ac$.	The discriminant of $2x^2 - 5x - 3 = 0$ is $(-5)^2 - 4(2)(-3) = 25 + 24 = 49$.
disjunction A compound statement that uses the word *or*.	**disyunción** Enunciado compuesto que contiene la palabra *o*.	John will walk to work OR he will stay home.
Distance Formula In a coordinate plane, the distance from (x_1, y_1) to (x_2, y_2) is $d = \sqrt{(x_2 - x_1)^2 + (y_2 - y_1)^2}$.	**Fórmula de distancia** En un plano cartesiano, la distancia desde (x_1, y_1) hasta (x_2, y_2) es $d = \sqrt{(x_2 - x_1)^2 + (y_2 - y_1)^2}$.	The distance from (2, 1) to (6, 4) is $d = \sqrt{(6-2)^2 + (4-1)^2}$ $= \sqrt{4^2 + 3^2} = \sqrt{9 + 16} = 5$.
diverge An infinite series diverges when the partial sums do not approach a fixed number.	**divergir** Una serie infinita diverge cuando las sumas parciales no se aproximan a un número fijo.	$1 + 2 + 4 + 8 + 16 + \ldots$ diverges.
domain The set of all possible input values of a relation or function.	**dominio** Conjunto de todos los posibles valores de entrada de una función o relación.	The domain of the function $f(x) = \sqrt{x}$ is $\{x \mid x \geq 0\}$.

E

ENGLISH	SPANISH	EXAMPLES
elementary row operations *See* row operations.	**operaciones elementales de fila** *Véase* operaciones de fila.	
elimination A method used to solve systems of equations in which one variable is eliminated by adding or subtracting two equations of the system.	**eliminación** Método utilizado para resolver sistemas de ecuaciones por el cual se elimina una variable sumando o restando dos ecuaciones del sistema.	
ellipse The set of all points *P* in a plane such that the sum of the distances from *P* to two fixed points F_1 and F_2, called the foci, is constant.	**elipse** Conjunto de todos los puntos *P* de un plano tal que la suma de las distancias desde *P* hasta los dos puntos fijos F_1 y F_2, denominados focos, es constante.	P, y, x, 0, Foci
empty set A set with no elements.	**conjunto vacío** Conjunto sin elementos.	The solution set of $\lvert x \rvert < 0$ is the empty set, $\{ \}$, or $\varnothing$.
end behavior The trends in the *y*-values of a function as the *x*-values approach positive and negative infinity.	**comportamiento extremo** Tendencia de los valores de y de una función a medida que los valores de *x* se aproximan al infinito positivo y negativo.	End behavior: $f(x) \to \infty$ as $x \to \infty$; $f(x) \to -\infty$ as $x \to -\infty$
entry Each value in a matrix; also called an element.	**entrada** Cada valor de una matriz; también denominado elemento.	3 is the entry in the first row and second column of $A = \begin{bmatrix} 2 & 3 \\ 0 & 1 \end{bmatrix}$, denoted a_{12}.

ENGLISH | **SPANISH** | **EXAMPLES**

equally likely outcomes Outcomes are equally likely if they have the same probability of occurring. If an experiment has n equally likely outcomes, then the probability of each outcome is $\frac{1}{n}$.

resultados igualmente probables Los resultados son igualmente probables si tienen la misma probabilidad de ocurrir. Si un experimento tiene n resultados igualmente probables, entonces la probabilidad de cada resultado es $\frac{1}{n}$.

If a coin is tossed, and heads and tails are equally likely, then $P(\text{heads}) = P(\text{tails}) = \frac{1}{2}$.

---

equation A mathematical statement that two expressions are equivalent.

ecuación Enunciado matemático que indica que dos expresiones son equivalentes.

$x + 4 = 7$
$2 + 3 = 6 - 1$
$(x - 1)^2 + (y + 2)^2 = 4$

---

even function A function in which $f(-x) = f(x)$ for all x in the domain of the function.

función par Función en la que para todos los valores de x dentro del dominio de la función.

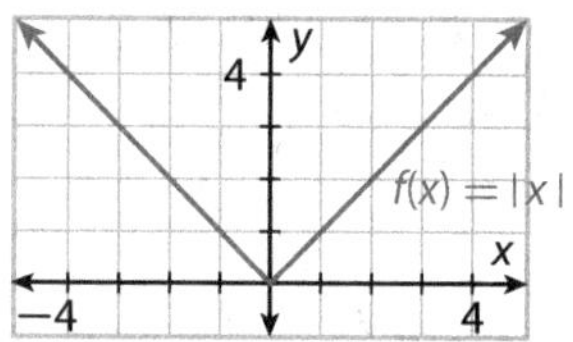

$f(x) = |x|$ is an even function.

---

event An outcome or set of outcomes in a probability experiment.

suceso Resultado o conjunto de resultados en un experimento de probabilidad.

In the experiment of rolling a number cube, the event "an odd number" consists of the outcomes 1, 3, and 5.

---

expected value The weighted average of the numerical outcomes of a probability experiment.

valor esperado Promedio ponderado de los resultados numéricos de un experimento de probabilidad.

The table shows the probability of getting a given score by guessing on a three-question quiz.

Score	0	1	2	3
Probability	0.42	0.42	0.14	0.02

The expected value is a score of
$0\,(0.42) + 1\,(0.42) + 2\,(0.14) + 3\,(0.02)$
$= 0.76$.

---

experiment An operation, process, or activity in which outcomes can be used to estimate probability.

experimento Una operación, proceso o actividad cuyo resultado se puede usar para estimar la probabilidad.

Tossing a coin 10 times and noting the number of heads.

---

experimental probability The ratio of the number of times an event occurs to the number of trials, or times, that an activity is performed.

probabilidad experimental Razón entre la cantidad de veces que ocurre un suceso y la cantidad de pruebas, o veces, que se realiza una actividad.

Kendra made 6 of 10 free throws. The experimental probability that she will make her next free throw is

$P(\text{free throw}) = \frac{\text{number made}}{\text{number attempted}} = \frac{6}{10}$.

---

explicit formula A formula that defines the nth term a_n, or general term, of a sequence as a function of n.

fórmula explícita Fórmula que define el enésimo término a_n, o término general, de una sucesión como una función de n.

Sequence: 4, 7, 10, 13, 16, 19, …
Explicit formula: $a_n = 1 + 3n$

---

exponent The number that indicates how many times the base in a power is used as a factor.

exponente Número que indica la cantidad de veces que la base de una potencia se utiliza como factor.

$3^4 = 3 \cdot 3 \cdot 3 \cdot 3 = 81$
(arrow pointing to the 4: exponent)

ENGLISH	SPANISH	EXAMPLES

exponential decay An exponential function of the form $f(x) = ab^x$ in which $0 < b < 1$. If r is the rate of decay, then the function can be written $y = a(1 - r)^t$, where a is the initial amount and t is the time.

decremento exponencial Función exponencial del tipo $f(x) = ab^x$ en la cual $0 < b < 1$. Si r es la tasa decremental, entonces la función se puede expresar como $y = a(1 - r)^t$, donde a es la cantidad inicial y t es el tiempo.

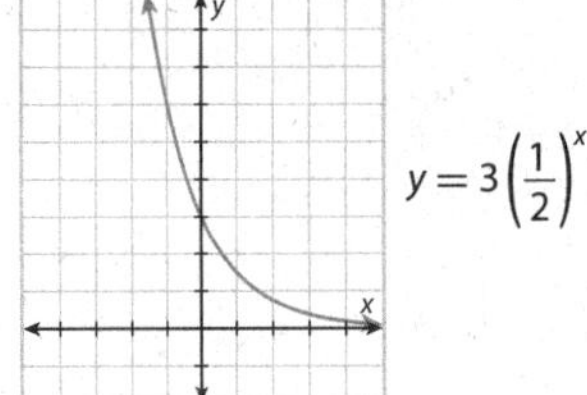

---

exponential equation An equation that contains one or more exponential expressions.

ecuación exponencial Ecuación que contiene una o más expresiones exponenciales.

$2^{x+1} = 8$

---

exponential function A function of the form $f(x) = ab^x$, where a and b are real numbers with $a \neq 0$, $b > 0$, and $b \neq 1$.

función exponencial Función del tipo $f(x) = ab^x$, donde a y b son números reales con $a \neq 0$, $b > 0$ y $b \neq 1$.

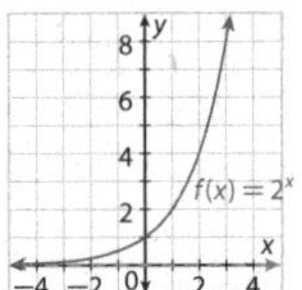

---

exponential growth An exponential function of the form $f(x) = ab^x$ in which $b > 1$. If r is the rate of growth, then the function can be written $y = a(1 + r)^t$, where a is the initial amount and t is the time.

crecimiento exponencial Función exponencial del tipo $f(x) = ab^x$ en la que $b > 1$. Si r es la tasa de crecimiento, entonces la función se puede expresar como $y = a(1 + r)^t$, donde a es la cantidad inicial y t es el tiempo.

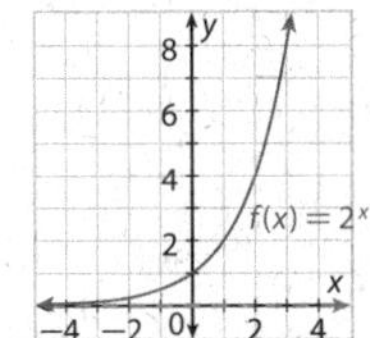

---

exponential regression A statistical method used to fit an exponential model to a given data set.

regresión exponencial Método estadístico utilizado para ajustar un modelo exponencial a un conjunto de datos determinado.

---

extraneous solution A solution of a derived equation that is not a solution of the original equation.

solución extraña Solución de una ecuación derivada que no es una solución de la ecuación original.

To solve $\sqrt{x} = -2$, square both sides; $x = 4$.
Check $\sqrt{4} = -2$ is false; so 4 is an extraneous solution.

F

Factor Theorem For any polynomial $P(x)$, $(x - a)$ is a factor of $P(x)$ if and only if $P(a) = 0$.

Teorema del factor Dado el polinomio $P(x)$, $(x - a)$ es un factor de $P(x)$ si y sólo si $P(a) = 0$.

$(x - 1)$ is a factor of $P(x) = x^2 - 1$ because $P(1) = 1^2 - 1 = 0$.

---

factorial If n is a positive integer, then n factorial, written $n!$, is $n \cdot (n - 1) \cdot (n - 2) \cdot \ldots \cdot 2 \cdot 1$. The factorial of 0 is defined to be 1.

factorial Si n es un entero positivo, entonces el factorial de n, expresado como n!, es $n \cdot (n - 1) \cdot (n - 2) \cdot \ldots \cdot 2 \cdot 1$ Por definición, el factorial de 0 será 1.

$7! = 7 \cdot 6 \cdot 5 \cdot 4 \cdot 3 \cdot 2 \cdot 1 = 5040$
$0! = 1$

---

factoring The process of writing a number or algebraic expression as a product.

factorización Proceso por el que se expresa un número o expresión algebraica como un producto.

$x^2 - 4x - 21 = (x - 7)(x + 3)$

ENGLISH	SPANISH	EXAMPLES
family of functions A set of functions whose graphs have basic characteristics in common. Functions in the same family are transformations of their parent function.	**familia de funciones** Conjunto de funciones cuyas gráficas tienen características básicas en común. Las funciones de la misma familia son transformaciones de su función madre.	Some members of the family of quadratic functions with the parent function $f(x) = x^2$ are: $f(x) = 3x^2$ $f(x) = x^2 + 1$ $f(x) = (x - 2)^2$ 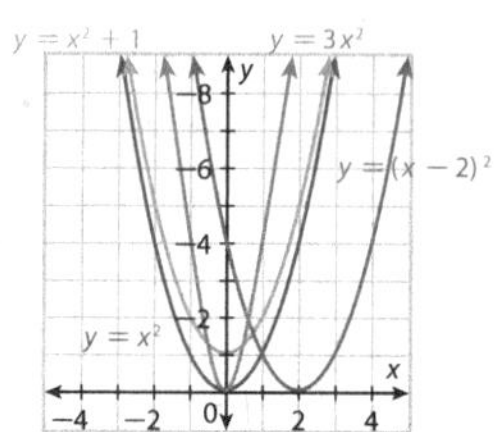
favorable outcome The occurrence of one of several possible outcomes of a specified event or probability experiment.	**resultado favorable** Cuando se produce uno de varios resultados posibles de un suceso específico o experimento de probabilidad.	In the experiment of rolling an odd number on a number cube, the favorable outcomes are 1, 3, and 5.
feasible region The set of points that satisfy the constraints in a linear-programming problem.	**región factible** Conjunto de puntos que cumplen con las restricciones de un problema de programación lineal.	Constraints: $x > 0$ $y > 0$ $x + y \leq 8$ $3x + 5y \leq 30$ Feasible region 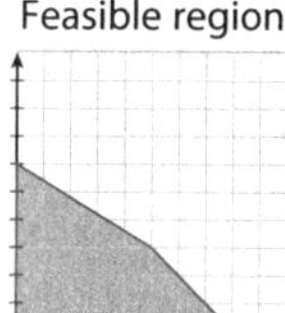
Fibonacci sequence The infinite sequence of numbers beginning with 1, 1 such that each term is the sum of the two previous terms.	**sucesión de Fibonacci** Sucesión infinita de números que comienza con 1, 1 de forma tal que cada término es la suma de los dos términos anteriores.	1, 1, 2, 3, 5, 8, 13, 21, …
finite sequence A sequence with a finite number of terms.	**sucesión finita** Sucesión con un número finito de términos.	1, 2, 3, 4, 5
finite set A set with a definite, or finite, number of elements.	**conjunto finito** Conjunto con un número de elementos definido o finito.	$\{2, 4, 6, 8, 10\}$
first differences The differences between *y*-values of a function for evenly spaced *x*-values.	**primeras diferencias** Diferencias entre los valores de y de una función para valores de *x* espaciados uniformemente.	x: 0, 1, 2, 3; y: 3, 7, 11, 15; first differences +4 +4 +4
first quartile The median of the lower half of a data set, denoted Q_1. Also called *lower quartile.*	**primer cuartil** Mediana de la mitad inferior de un conjunto de datos, expresada como Q_1. También se llama *cuartil inferior.*	Lower half: 18, (23,) 28, Upper half: 36, 42, 49, — First quartile
focus (pl. foci) of a hyperbola One of two fixed points F_1 and F_2 that are used to define a hyperbola. For every point P on the hyperbola, $PF_1 - PF_2$ is constant.	**foco de una hipérbola** Uno de los dos puntos fijos F_1 y F_2 utilizados para definir una hipérbola, Para cada punto P de la hipérbola, $PF_1 - PF_2$ es constante.	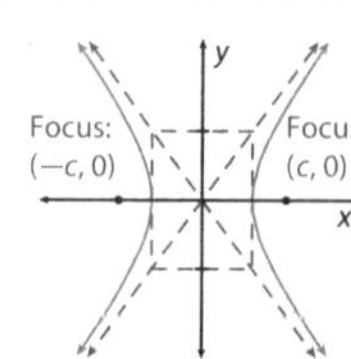

Glossary/Glosario

ENGLISH	SPANISH	EXAMPLES
focus (pl. foci) of an ellipse One of two fixed points F_1 and F_2 that are used to define an ellipse. For every point P on the ellipse, $PF_1 + PF_2$ is constant.	**foco de una elipse** Uno de los dos puntos fijos F_1 y F_2 utilizados para definir una elipse. Para cada punto P de la elipse, $PF_1 + PF_2$ es constante.	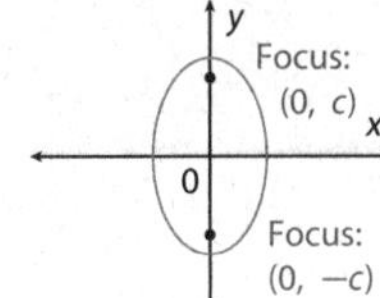
focus (pl. foci) of a parabola A fixed point F used with a *directrix* to define a *parabola*.	**foco de una parábola** Punto fijo F utilizado con una *directriz* para definir una *parábola*.	Focus, F
frequency of a data value The number of times the value appears in the data set.	**frecuencia de un valor de datos** Cantidad de veces que aparece el valor en un conjunto de datos.	In the data set 5, 6, 6, 6, 8, 9, the data value 6 has a frequency of 3.
frequency of a periodic function The number of cycles per unit of time. Also the reciprocal of the period.	**frecuencia de una función periódica** Cantidad de ciclos por unidad de tiempo. También es la inversa del periodo.	The function $y = \sin(2x)$ has a period of π and a frequency of $\frac{1}{\pi}$.
function A relation in which every input is paired with exactly one output.	**función** Una relación en la que cada entrada corresponde exactamente a una salida.	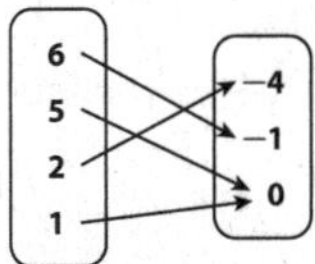
function notation If x is the independent variable and y is the dependent variable, then the function notation for y is $f(x)$, read "f of x," where f names the function.	**notación de función** Si x es la variable independiente e y es la variable dependiente, entonces la notación de función para y es $f(x)$, que se lee "f de x", donde f nombra la función.	equation: $y = 2x$ function notation: $f(x) = 2x$
function rule An algebraic expression that defines a function.	**regla de función** Expresión algebraica que define una función.	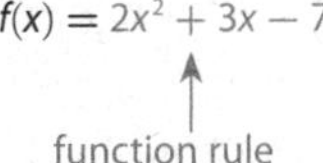
Fundamental Counting Principle For n items, if there are m_1 ways to choose a first item, m_2 ways to choose a second item after the first item has been chosen, and so on, then there are $m_1 \cdot m_2 \cdot \ldots \cdot m_n$ ways to choose n items.	**Principio fundamental de conteo** Dados n elementos, si existen m_1 formas de elegir un primer elemento, m_2 formas de elegir un segundo elemento después de haber elegido el primero, y así sucesivamente, entonces existen $m_1 \cdot m_2 \cdot \ldots \cdot m_n$ formas de elegir n elementos.	If there are 4 colors of shirts, 3 colors of pants, and 2 colors of shoes, then there are $4 \cdot 3 \cdot 2 = 24$ possible outfits.

G

Gaussian Elimination An algorithm for solving systems of equations using matrices and row operations to eliminate variables in each equation in the system.	**Eliminación Gaussiana** Algoritmo para resolver sistemas de ecuaciones mediante matrices y operaciones de fila con el fin de eliminar variables en cada ecuación del sistema.	

ENGLISH	SPANISH	EXAMPLES
general form of a conic section $Ax^2 + Bxy + Cy^2 + Dx + Ey + F = 0$, where A and B are not both 0.	**forma general de una sección cónica** $Ax^2 + Bxy + Cy^2 + Dx + Ey + F = 0$, donde A y B no son los dos 0.	A circle with a vertex at $(1, 2)$ and radius 3 has the general form $x^2 + y^2 - 2x - 4y - 4 = 0$.
geometric mean In a geometric sequence, a term that comes between two given nonconsecutive terms of the sequence. For positive numbers a and b, the geometric mean is $\sqrt{ab}$.	**media geométrica** En una sucesión geométrica, un término que se encuentra entre dos términos no consecutivos dados de la sucesión. Dados los números positivos a y b, la media geométrica es $\sqrt{ab}$.	The geometric mean of 4 and 9 is $\sqrt{4(9)} = \sqrt{36} = 6$.
geometric probability A form of theoretical probability determined by a ratio of geometric measures such as lengths, areas, or volumes.	**probabilidad geométrica** Una forma de la probabilidad teórica determinada por una razón de medidas geométricas, como longitud, área o volumen.	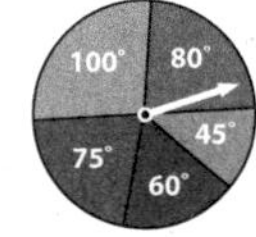 The probability of the pointer landing on 80° is $\frac{2}{9}$.
geometric sequence A sequence in which the ratio of successive terms is a constant r, called the common ratio, where $r \neq 0$ and $r \neq 1$.	**sucesión geométrica** Sucesión en la que la razón de los términos sucesivos es una constante r, denominada razón común, donde $r \neq 0$ y $r \neq 1$.	1, 2, 4, 8, 16, ... $\cdot 2 \quad \cdot 2 \quad \cdot 2 \quad \cdot 2 \qquad r = 2$
geometric series The indicated sum of the terms of a geometric sequence.	**serie geométrica** Suma indicada de los términos de una sucesión geométrica.	$1 + 2 + 4 + 8 + 16 + \ldots$
grade A measure of the steepness of surfaces, expressed as a percent.	**grado** Medida de la inclinación de las superficies, expresada como un porcentaje.	A ramp that rises 1 foot for every 5 feet of the horizontal distance has a grade of 20%.
greatest common factor (GCF) The product of the greatest integer and the greatest power of each variable that divides evenly into each term.	**máximo común divisor (MCD)** Dado Producto del entero mayor y la potencia mayor de cada variable que divide exactamente cada término.	The GCF of $4x^3y$ and $6x^2y$ is $2x^2y$. The GCF of 27 and 45 is 9.
greatest-integer function A function denoted by $f(x) = [x]$ or $f(x) = \lfloor x \rfloor$ in which the number x is rounded down to the greatest integer that is less than or equal to x.	**función de entero mayor** Función expresada como $f(x) = [x]$ o $f(x) = \lfloor x \rfloor$ en la cual el número x se redondea hacia abajo hasta el entero mayor que sea menor que o igual a x.	$\lfloor 4.98 \rfloor = 4$ $\lfloor -2.1 \rfloor = -3$
growth factor The base $1 + r$ in an exponential expression.	**factor de crecimiento** La base $1 + r$ en una expresión exponencial.	$12{,}000(1 + 0.14)^t$ ↑ growth factor
growth rate The constant percent increase, in decimal form, in an exponential growth function.	**tasa de crecimiento** Aumento porcentual constante, en forma decimal, en una función de crecimiento exponencial.	In the function $f(t) = a(1 + 0.3)^t$, 0.3 is the growth rate.

H

half-life The half-life of a substance is the time it takes for one-half of the substance to decay into another substance.

vida media La vida media de una sustancia es el tiempo que tarda la mitad de la sustancia en desintegrarse y transformarse en otra sustancia.

Carbon-14 has a half-life of 5730 years, so 5 g of an initial amount of 10 g will remain after 5730 years.

---

half-plane The part of the coordinate plane on one side of a line, which may include the line.

semiplano Parte del plano cartesiano de un lado de una línea, que puede incluir la línea.

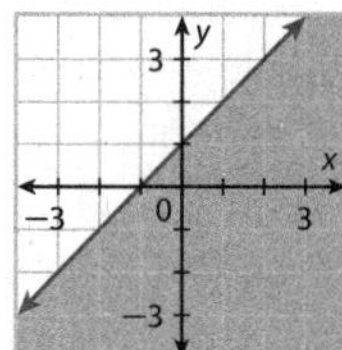

Heron's Formula A triangle with side lengths a, b, and c has area $A=\sqrt{s(s-a)(s-b)(s-c)}$, where s is one-half the perimeter, or $s=\frac{1}{2}(a+b+c)$.

fórmula de Herón Un triángulo con longitudes de lado a, b y c tiene un área $A=\sqrt{s(s-a)(s-b)(s-c)}$, donde s es la mitad del perímetro ó $s=\frac{1}{2}(a+b+c)$.

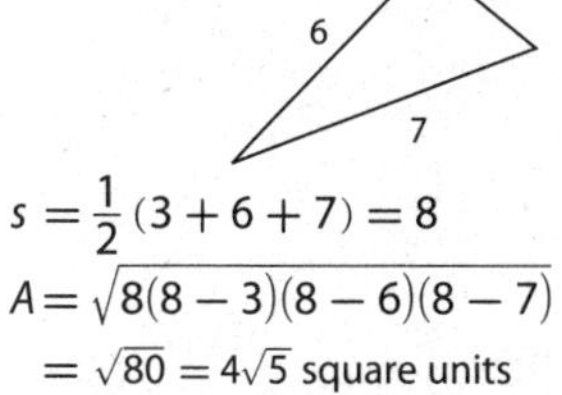

$s=\frac{1}{2}(3+6+7)=8$

$A=\sqrt{8(8-3)(8-6)(8-7)}$

$=\sqrt{80}=4\sqrt{5}$ square units

---

hole (in a graph) An omitted point on a graph. If a rational function has the same factor $x-b$ in both the numerator and the denominator, and the line $x=b$ is not a vertical asymptote, then there is a hole in the graph at the point where $x=b$.

hoyo (en una gráfica) Punto omitido en una gráfica. Si una función racional tiene el mismo factor $x-b$ tanto en el numerador como en el denominador, y la línea $x=b$ no es una asíntota vertical, entonces hay un hoyo en la gráfica en el punto donde $x=b$.

$f(x)=\frac{(x-2)(x+2)}{(x+2)}$ has a hole at $x=-2$.

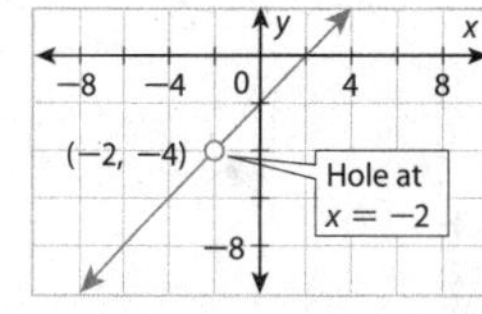

---

horizontal line A line described by the equation $y=b$, where b is the y-intercept.

línea horizontal Línea descrita por la ecuación $y=b$, donde b es la intersección con el eje y.

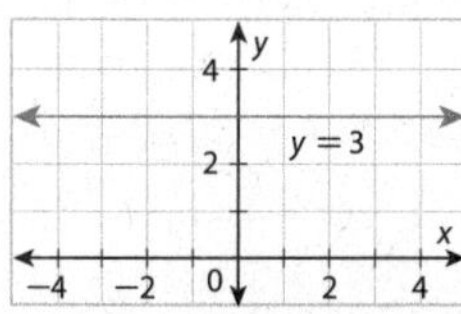

---

horizontal line test If a horizontal line crosses the graph of a function f at more than one point, then the inverse is not a function.

prueba de la línea horizontal Si una línea horizontal cruza la gráfica de una función f en más de un punto, entonces la inversa no es una función.

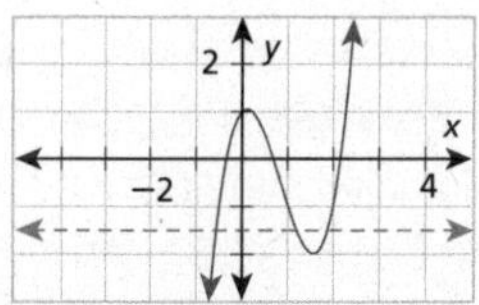

The inverse is not a function.

---

hyperbola The set of all points P in a plane such that the difference of the distances from P to two fixed points F_1 and F_2, called the foci, is a constant $d=|PF_1-PF_2|$.

hipérbola Conjunto de todos los puntos P en un plano tal que la diferencia de las distancias de P a dos puntos fijos F_1 y F_2, llamados focos, es una constante $d=|PF_1-PF_2|$.

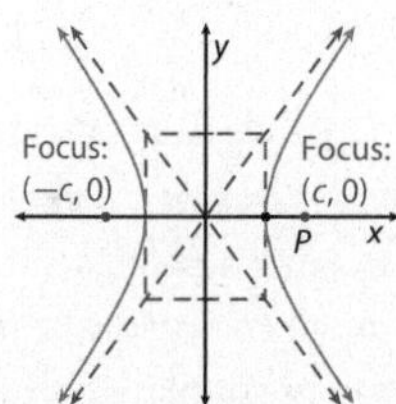

Glossary/Glosario

ENGLISH	SPANISH	EXAMPLES
hypothesis testing A type of testing used to determine whether the difference in two groups is likely to be caused by chance.	**comprobación de hipótesis** Tipo de comprobación que sirve para determinar si el azar es la causa probable de la diferencia entre dos grupos.	

I

ENGLISH	SPANISH	EXAMPLES
imaginary axis The vertical axis in the complex plane, it graphically represents the purely imaginary part of complex numbers.	**eje imaginario** Eje vertical de un plano complejo. Representa gráficamente la parte puramente imaginaria de los números complejos.	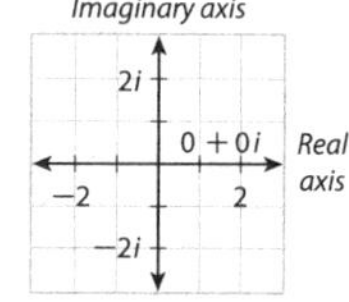
imaginary number The square root of a negative number, written in the form bi, where b is a real number and i is the imaginary unit, $\sqrt{-1}$. Also called a *pure imaginary number*.	**número imaginario** Raíz cuadrada de un número negativo, expresado como bi, donde b es un número real e i es la unidad imaginaria, $\sqrt{-1}$. También se denomina *número imaginario puro*.	$\sqrt{-16} = \sqrt{16} \cdot \sqrt{-1} = 4i$
imaginary part of a complex number For a complex number of the form $a + bi$, the real number b is called the imaginary part, represented graphically as b units on the imaginary axis of a complex plane.	**parte imaginaria de un número complejo** Dado un número complejo del tipo $a + bi$, el número real b se denomina parte imaginaria y se representa gráficamente como b unidades en el eje imaginario de un plano complejo.	
imaginary unit The unit in the imaginary number system, $\sqrt{-1}$.	**unidad imaginaria** Unidad del sistema de números imaginarios, $\sqrt{-1}$.	$\sqrt{-1} = i$
inconsistent system A system of equations or inequalities that has no solution.	**sistema inconsistente** Sistema de ecuaciones o desigualdades que no tiene solución.	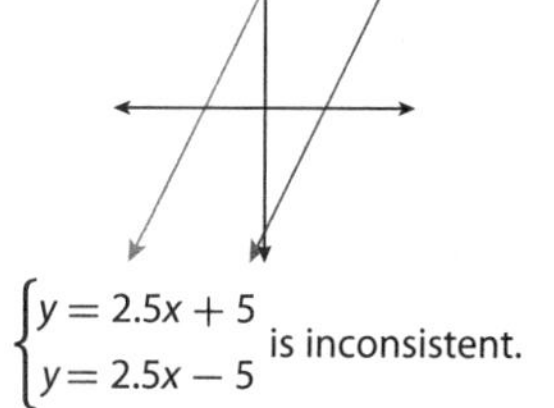 $\begin{cases} y = 2.5x + 5 \\ y = 2.5x - 5 \end{cases}$ is inconsistent.
Increasing A function is increasing on an interval if $f(x_1) < f(x_2)$ when $x_1 < x_2$ for any x-values x_1 and x_2 from the interval.	**creciente** Una función es creciente en un intervalo si $f(x_1) < f(x_2)$ cuando $x_1 < x_2$ dados los valores de x, x_1 y x_2, pertenecienlos al intervalo.	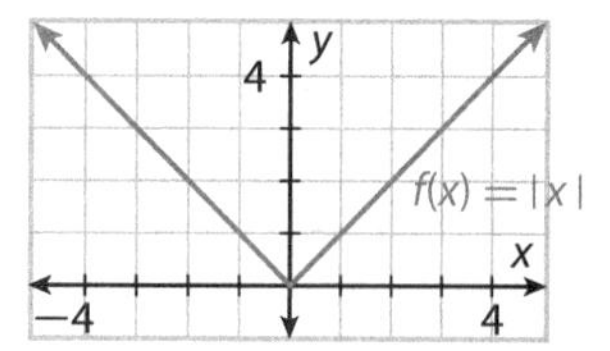 $f(x)$ is increasing on the interval $x > 0$.
independent events Events for which the occurrence or non-occurrence of one event does not affect the probability of the other event.	**sucesos independientes** Dos sucesos son independientes si el hecho de que se produzca o no uno de ellos no afecta la probabilidad del otro suceso.	From a bag containing 3 red marbles and 2 blue marbles, drawing a red marble, replacing it, and then drawing a blue marble.

Glossary/Glosario

ENGLISH	SPANISH	EXAMPLES
independent system A system of equations that has exactly one solution.	**sistema independiente** Sistema de ecuaciones que tiene exactamente una solución.	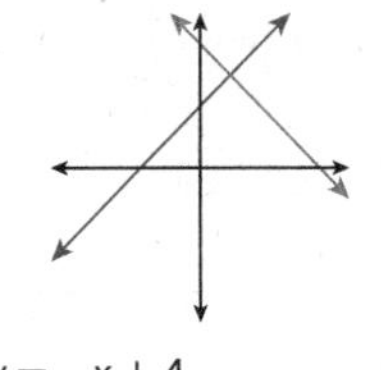 $\begin{cases} y = -x + 4 \\ y = x + 2 \end{cases}$ Solution: (1, 3)
independent variable The input of a function; a variable whose value determines the value of the output, or dependent variable.	**variable independiente** Entrada de una función; variable cuyo valor determina el valor de la salida, o variable dependiente.	$y = 2x + 1$ (arrow to x: independent variable)
index In the radical $\sqrt[n]{x}$ which represents the *n*th root of *x*, *n* is the index. In the radical $\sqrt{x}$, the index is understood to be 2.`	**index** En el radical $\sqrt[n]{x}$, que representa la enésima raíz de *x*, *n* es el índice. En el radical $\sqrt{x}$, se da por sentado que el índice es 2.	The radical $\sqrt[3]{8}$ has an index of 3.
indirect measurement A method of measurement that uses formulas, similar figures, and/or proportions.	**medición indirecta** Método para medir objetos mediante fórmulas, figuras semejantes y/o proporciones.	
inequality A statement that compares two expressions by using one of the following signs: $<, >, \leq, \geq$ or $\neq$.	**desigualdad** Enunciado que compara dos expresiones utilizando uno de los siguientes signos: $<, >, \leq, \geq$, ó, $\neq$.	−4 −3 −2 −1 0 1 2 $x \geq -2$
infinite geometric series A geometric series with infinitely many terms.	**serie geométrica infinita** Serie geométrica con una cantidad infinita de términos.	$\frac{1}{10} + \frac{1}{100} + \frac{1}{1000} + \frac{1}{10{,}000} + \ldots$
infinite sequence A sequence with infinitely many terms.	**sucesión infinita** Sucesión con infinitos términos.	1, 3, 5, 7, 9, 11 ...
infinite set A set with an unlimited, or infinite, number of elements.	**conjunto infinito** Conjunto con un número de elementos ilimitado o infinito.	The set of all integers is an infinite set.
initial side The ray that lies on the positive *x*-axis when an angle is drawn in standard position.	**lado inicial** El rayo que se encuentra en el eje positivo *x* cuando se traza un ángulo en la posición estándar.	Terminal side, 135°, 45°, *y*, *x*, 0, Initial side
integer A member of the set of whole numbers and their opposites.	**entero** Miembro del conjunto de números cabales y sus opuestos.	... −3, −2, −1, 0, 1, 2, 3 ...
interquartile range (IQR) The difference of the third (upper) and first (lower) quartiles in a data set, representing the middle half of the data.	**rango entre cuartiles** Diferencia entre el tercer cuartil (superior) y el primer cuartil (inferior) de un conjunto de datos, que representa la mitad central de los datos.	Lower half: 18, (23), 28, — Upper half: 29, (36), 42 First quartile — Third quartile Interquartile range: 36 − 23 = 13.

Glossary/Glosario

Glossary/Glosario

ENGLISH	SPANISH	EXAMPLES
interval notation A way of writing the set of all real numbers between two endpoints. The symbols [and] are used to include an endpoint in an interval, and the symbols (and) are used to exclude an endpoint from an interval.	**notación de intervalo** Forma de expresar el conjunto de todos los números reales entre dos extremos. Los símbolos [y] se utilizan para incluir un extremo en un intervalo y los símbolos (y) se utilizan para excluir un extremo de un intervalo.	Interval notation / Set-builder notation (a, b) $\{x \mid a < x < b\}$ $(a, b]$ $\{x \mid a < x \leq b\}$ $[a, b)$ $\{x \mid a \leq x < b\}$ $[a, b]$ $\{x \mid a \leq x \leq b\}$
inverse cosine function If the domain of the cosine function is restricted to $[0, \pi]$, then the function Cos $\theta = a$ has an inverse function $\text{Cos}^{-1}\, a = \theta$, also called *arccosine*.	**función coseno inverso** Si el dominio de la función coseno se restringe a $[0, \pi]$, entonces la función Cos $\theta = a$ tiene una función inversa $\text{Cos}^{-1}\, a = \theta$, también llamada *arco coseno*.	$\text{Cos}^{-1}\frac{1}{2} = \frac{\pi}{3}$
inverse function The function that results from exchanging the input and output values of a one-to-one function. The inverse of $f(x)$ is denoted $f^{-1}(x)$.	**función inversa** Función que resulta de intercambiar los valores de entrada y salida de una función uno a uno. La función inversa de $f(x)$ se expresa $f^{-1}(x)$	$f(x)=x^3$ $f^{-1}(x)=\sqrt[3]{x}$ $y = x$
inverse relation The inverse of the relation consisting of all ordered pairs (x, y) is the set of all ordered pairs (y, x). The graph of an inverse relation is the reflection of the graph of the relation across the line $y = x$.	**relación inversa** La inversa de la relación que consta de todos los pares ordenados (x, y) es el conjunto de todos los pares ordenados (y, x). La gráfica de una relación inversa es el reflejo de la gráfica de la relación sobre la línea $y = x$.	$y = x$
inverse sine function If the domain of the sine function is restricted to $\left[-\frac{\pi}{2}, \frac{\pi}{2}\right]$ then the function Sin $\theta =$ a has an inverse function, $\text{Sin}^{-1}\, a = \theta$, also called *arcsine*.	**función seno inverso** Si el dominio de la función seno se restringe a $\left[-\frac{\pi}{2}, \frac{\pi}{2}\right]$, entonces la función Sen $\theta = a$ tiene una función inversa, $\text{Sen}^{-1}\, a = \theta\theta$, también llamada *arco seno*.	$\text{Sin}^{-1}\frac{\sqrt{3}}{2} = \frac{\pi}{3}$
inverse tangent function If the domain of the tangent function is restricted to $\left(-\frac{\pi}{2}, \frac{\pi}{2}\right)$, then the function Tan$\theta = a$ has an inverse function, $\text{Tan}^{-1}\, a = \theta$, also called *arctangent*.	**función tangente inversa** Si el dominio de la función tangente se restringe a $\left(-\frac{\pi}{2}, \frac{\pi}{2}\right)$ entonces la función Tan$\theta =$ a tiene una función inversa, $\text{Tan}^{-1}\, a = \theta$, también llamada *arco tangente*.	$\text{Tan}^{-1}\sqrt{3} = \frac{\pi}{3}$
inverse variation A relationship between two variables, x and y, that can be written in the form $y = \frac{k}{x}$, where k is a nonzero constant and $x \neq 0$.	**variación inversa** Relación entre dos x e y, que puede expresarse en la forma $y = \frac{k}{x}$ donde k es una constante distinta de cero y $x \neq 0$.	$y = \frac{24}{x}$
irrational number A real number that cannot be expressed as the ratio of two integers.	**número irracional** Número real que no se puede expresar como una razón de enteros.	$\sqrt{2}, \pi, e$

ENGLISH	SPANISH	EXAMPLES

irreducible factor A factor of degree 2 or greater that cannot be factored further.

factor irreductible Factor de grado 2 o mayor que no se puede seguir factorizando.

$x^2 + 7x + 1$

iteration The repetitive application of the same rule.

iteración Aplicación repetitiva de la misma regla.

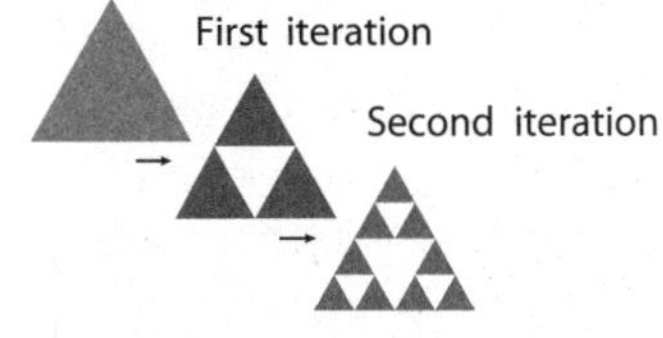

J

joint relative frequency The ratio of the frequency in a particular category divided by the total number of data values.

frecuencia relativa conjunta La razón de la frecuencia en una determinada categoría dividida entre el número total de valores.

joint variation A relationship among three variables that can be written in the form $y = kxz$, where k is a nonzero constant.

variación conjunta Relación entre tres variables que se puede expresar de la forma $y = kxz$, donde k es una constante distinta de cero.

$y = 3xz$

L

Law of Cosines For $\triangle ABC$ with side lengths a, b, and c,
$a^2 = b^2 + c^2 - 2bc \cos A$
$b^2 = a^2 + c^2 - 2ac \cos B$
$c^2 = a^2 + b^2 - 2ab \cos C$.

Ley de cosenos Dado $\triangle ABC$ con longitudes de lado a, b y c,
$a^2 = b^2 + c^2 - 2bc \cos A$
$b^2 = a^2 + c^2 - 2ac \cos B$
$c^2 = a^2 + b^2 - 2ab \cos C$

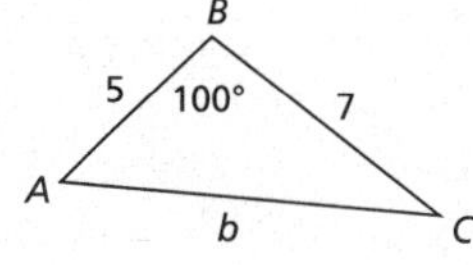

$b^2 = 7^2 + 5^2 - 2(7)(5)\cos 100°$
$b^2 \approx 86.2$
$b \approx 9.3$

law of large numbers The tendency of experimental probability to approach theoretical probability as the number of trials gets very large.

Ley de los números grandes Tendencia de la probabilidad experimental a acercarse a la probabilidad teórica cuando el número de pruebas es muy grande.

The more times you toss a coin, the closer the experimental probability will be to $\frac{1}{2}$.

Law of Sines For $\triangle ABC$ with side lengths a, b, and c,
$\frac{\sin A}{a} = \frac{\sin B}{b} = \frac{\sin C}{c}$.

Ley de senos Dado $\triangle ABC$ con longitudes de lado a, b y c,
$\frac{\text{sen} A}{a} = \frac{\text{sen} B}{b} = \frac{\text{sen} C}{c}$.

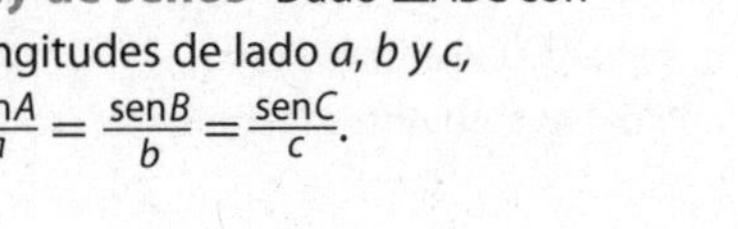

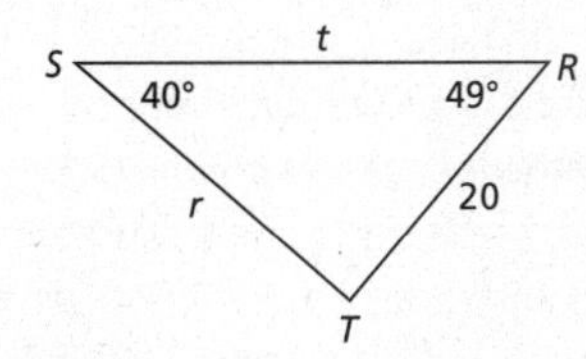

$\frac{\sin 49°}{r} = \frac{\sin 40°}{20}$

$r = \frac{20 \sin 49°}{\sin 40°} \approx 23.5$

leading coefficient The coefficient of the first term of a polynomial in standard form.

coeficiente principal Coeficiente del primer termino de un polinomio en forma estandar

$3x^2 + 7x - 2$
↑
Leading coefficient

ENGLISH	SPANISH	EXAMPLES
least common denominator (LCD) The least common multiple of two or more given denominators.	**minimo comun denominador (mcd)** Minimo comun multiplo de dos o mas denominadores dados.	The LCD of $\frac{3}{4}$ and $\frac{5}{6}$ is 12.
least common multiple (LCM) The product of the smallest positive number and the lowest power of each variable that divides evenly into each term.	**mínimo común múltiplo (mcm)** El producto del número positivo más pequeño y la potencia más baja de cada variable que divide exactamente cada término.	The LCM of 10 and 18 is 90. The LCM of $2x^2$ and $5x^3$ is $10x^3$.
least-squares line The line of fit for which the sum of the squares of the residuals is as small as possible.	**recta de mínimos cuadrados** Recta de ajuste para la cual la suma de los cuadrados de los residuos es la menor posible.	
limit For an infinite arithmetic series that converges, the number that the partial sums approach.	**límite** Para un serie que coverge, el número que se aproximan las sumas.	The series $\frac{1}{2}+\frac{1}{4}+\frac{1}{8}+\frac{1}{16}+\ldots$ has a limit of 1.
line of best fit The line that comes closest to all of the points in a data set.	**línea de mejor ajuste** Línea que más se acerca a todos los puntos de un conjunto de datos.	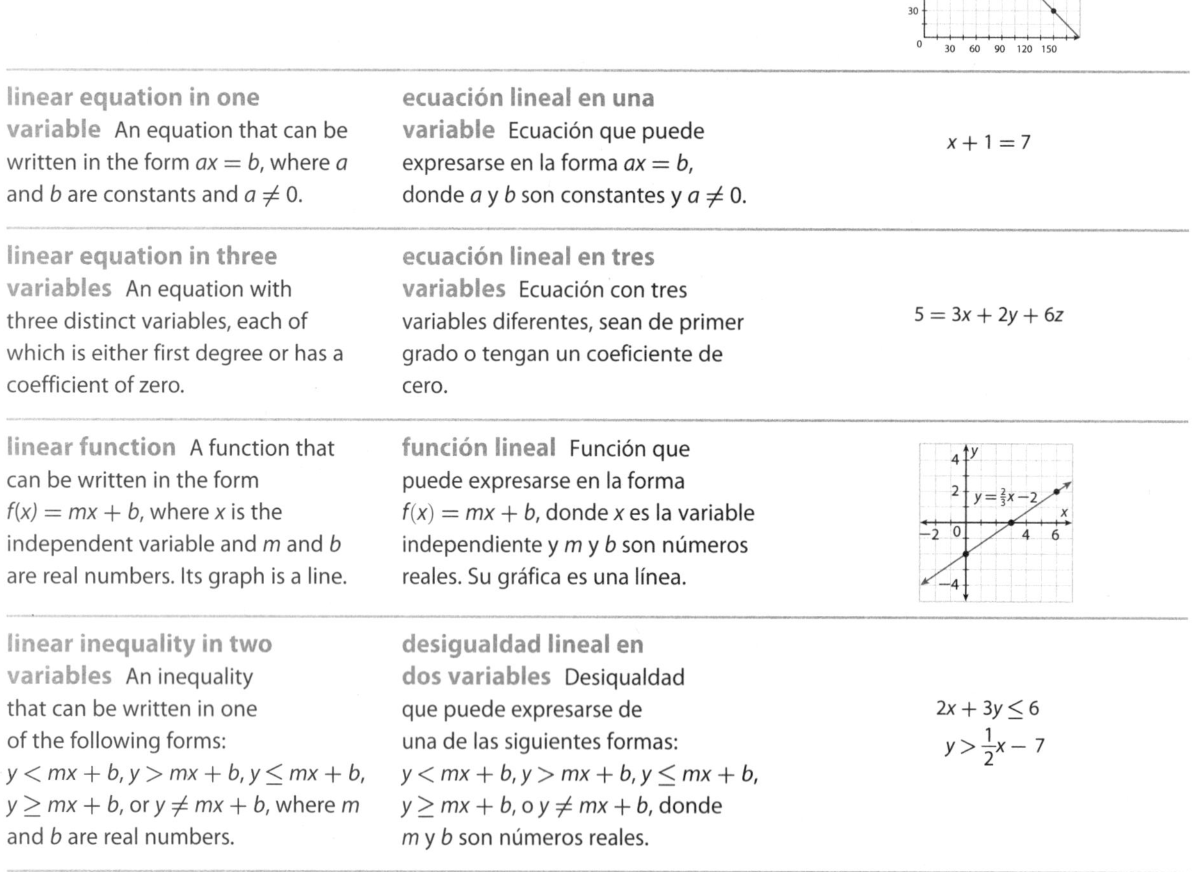
linear equation in one variable An equation that can be written in the form $ax = b$, where a and b are constants and $a \neq 0$.	**ecuación lineal en una variable** Ecuación que puede expresarse en la forma $ax = b$, donde a y b son constantes y $a \neq 0$.	$x + 1 = 7$
linear equation in three variables An equation with three distinct variables, each of which is either first degree or has a coefficient of zero.	**ecuación lineal en tres variables** Ecuación con tres variables diferentes, sean de primer grado o tengan un coeficiente de cero.	$5 = 3x + 2y + 6z$
linear function A function that can be written in the form $f(x) = mx + b$, where x is the independent variable and m and b are real numbers. Its graph is a line.	**función lineal** Función que puede expresarse en la forma $f(x) = mx + b$, donde x es la variable independiente y m y b son números reales. Su gráfica es una línea.	$y = \frac{2}{3}x - 2$
linear inequality in two variables An inequality that can be written in one of the following forms: $y < mx + b$, $y > mx + b$, $y \leq mx + b$, $y \geq mx + b$, or $y \neq mx + b$, where m and b are real numbers.	**desigualdad lineal en dos variables** Desigualdad que puede expresarse de una de las siguientes formas: $y < mx + b$, $y > mx + b$, $y \leq mx + b$, $y \geq mx + b$, o $y \neq mx + b$, donde m y b son números reales.	$2x + 3y \leq 6$ $y > \frac{1}{2}x - 7$
linear regression A statistical method used to fit a linear model to a given data set.	**regresión lineal** Método estadístico utilizado para ajustar un modelo lineal a un conjunto de datos determinado.	LinReg y=ax+b

ENGLISH	SPANISH	EXAMPLES
linear system A system of equations containing only linear equations.	**sistema lineal** Sistema de ecuaciones que contiene sólo ecuaciones lineales.	$\begin{cases} y = 2x + 1 \\ x + y = 8 \end{cases}$
local maximum For a function f, $f(a)$ is a local maximum if there is an interval around a such that $f(x) < f(a)$ for every x-value in the interval except a.	**máximo local** Dada una función f, $f(a)$ es el máximo local si hay un intervalo en a tal que $f(x) < f(a)$ para cada valor de x en el intervalo excepto a.	Local maximum
local minimum For a function f, $f(a)$ is a local minimum if there is an interval around a such that $f(x) > f(a)$ for every x-value in the interval except a.	**mínimo local** Dada una función f, $f(a)$ es el mínimo local si hay un intervalo en a tal que $f(x) > f(a)$ para cada valor de x en el intervalo excepto a.	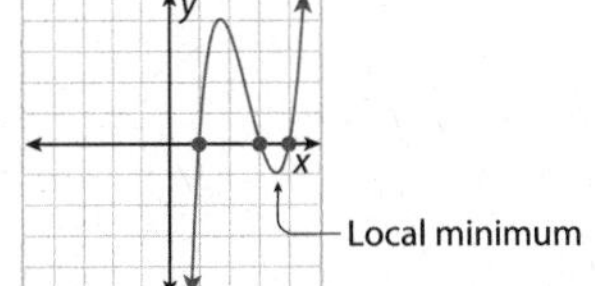
logarithm The exponent that a specified base must be raised to in order to get a certain value.	**logaritmo** Exponente al cual debe elevarse una base determinada a fin de obtener cierto valor.	$\log_2 8 = 3$, because 3 is the power that 2 is raised to in order to get 8; or $2^3 = 8$.
logarithmic equation An equation that contains a logarithm of a variable.	**ecuación logarítmica** Ecuación que contiene un logaritmo de una variable.	$\log x + 3 = 7$
logarithmic function A function of the form $f(x) = \log_b x$, where $b \neq 1$ and $b > 0$, which is the inverse of the exponential function $f(x) = b^x$.	**función logarítmica** Función del tipo $f(x) = \log_b x$, donde $b \neq 1$ y $b > 0$, que es la inversa de la función exponencial $f(x) = b^x$.	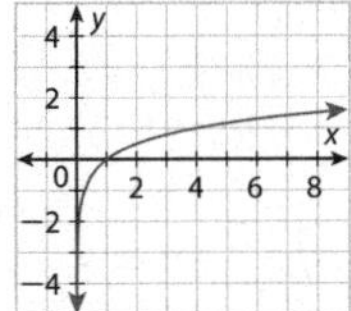 $f(x) = \log_4 x$
logarithmic regression A statistical method used to fit a logarithmic model to a given data set.	**regresión logarítmica** Método estadístico utilizado para ajustar un modelo logarítmico a un conjunto de datos determinado.	
logistic function An exponential growth function that tapers off at an asymptote.	**función logística** Función de crecimiento exponencial que disminuye en una asíntota.	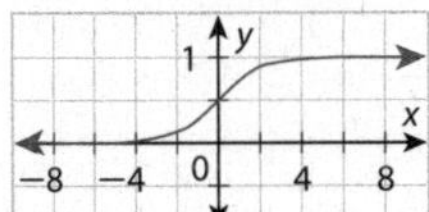

M

ENGLISH	SPANISH	EXAMPLES
major axis The longer axis of an ellipse. The foci of the ellipse are located on the major axis, and its endpoints are the *vertices of the ellipse*.	**eje mayor** El eje más largo de una elipse. Los focos de la elipse se encuentran sobre el eje mayor y sus extremos son los *vértices de la elipse*.	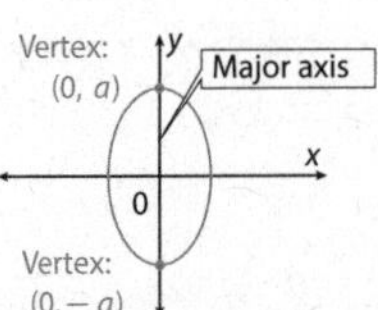
margin of error In a random sample, it defines an interval, centered on the sample percent, in which the population percent is most likely to lie.	**margen de error** En una muestra aleatoria, define un intervalo, centrado en el porcentaje de muestra, en el que es más probable que se encuentre el porcentaje de población.	

ENGLISH	SPANISH	EXAMPLES
marginal relative frequency The sum of the joint relative frequencies in a row or column of a two-way table.	**frecuencia relativa marginal** La suma de las frecuencias relativas conjuntas en una fila o columna de una tabla de doble entrada.	
mathematical induction A type of mathematical proof. To prove that a statement is true for all natural numbers *n*, first show that the statement is true for $n = 1$; then assume it is true for some number *k* and prove that it is true for $k + 1$. It follows that the statement is true for all values of *n*.	**inducción matemática** Tipo de demostración matemática. Para demostrar que un enunciado se cumple para todos los números naturales *n*, primero se demuestra que el enunciado se cumple para $n = 1$; luego se supone que se cumple para un número *k* y se demuestra que se cumple para $k + 1$. Por lo tanto, el enunciado se cumplirá para todos los valores de *n*.	
matrix A rectangular array of numbers.	**matriz** Arreglo rectangular de números.	$\begin{bmatrix} 1 & 0 & 3 \\ -2 & 2 & -5 \\ 7 & -6 & 3 \end{bmatrix}$
matrix equation An equation of the form $AX = B$, where A is the coefficient matrix, *X* is the variable matrix, and *B* is the constant matrix of a system of equations.	**ecuación matricial** Ecuación del tipo $AX = B$, donde *A* es la matrizde coeficientes, *X* es la matriz de variables y *B* es la matriz de constantes de un sistema de ecuaciones.	System of equations: $\begin{matrix} 2x + 3y = 7 \\ 4x - 6y = 5 \end{matrix}$ Matrix equation: $\begin{bmatrix} 2 & 3 \\ 4 & -6 \end{bmatrix}\begin{bmatrix} x \\ y \end{bmatrix} = \begin{bmatrix} 7 \\ 5 \end{bmatrix}$
matrix product The product of two matrices, where each entry in P_{ij} is the sum of the products of consecutive entries in row *i* in matrix *A* and column *j* in matrix *B*.	**producto matricial** Producto de dos matrices, donde cada entrada de P_{ij} es la suma de los productos de las entradas consecutivas de la fila *i* de la matriz *A* y de la columna *j* de la matriz *B*.	$\begin{bmatrix} 1 & 2 \\ 3 & 4 \end{bmatrix}\begin{bmatrix} 5 & 6 \\ 7 & 8 \end{bmatrix}$ $= \begin{bmatrix} 1(5) + 2(7) & 1(6) + 2(8) \\ 3(5) + 4(7) & 3(6) + 4(8) \end{bmatrix}$ $= \begin{bmatrix} 19 & 22 \\ 43 & 50 \end{bmatrix}$
maximum value of a function The *y*-value of the highest point on the graph of the function.	**máximo de una función** Valor de *y* del punto más alto en la gráfica de la función.	Maximum value
mean The sum of all the values in a data set divided by the number of data values. Also called the *average*.	**media** Suma de todos los valores de un conjunto de datos dividida entre el número de valores de datos. También llamada *promedio*.	Data set: 4, 6, 7, 8, 10 Mean: $\frac{4 + 6 + 7 + 8 + 10}{5} = \frac{35}{5} = 7$
measure of central tendency A measure that describes the center of a data set.	**medida de tendencia dominante** Medida que describe el centro de un conjunto de datos.	the mean, median, or mode
measure of variation A measure that describes the spread of a data set.	**medida de variación** Medida que describe la amplitud de un conjunto de datos.	the range, variance, standard deviation, or interquartile range

ENGLISH	SPANISH	EXAMPLES
median of a data set For an ordered data set with an odd number of values, the median is the middle value. For an ordered data set with an even number of values, the median is the average of the two middle values.	**mediana de un conjunto de datos** Dado un conjunto de datos ordenados con un número impar de valores, la mediana es el valor del medio. Dado un conjunto de datos ordenados con un número par de valores, la mediana es el promedio de los dos valores del medio	8, 9, (9), 12, 15 median: 9 4, 6, (7, 10), 10, 12 median: $\frac{7+10}{2} = 8.5$
midpoint The point that divides a segment into two congruent segments.	**punto medio** Punto que divide un segmento en dos segmentos congruentes.	A B C Point B is the midpoint of $\overline{AC}$.
minimum value of a function The *y*-value of the lowest point on the graph of the function.	**mínimo de una función** Valor de *y* del punto más bajo en la gráfica de la función.	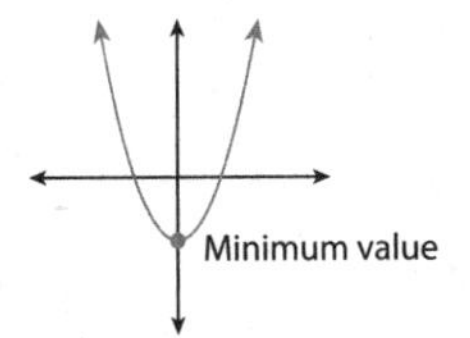
minor axis The shorter axis of an ellipse. Its endpoints are the *co-vertices of the ellipse.*	**eje menor** El eje más corto de una elipse. Sus extremos son los *co-vértices de la elipse.*	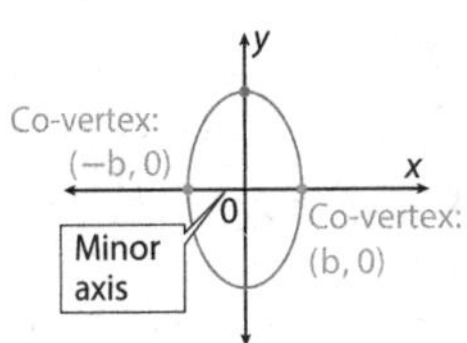
mode The value or values that occur most frequently in a data set; if all values occur only once, the data set is said to have no mode.	**moda** El valor o los valores que se presentan con mayor frecuencia en un conjunto de datos. Si todos los valores se presentan con la misma frecuencia, se dice que el conjunto de datos no tiene moda.	Data set: 3, 6, (8, 8), 10 mode: 8 Data set: 2, (5, 5), (7, 7) Mode: 5 and 7 Data set: 2, 3, 6, 9, 11 No mode
monomial A number or a product of numbers and variables with whole-number exponents, or a polynomial with one term.	**monomio** Número o producto de números y variables con exponentes de números cabales, o polinomio con un término.	$8x, 9, 3x^2y^4$
multiple root A root *r* is a multiple root when the factor $(x - r)$ appears in the equation more than once.	**raíz múltiple** Una raíz r es una raíz múltiple cuando el factor $(x - r)$ aparece en la ecuación más de una vez.	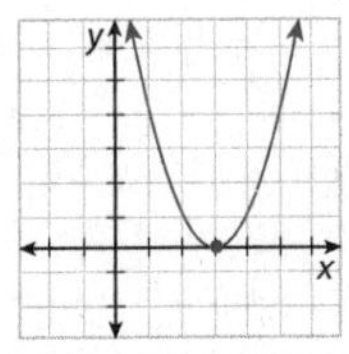3 is a multiple root of $P(x) = (x - 3)^2$.
multiplicative identity matrix A square matrix with 1 in every entry of the main diagonal and 0 in every other entry.	**matriz de identidad multiplicativa** Una matriz cuadrada que contiene 1 en cada entrada de la diagonal principal y 0 en las demás entradas.	$\begin{bmatrix} 1 & 0 \\ 0 & 1 \end{bmatrix}, \begin{bmatrix} 1 & 0 & 0 \\ 0 & 1 & 0 \\ 0 & 0 & 1 \end{bmatrix}$

ENGLISH	SPANISH	EXAMPLES
multiplicative inverse of a square matrix The multiplicative inverse of square matrix A, if it exists, is notated A^{-1}, where the product of A and A^{-1} is the identity matrix.	**inverso multiplicativo de una matriz cuadrada** El inverso multiplicativo de una matriz cuadrada A, si existe, se escribe A^{-1}, donde el producto de A y A^{-1} es la matriz de identidad.	The multiplicative inverse of $A = \begin{bmatrix} -2 & 5 \\ 1 & -3 \end{bmatrix}$ is $A^{-1} = \begin{bmatrix} -3 & -5 \\ -1 & -2 \end{bmatrix}$ because $AA^{-1} = A^{-1}A = \begin{bmatrix} 1 & 0 \\ 0 & 1 \end{bmatrix}$.
multiplicity If a polynomial $P(x)$ has a multiple root at r, the multiplicity of r is the number of times $(x - r)$ appears as a factor in $P(x)$.	**multiplicidad** Si un polinomio $P(x)$ tiene una raíz múltiple en r, la multiplicidad de r es la cantidad de veces que $(x - r)$ aparece como factor en $P(x)$.	For $P(x) = (x - 3)^2$, the root 3 has a multiplicity of 2.
mutually exclusive events Two events are mutually exclusive if they cannot both occur in the same trial of an experiment.	**sucesos mutuamente excluyentes** Dos sucesos son mutuamente excluyentes si ambos no pueden ocurrir en la misma prueba de un experimento.	In the experiment of rolling a number cube, rolling a 3 and rolling an even number are mutually exclusive events.

N

ENGLISH	SPANISH	EXAMPLES
natural logarithm A logarithm with base e, written as ln.	**logaritmo natural** Logaritmo con base e, que se escribe ln.	$\ln 5 = \log_e 5 \approx 1.6$
natural logarithmic function The function $f(x) = \ln x$, which is the inverse of the natural exponential function $f(x) = e^x$. Domain is $\{x \mid x > 0\}$; range is all real numbers	**función logarítmica natural** Función $f(x) = \ln x$, que es la inversa de la función exponencial natural $f(x) = e^x$. El dominio es $\{x \mid x > 0\}$; el rango es todos los números reales.	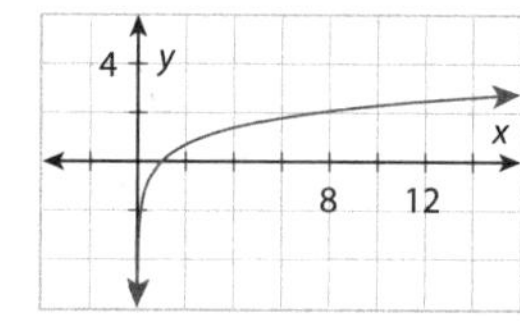
natural number A counting number.	**número natural** Número que sirve para contar.	1, 2, 3, 4, 5, 6, . . .
negative exponent A base raised to a negative exponent is equal to the reciprocal of that base raised to the opposite exponent: $b^{-n} = \frac{1}{b^n}$.	**exponente negativo** Una base elevada a un exponente negativo es igual al recíproco de dicha base elevado al exponente opuesto: $b^{-n} = \frac{1}{b^n}$.	$5^{-3} = \frac{1}{5^3} = \frac{1}{125}$
net A diagram of the faces of a three-dimensional figure arranged in such a way that the diagram can be folded to form the three-dimensional figure.	**plantilla** Diagrama de las caras de una figura tridimensional que se puede plegar para formar la figura tridimensional.	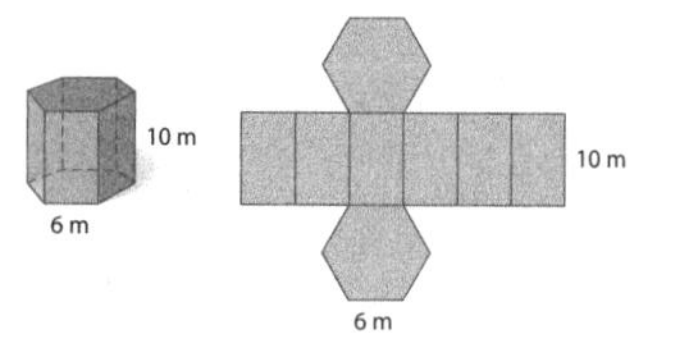
nonlinear system of equations A system in which at least one of the equations is not linear.	**sistema no lineal de ecuaciones** Sistema en el cual por lo menos una de las ecuaciones no es lineal.	$\begin{cases} y = 2x^2 \\ y = -3^2 + 5 \end{cases}$

ENGLISH	SPANISH	EXAMPLES
***n*th root** The *n*th root of a number a, written as $\sqrt[n]{a}$ or $a^{\frac{1}{n}}$, is a number that is equal to a when it is raised to the *n*th power.	**enésima raíz** La enésima raíz de un número a, que se escribe como $\sqrt[n]{a}$ or $a^{\frac{1}{n}}$, es un número igual a a cuando se eleva a la enésima potencia.	$\sqrt[5]{32} = 2$, because $2^5 = 32$.

ENGLISH	SPANISH	EXAMPLES
objective function The function to be maximized or minimized in a linear programming problem.	**función objetiva** Función que se debe maximizar o minimizar en un problema de programación lineal.	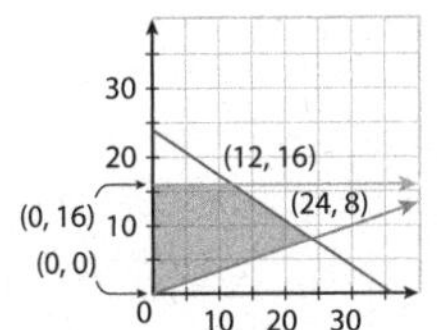 The objective function $P = 18x + 25y$ is maximized at (24, 8).
observational study A study that observes individuals and measures variables without controlling the individuals or their environment in any way.	**estudio de observación** Estudio que permite observar a individuos y medir variables sin controlar a los individuos ni su ambiente.	
odd function A function in which $f(-x) = -f(x)$ for all x in the domain of the function.	**función impar** Función en la que $f(-x) = -f(x)$ para todos los valores de x dentro del dominio de la función	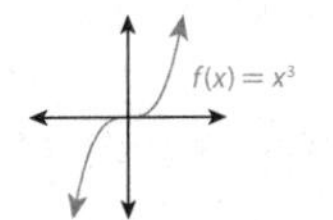 $f(x) = x^3$ is an odd function.
one-to-one function A function in which each *y*-value corresponds to only one *x*-value. The inverse of a one-to-one function is also a function.	**función uno a uno** Función en la que cada valor de *y* corresponde a sólo un valor de *x*. La inversa de una función uno a uno es también una función.	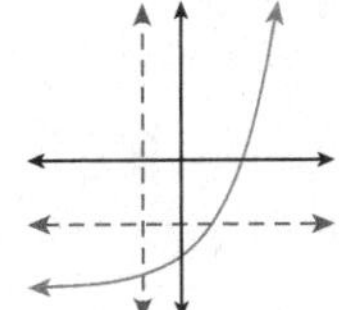
order of operations A process for evaluating expressions: First, perform operations in parentheses or other grouping symbols. Second, evaluate powers and roots. Third, perform all multiplication and division from left to right. Fourth, perform all addition and subtraction from left to right.	**orden de las operaciones** Proceso para evaluar las expresiones: Primero, realizar las operaciones entre paréntesis u otros símbolos de agrupación. Segundo, evaluar las potencias y las raíces. Tercero, realizar todas las multiplicaciones y divisiones de izquierda a derecha. Cuarto, realizar todas las sumas y restas de izquierda a derecha.	$2 + 3^2 - (7 + 5) \div 4 \cdot 3$ $2 + 3^2 - 12 \div 4 \cdot 3$ Add inside parentheses. $2 + 9 - 12 \div 4 \cdot 3$ Evaluate the power. $2 + 9 - 3 \cdot 3$ Divide. $2 + 9 - 9$ Multiply. $11 - 9$ Add. 2 Subtract.
ordered triple A set of three numbers that can be used to locate a point (x, y, z) in a three-dimensional coordinate system.	**tripleta ordenada** Conjunto de tres números que se pueden utilizar para ubicar un punto (x, y, z) en un sistema de coordenadas tridimensional.	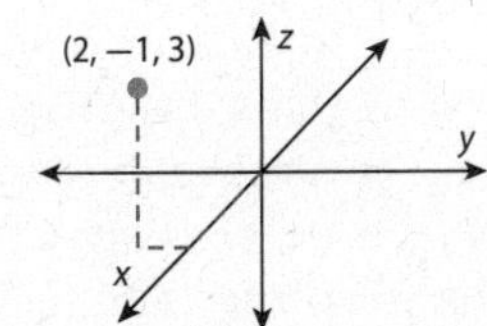

ENGLISH	SPANISH	EXAMPLES
origin The intersection of the x- and y-axes in a coordinate plane. The coordinates of the origin are $(0, 0)$.	**origen** Intersección de los ejes x e y en un plano cartesiano. Las coordenadas de origen son $(0, 0)$.	Origin; 0

P

ENGLISH	SPANISH	EXAMPLES
parabola The shape of the graph of a quadratic function. Also, the set of points equidistant from a point F, called the focus, and a line d, called the *directrix*.	**parábola** Forma de la gráfica de una función cuadrática. También, conjunto de puntos equidistantes de un punto F, denominado *foco*, y una línea d, denominada *directriz*.	Focus; Directrix
parameter One of the constants in a function or equation that may be changed. Also the third variable in a set of parametric equations.	**parámetro** Una de las constantes en una función o ecuación que se puede cambiar. También es la tercera variable en un conjunto de ecuaciones paramétricas.	$y = (x - h)^2 + k$ — parameters
parametric equations A pair of equations that define the x- and y-coordinates of a point in terms of a third variable called a parameter.	**ecuaciones paramétricas** Par de ecuaciones que definen las coordenadas x e y de un punto en función de una tercera variable denominada parámetro.	$x(t) = t + 1$ $y(t) = -2t$
parent cube root function The function $f(x) = \sqrt[3]{x}$.	**función madre de la raíz cúbica** Función del tipo $f(x) = \sqrt[3]{x}$.	$f(x) = \sqrt[3]{x}$
parent function The simplest function with the defining characteristics of the family. Functions in the same family are transformations of their parent function.	**función madre** La función más básica con las características de la familia. Las funciones de la misma familia son transformaciones de su función madre.	$f(x) = x^2$ is the parent function for $g(x) = x^2 + 4$ and $h(x) = 5(x + 2)^2 - 3$.
parent square root function The function $f(x) = \sqrt{x}$, where $x \geq 0$.	**función madre de la raíz cuadrada** Función del tipo $f(x) = \sqrt{x}$, donde $x \geq 0$.	$f(x) = \sqrt{x}$
partial sum Indicated by $S_n = \sum_{i=1}^{n} a_i$, the sum of a specified number of terms n of a sequence whose total number of terms is greater than n.	**suma parcial** Expresada por $S_n = \sum_{i=1}^{n} a_i$, la suma de un número específico n de términos de una sucesión cuyo número total de términos es mayor que n.	For the sequence $a_n = n^2$, the fourth partial sum of the infinite series $\sum_{k=1}^{\infty} k^2$ is $\sum_{k=1}^{4} k^2 = 1^2 + 2^2 + 3^2 + 4^2 = 30$.
Pascal's triangle A triangular arrangement of numbers in which every row starts and ends with 1 and each other number is the sum of the two numbers above it.	**triángulo de Pascal** Arreglo triangular de números en el cual cada fila comienza y termina con 1 y cada uno de los demás números es la suma de los dos números que están encima de él.	1 1 1 1 2 1 1 3 3 1 1 4 6 4 1

ENGLISH	SPANISH	EXAMPLES

perfect square A number whose positive square root is a whole number.

cuadrado perfecto Número cuya raíz cuadrada positiva es un número cabal.

36 is a perfect square because $\sqrt{36} = 6$.

---

perfect-square trinomial A trinomial whose factored form is the square of a binomial. A perfect-square trinomial has the form $a^2 - 2ab + b^2 = (a - b)^2$ or $a^2 + 2ab + b^2 = (a + b)^2$.

trinomio cuadrado perfecto Trinomio cuya forma factorizada es el cuadrado de un binomio. Un trinomio cuadrado perfecto tiene la forma $a^2 - 2ab + b^2 = (a - b)^2$ o $a^2 + 2ab + b^2 = (a + b)^2$.

$x^2 + 6x + 9$ is a perfectsquare trinomial, because $x^2 + 6x + 9 = (x + 3)^2$.

---

period of a periodic function The length of a cycle measured in units of the independent variable (usually time in seconds). Also the reciprocal of the frequency.

periodo de una función periódica Longitud de un ciclo medido en unidades de la variable independiente (generalmente el tiempo en segundos). También es la inversa de la frecuencia.

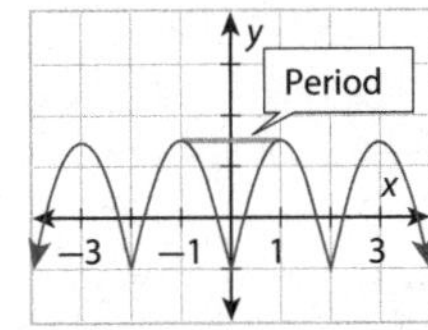

---

periodic function A function that repeats exactly in regular intervals, called *periods*.

función periódica Función que se repite exactamente a intervalos regulares denominados *periodos*.

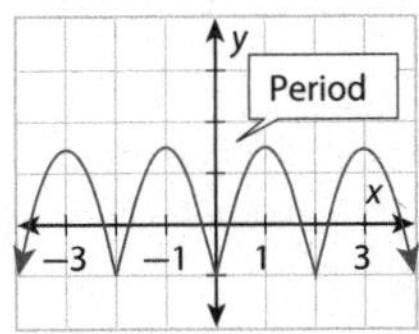

---

permutation An arrangement of a group of objects in which order is important. The number of permutations of r objects from a group of n objects is denoted ${}_nP_r$.

permutación Arreglo de un grupo de objetos en el cual el orden es importante. El número de permutaciones de r objetos de un grupo de n objetos se expresa ${}_nP_r$.

For 4 objects A, B, C, and D, there are ${}_4P_2 = 12$ different permutations of 2 objects: *AB*, *AC*, *AD*, *BC*, *BD*, *CD*, *BA*, *CA*, *DA*, *CB*, *DB*, and *DC*.

---

phase shift A horizontal translation of a periodic function.

cambio de fase Traslación horizontal de una función periódica.

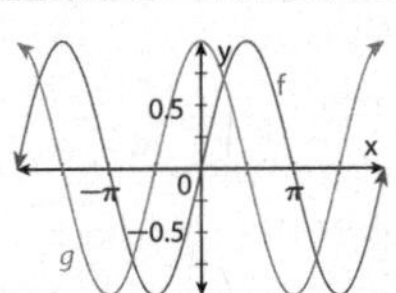

g is a phase shift of f $\frac{\pi}{2}$ units left.

---

piecewise function A function that is a combination of one or more functions.

función a trozos Función que es una combinación de una o más funciones.

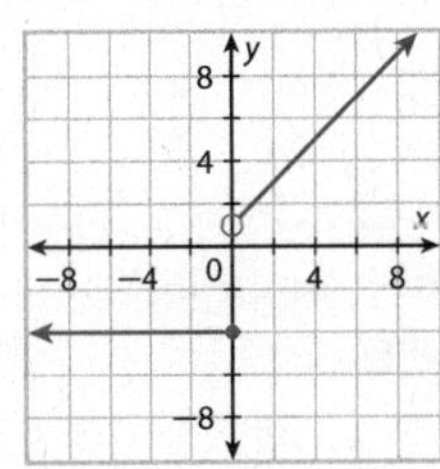

$$f(x) = \begin{cases} -4 & \text{if } x \leq 0 \\ x + 1 & \text{if } x > 0 \end{cases}$$

---

point-slope form The point-slope form of a linear equation is $y - y_1 = m(x - x_1)$, where m is the slope and (x_1, y_1) is a point on the line.

forma de punto y pendiente La forma de punto y pendiente de una ecuación lineal es $y - y_1 = m(x - x_1)$, donde m es la pendiente y (x_1, y_1) es un punto en la línea.

The equation of the line through $(2, 1)$ with slope 3 is $y - 1 = 3(x - 2)$.

ENGLISH | SPANISH | EXAMPLES

polynomial A monomial or a sum or difference of monomials.

polinomio Monomio o suma o diferencia de monomios.

$2x^2 + 3x - 7$

---

polynomial function A function whose rule is a polynomial.

función polinomial Función cuya regla es un polinomio.

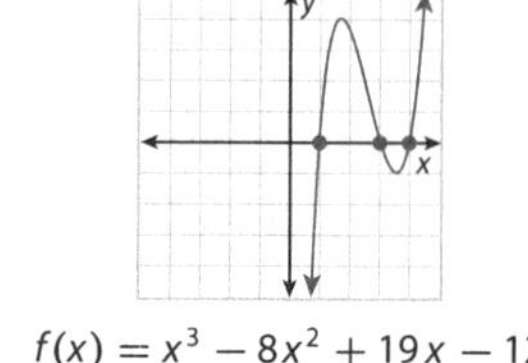

$f(x) = x^3 - 8x^2 + 19x - 12$

---

polynomial identity A mathematical relationship equating one polynomial quantity to another.

identidad de polinomios Relación matemática que iguala una cantidad polinomial con otra.

$(x^4 - y^4) = (x^2 + y^2)(x^2 - y^2)$

---

population The entire group of objects or individuals considered for a survey.

población Grupo completo de objetos o individuos que se desea estudiar.

In a survey about the study habits of high school students, the population is all high school students.

---

probability A number from 0 to 1 (or 0% to 100%) that is the measure of how likely an event is to occur.

probabilidad Número entre 0 y 1 (o entre 0% y 100%) que describe cuán probable es que ocurra un suceso.

A bag contains 3 red marbles and 4 blue marbles. The probability of choosing a red marble is $\frac{3}{7}$.

---

probability distribution for an experiment The function that pairs each outcome with its probability.

distribución de probabilidad para un experimento Función que asigna a cada resultado su probabilidad.

A number cube is rolled 10 times. The results are shown in the table.

Outcome	1	2	3	4	5	6
Probability	$\frac{1}{10}$	$\frac{1}{5}$	$\frac{1}{5}$	0	$\frac{3}{10}$	$\frac{1}{5}$

---

probability sample A sample in which every member of the population being sampled has a nonzero probability of being selected.

muestra de probabilidad Muestra en la que cada miembro de la población que se estudia tiene una probabilidad distinta de cero de ser elegido.

---

proportion A statement that two ratios are equal; $\frac{a}{b} = \frac{c}{d}$.

proporción Enunciado que establece que dos razones son iguales; $\frac{a}{b} = \frac{c}{d}$.

$\frac{2}{3} = \frac{4}{6}$

---

pure imaginary number *See* imaginary number.

número imaginario puro Ver número imaginario.

$3i$

Q

quadratic equation An equation that can be written in the form $ax^2 + bx + c = 0$, where a, b, and c are real numbers and $a \neq 0$.

ecuación cuadrática Ecuación que se puede expresar como $ax^2 + bx + c = 0$, donde a, b y c son números reales y $a \neq 0$.

$x^2 + 3x - 4 = 0$
$x^2 - 9 = 0$

ENGLISH	SPANISH	EXAMPLES

Quadratic Formula The formula $x = \frac{-b \pm \sqrt{b^2 - 4ac}}{2a}$, which gives solutions, or roots, of equations in the form $ax^2 + bx + c = 0$, where $a \neq 0$.

fórmula cuadrática La fórmula $x = \frac{-b \pm \sqrt{b^2 - 4ac}}{2a}$, que da soluciones, o raíces, para las ecuaciones del tipo $ax^2 + bx + c = 0$, donde $a \neq 0$.

The solutions of $2x^2 - 5x - 3 = 0$ are given by

$$x = \frac{-(-5) \pm \sqrt{(-5)^2 - 4(2)(-3)}}{2(2)}$$

$$= \frac{5 \pm \sqrt{25 + 24}}{4} = \frac{5 \pm 7}{4};$$

$x = 3$ or $x = -\frac{1}{2}$.

---

quadratic function A function that can be written in the form $f(x) = ax^2 + bx + c$, where a, b, and c are real numbers and $a \neq 0$, or in the form $f(x) = a(x - h)^2 + k$, where a, h, and k are real numbers and $a \neq 0$.

función cuadrática Función que se puede expresar como $f(x) = ax^2 + bx + c$, donde a, b y c son números reales y $a \neq 0$, o como $f(x) = a(x - h)^2 + k$, donde a, h y k son números reales y $a \neq 0$.

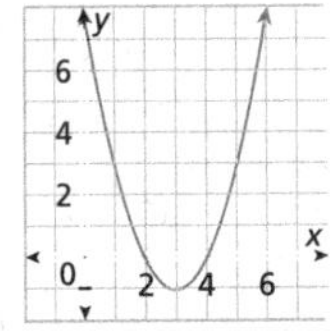

$f(x) = x^2 - 6x + 8$

---

quadratic inequality in two variables An inequality that can be written in one of the following forms:
$y < ax^2 + bx + c$,
$y > ax^2 + bx + c$,
$y \leq ax^2 + bx + c$,
$y \geq ax^2 + bx + c$,
or $y \neq ax^2 + bx + c$,
where a, b, and c are real numbers and $a \neq 0$.

desigualdad cuadrática en dos variables Desigualdad que puede expresarse de una de las siguientes formas:
$y < ax^2 + bx + c$,
$y > ax^2 + bx + c$,
$y \leq ax^2 + bx + c$,
$y \geq ax^2 + bx + c$,
o $y \neq ax^2 + bx + c$, donde a, b y c son números reales y $a \neq 0$.

$y > -x^2 - 2x + 3$

---

quadratic model A quadratic function used to represent a set of data.

modelo cuadrático Función cuadrática que se utiliza para representar un conjunto de datos.

x	4	6	8	10
$f(x)$	27	52	89	130

A quadratic model for the data is $f(x) = x^2 + 3.3x - 2.6$.

---

quadratic regression A statistical method used to fit a quadratic model to a given data set.

regresión cuadrática Método estadístico utilizado para ajustar un modelo cuadrático a un conjunto de datos determinado.

R

radian A unit of angle measure based on arc length. In a circle of radius r, if a central angle has a measure of 1 radian, then the length of the intercepted arc is r units.

2π radians $= 360°$
1 radian $\approx 57°$

radián Unidad de medida de un ángulo basada en la longitud del arco. En un círculo de radio r, si un ángulo central mide 1 radián, entonces la longitud del arco abarcado es r unidades.

2π radianes $= 360°$
1 radián $\approx 57°$

---

radical An indicated root of a quantity.

radical Raíz indicada de una cantidad.

$\sqrt{36} = 6$, $\sqrt[3]{27} = 3$

Glossary/Glosario

ENGLISH	SPANISH	EXAMPLES
radical equation An equation that contains a variable within a radical.	**ecuación radical** Ecuación que contiene una variable dentro de un radical.	$\sqrt{x+3}+4=7$
radical function A function whose rule contains a variable within a radical.	**función radical** Función cuya regla contiene una variable dentro de un radical.	(graph with points (0, 0), (1, 1), (4, 2), (9, 3)) $f(x)=\sqrt{x}$
radical inequality An inequality that contains a variable within a radical.	**desigualdad radical** Desigualdad que contiene una variable dentro de un radical.	$\sqrt{x+3}\leq 7$
radical symbol The symbol $\sqrt{}$ used to denote a root. The symbol is used alone to indicate a square root or with an index, $\sqrt[n]{}$, to indicate the *n*th root.	**símbolo de radical** Símbolo $\sqrt{}$ que se utiliza para expresar una raíz. Puede utilizarse solo para indicar una raíz cuadrada, o con uníndice, $\sqrt[n]{}$, para indicar la enésima raíz.	$\sqrt{36}=6$, $\sqrt[3]{27}=3$
radicand The expression under a radical sign.	**radicando** Número o expresión debajo del signo de radical.	$\sqrt{x+3}-2$ ↑ Radicand
randomized comparative experiment An experiment in which the individuals are assigned to the control group or the treatment group at random, in order to minimize bias.	**experimento comparativo aleatorizado** Experimento en el que se elige al azar a los individuos para el grupo de control o para el grupo experimental, a fin de minimizar el sesgo.	
range of a data set The difference of the greatest and least values in the data set.	**rango de un conjunto de datos** La diferencia del mayor y menor valor en un conjunto de datos.	The data set $\{3, 3, 5, 7, 8, 10, 11, 11, 12\}$ has a range of $12-3=9$.
range of a function or relation The set of output values of a function or relation.	**rango de una función o relación** Conjunto de los valores desalida de una función o relación.	The range of $y=x^2$ is $\{y \mid y\geq 0\}$.
rate A ratio that compares two quantities measured in different units.	**tasa** Razón que compara doscantidades medidas en diferentes unidades.	$\frac{55 \text{ miles}}{1 \text{ hour}}=55\text{mi/hr}$
ratio A comparison of two quantities by division.	**razón** Comparación de dos cantidades mediante una división.	$\frac{1}{2}$ or 1:2
rational equation An equation that contains one or more rational expressions.	**ecuación racional** Ecuación que contiene una o más expresiones racionales.	$\frac{x+2}{x^2+3x-1}=6$

ENGLISH	SPANISH	EXAMPLES
rational exponent An exponent that can be expressed as $\frac{m}{n}$ such that if m and n are integers, then $b^{\frac{m}{n}} = \sqrt[n]{b^m} = \left(\sqrt[n]{b}\right)^m$.	**exponente racional** Exponente quese puede expresar como $\frac{m}{n}$ tal que, si m y n son números enteros, entonces $b^{\frac{m}{n}} = \sqrt[n]{b^m} = \left(\sqrt[n]{b}\right)^m$	$4^{\frac{3}{2}} = \sqrt{4^3} = \sqrt{64} = 8$ $4^{\frac{3}{2}} = \left(\sqrt{4}\right)^3 = 2^3 = 8$
rational expression An algebraic expression whose numerator and denominator are polynomials and whose denominator has a degree ≥ 1.	**expresión racional** Expresión algebraica cuyo numerador y denominador son polinomios y cuyo denominador tiene un grado ≥ 1.	$\frac{x+2}{x^2+3x-1}$
rational function A function whose rule can be written as a rational expression.	**función racional** Función cuya regla se puede expresar como una expresión racional.	$f(x) = \frac{x+2}{x^2+3x-1}$
rational inequality An inequality that contains one or more rational expressions.	**desigualdad racional** Desigualdad que contiene una o más expresiones racionales.	$\frac{x+2}{x^2+3x-1} \geq 6$
rational number A number that can be written in the form $\frac{a}{b}$, where a and b are integers and $b \neq 0$.	**número racional** Número que se puede expresar como $\frac{a}{b}$, donde a y b son números enteros y $b \neq 0$.	$3, 1.75, 0.\overline{3}, -\frac{2}{3}, 0$
rationalizing the denominator A method of rewriting a fraction by multiplying by another fraction that is equivalent to 1 in order to remove radical terms from the denominator.	**racionalizar el denominador** Método que consiste en escribir nuevamente una fracción multiplicándola por otra fracción equivalente a 1 a fin de eliminar los términos radicales del denominador.	$\frac{1}{\sqrt{2}}\left(\frac{\sqrt{2}}{\sqrt{2}}\right) = \frac{\sqrt{2}}{2}$
real axis The horizontal axis in the complex plane; it graphically represents the real part of complex numbers.	**eje real** Eje horizontal de un plano complejo. Representa gráficamente la parte real de los números complejos.	Imaginary axis; $2i$; $0 + 0i$; Real axis; -2; 2; $-2i$
real number A rational or irrational number. Every point on the number line represents a real number.	**número real** Número racional o irracional. Cada punto de la recta numérica representa un número real.	$-5, 0, \frac{2}{3}, \sqrt{2}, 3.1, \pi$
real part of a complex number For a complex number of the form $a + bi$, a is the real part.	**parte real de un número complejo** Dado un número complejo del tipo $a + bi$, a es la parte real.	$5 + 6i$ Real part (5) Imaginary part (6i)
reciprocal For a real number $a \neq 0$, the reciprocal of a is $\frac{1}{a}$. The product of reciprocals is 1.	**recíproco** Dado el número real $a \neq 0$, el recíproco de a es $\frac{1}{a}$. El producto de los recíprocos es 1.	$\frac{1}{2}$ is the reciprocal of 2. $\frac{5}{3}$ is the reciprocal of $\frac{3}{5}$.
recursive rule A rule for a sequence in which one or more previous terms are used to generate the next term.	**Regla recurrente** Regla para una sucesión en la cual uno o más términos anteriores se utilizan para generar el término siguiente.	For the sequence 5, 7, 9, 11, ..., a recursive rule is $a_1 = 5$ and $a_n = a_{n-1} + 2$.

Glossary/Glosario

ENGLISH	SPANISH	EXAMPLES
reduced row-echelon form A form of an augmented matrix in which the coefficient columns form an identity matrix.	**forma escalonada reducida por filas** Forma de matriz aumentada en la que las columnas de coeficientes forman una matriz de identidad.	$\left[\begin{array}{cc:c} 1 & 0 & -1 \\ 0 & 1 & 3 \end{array}\right]$
reference angle For an angle in standard position, the reference angle is the positive acute angle formed by the terminal side of the angle and the x-axis.	**ángulo de referencia** Dado un ángulo en posición estándar, el ángulo de referencia es el ángulo agudo positivo formado por el lado terminal del ángulo y el eje x.	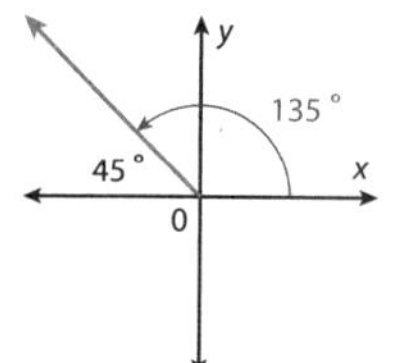
reflection A transformation that reflects, or "flips," a graph or figure across a line, called the line of reflection, such that each reflected point is the same distance from the line of reflection but is on the opposite side of the line.	**reflexión** Transformación que refleja, o invierte, una gráfica o figura sobre una línea, llamada la línea de reflexión, de manera tal que cada punto reflejado esté a la misma distancia de la línea de reflexión pero que se encuentre en el lado opuesto de la línea.	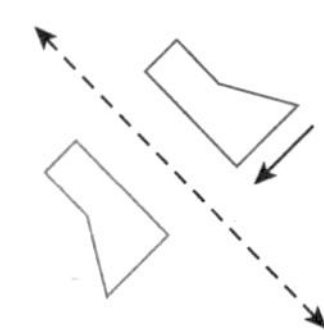
regression The statistical study of the relationship between variables.	**regresión** Estudio estadístico de la relación entre variables.	
relation A set of ordered pairs.	**relación** Conjunto de pares ordenados.	$\{(0, 5), (0, 4), (2, 3), (4, 0)\}$
Remainder Theorem If the polynomial function $P(x)$ is divided by $x - a$, then the remainder r is $P(a)$.	**Teorema del resto** Si la función polinomial $P(x)$ se divide entre $x - a$, entonces, el residuo r será $P(a)$.	
replacement set A set of numbers that can be substituted for a variable.	**conjunto de reemplazo** Conjunto de números que pueden sustituir una variable.	The solution set of $y = x + 3$ for the replacement set $\{1, 2, 3\}$ is $\{4, 5, 6\}$.
right angle An angle that measures 90°.	**ángulo recto** Ángulo que mide 90°.	
right triangle A triangle with one right angle.	**triángulo rectángulo** Triángulo con un ángulo recto.	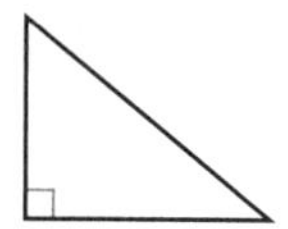
rigid transformation A transformation that does not change the size or shape of a figure.	**transformación rígida** Transformación que no cambia el tamaño o la forma de una figura.	Reflection, rotations, and translations are rigid transformations.
root of an equation Any value of the variable that makes the equation true.	**raíz de una ecuación** Cualquier valor de la variable que transforme la ecuación en verdadera.	The roots of $(x - 2)(x + 1) = 0$ are 2 and -1.

ENGLISH	SPANISH	EXAMPLES
roster notation A way of representing a set by listing the elements between braces, { }.	**notación de lista** Forma de representar un conjunto enumerando los elementos entre llaves, { }.	The first 5 positive odd numbers are $\left\{1, 3, 5, 7, 9\right\}$.
rotation A transformation that rotates or turns a figure about a point called the center of rotation.	**rotación** Transformación que hace rotar o girar una figura sobre un punto llamado centro de rotación.	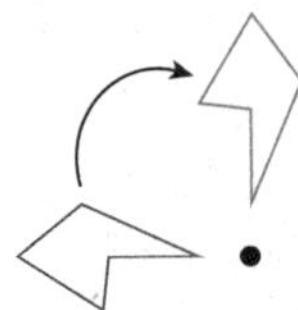
rotation transformation graph A transformation used to rotate a figure about the origin.	**Transformatión de rotación** transformatión utilizada para rotar una figura sobre el origen.	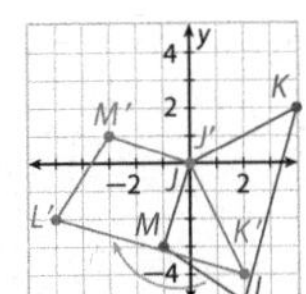 The system $\begin{cases} x^1 = x\cos 90 - y\sin 90 \\ y^1 = x\sin 90 + y\cos 90 \end{cases}$ was used to rotate the 90° clockwise.
row operation An operation performed on a row of an augmented matrix that creates an equivalent matrix.	**operación por filas** Operación realizada en una fila de una matriz aumentada que crea una matriz equivalente.	$\left[\begin{array}{cc\|c} 2 & 0 & -2 \\ 0 & 1 & 3 \end{array}\right] = \left[\begin{array}{cc\|c} \frac{1}{2}(2) & \frac{1}{2}(0) & \frac{1}{2}(-2) \\ 0 & 1 & 3 \end{array}\right] = \left[\begin{array}{cc\|c} 1 & 0 & -1 \\ 0 & 1 & 3 \end{array}\right]$
row-reduction method The process of performing elementary row operations on an augmented matrix to transform the matrix to reduced row echelon form.	**método de reducción por filas** Proceso por el cual se realizan operaciones elementales de filas en una matriz aumentada para transformar la matriz en una forma reducida de filas escalonadas.	$\left[\begin{array}{cc\|c} 2 & 0 & -2 \\ 0 & 1 & 3 \end{array}\right] = \left[\begin{array}{cc\|c} \frac{1}{2}(2) & \frac{1}{2}(0) & \frac{1}{2}(-2) \\ 0 & 1 & 3 \end{array}\right] = \left[\begin{array}{cc\|c} 1 & 0 & -1 \\ 0 & 1 & 3 \end{array}\right]$

S

ENGLISH	SPANISH	EXAMPLES
sample A part of the population.	**muestra** Una parte de la población.	In a survey about the study habits of high school students, a sample is a survey of 100 students.
sample space The set of all possible outcomes of a probability experiment.	**espacio muestral** Conjunto de todos los resultados posibles en un experimento de probabilidades.	In the experiment of rolling a number cube, the sample space is 1, 2. 3, 4, 5, 6.
scalar A number that is multiplied by a matrix.	**escalar** Número que se multiplica por una matriz.	$3\begin{bmatrix} 1 & -2 \\ 2 & 3 \end{bmatrix} = \begin{bmatrix} 3 & -6 \\ 6 & 9 \end{bmatrix}$ (3 = scalar)
scatter plot A graph with points plotted to show a possible relationship between two sets of data.	**diagrama de dispersión** Gráfica con puntos que se usa para demostrar una relación posible entre dos conjuntos de datos.	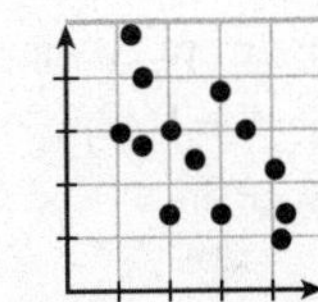

ENGLISH	SPANISH	EXAMPLES
second-degree equation in two variables An equation constructed by adding terms in two variables with powers no higher than 2.	**ecuación de segundo grado en dos variables** Ecuación compuesta por la suma de términos en dos variables con potencias no mayores a 2.	$ax^2 + by^2 + cx + dy + e = 0$
second differences Differences between first differences of a function.	**segundas diferencias** Diferencias entre las primerasdiferencias de una función.	x: 0, 1, 2, 3; y: 1, 4, 9, 16 first differences +3 +5 +7 second differences +2 +2
self-selected sample A sample in which members volunteer to participate.	**muestra de voluntarios** Muestra en la que los miembros se ofrecen voluntariamente para participar.	
sequence A list of numbers that often form a pattern.	**sucesión** Lista de números que generalmente forman un patrón.	1, 2, 4, 8, 16, ...
series The indicated sum of the terms of a sequence.	**serie** Suma indicada de los términos de una sucesión.	1 + 2 + 4 + 8 + 16 + ...
set A collection of items called elements.	**conjunto** Grupo de componentes denominados elementos.	$\{1, 2, 3\}$
set-builder notation A notation for a set that uses a rule to describe the properties of the elements of the set.	**notación de conjuntos** Notación para un conjunto que se vale de una regla para describir las propiedades de los elementos del conjunto.	$\{x \mid x > 3\}$ read, "The set of all x such that x is greater than 3."
Sierpinski triangle A fractal formed from a triangle by removing triangles with vertices at the midpoints of the sides of each remaining triangle.	**triángulo de Sierpinski** Fractal formado a partir de un triángulo al cual se le recortan triángulos cuyos vértices se encuentran en los puntos medios de los lados de cada triángulo restante.	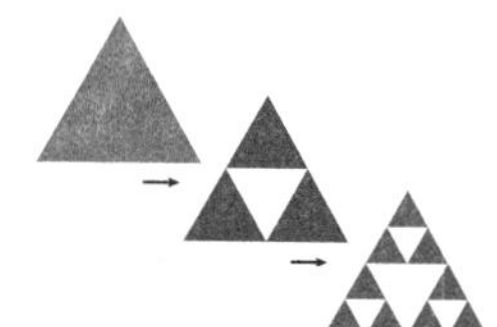
simple event An event consisting of only one outcome.	**suceso simple** Suceso que contiene sólo un resultado.	In the experiment of rolling a number cube, the event consisting of the outcome 3 is a simple event.
simple random sample A sample selected from a population so that each member of the population has an equal chance of being selected.	**muestra aleatoria simple** Muestra seleccionada de una población tal que cada miembro de ésta tenga igual probabilidad de ser seleccionada.	Mr. Hansen chose a random sample of the class by writing each student's name on a slip of paper, mixing up the slips, and drawing five slips without looking.

ENGLISH	SPANISH	EXAMPLES
simulation A model of an experiment, often one that would be too difficult or time-consuming to actually perform.	**simulación** Modelo de un experimento; generalmente se recurre a la simulación cuando realizar dicho experimento sería demasiado difícil o llevaría mucho tiempo.	A random number generator is used to simulate the roll of a number cube.
sine In a right triangle, the ratio of the length of the side opposite $\angle A$ to the length of the hypotenuse.	**seno** En un triángulo rectángulo, razón entre la longitud del cateto opuesto a $\angle A$ y la longitud de la hipotenusa.	opposite hypotenuse A $\sin A = \dfrac{\text{opposite}}{\text{hypotenuse}}$.
slope A measure of the steepness of a line. If (x_1, y_1) and (x_2, y_2) are any two points on the line, the slope of the line, known as m, is represented by the equation $m = \frac{y_2 - y_1}{x_2 - x_1}$.	**pendiente** Medida de la inclinación de una línea. Dados dos puntos (x_1, y_1) y (x_2, y_2) en una línea, la pendiente de la línea, denominada m, se representa con la ecuación $m = \frac{y_2 - y_1}{x_2 - x_1}$.	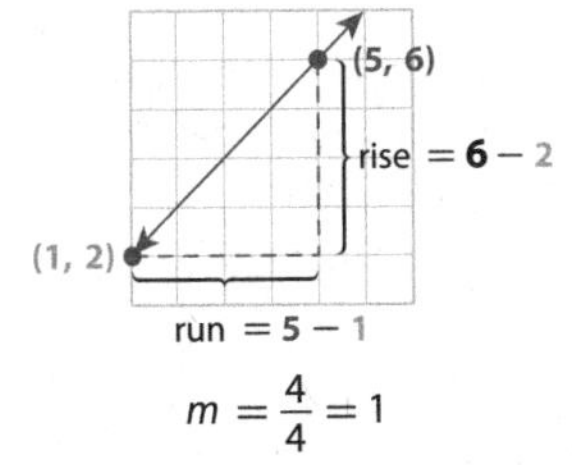
slope-intercept form The slope intercept form of a linear equation is $y = mx + b$, where m is the slope and b is the y-intercept.	**forma de pendiente-intersección** La forma de pendiente-intersección de una ecuación lineal es $y = mx + b$, donde m es la pendiente y b es la intersección y.	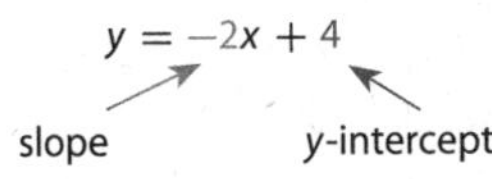
solution set of an equation The set of values that make an equation true.	**conjunto solución de una ecuación** Conjunto de valores que hacen verdadero un enunciado.	The solution set of $x^2 = 9$ is $\{-3, 3\}$.
solving a triangle Using given measures to find unknown angle measures or side lengths of a triangle.	**resolución de un triángulo** Utilizar medidas dadas para hallar las medidas desconocidas de los ángulos o las longitudes de los lados de un triángulo.	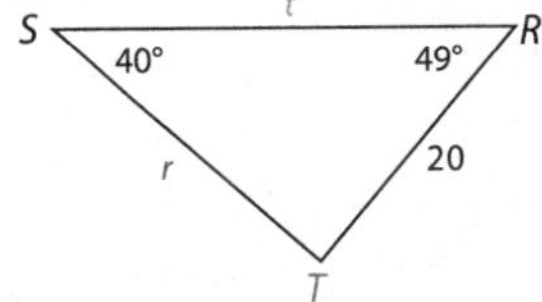 $49° + 40° + m\angle T = 180°$ $m\angle T = 91°$ $\frac{\sin 49°}{r} = \frac{\sin 40°}{20}$ $\quad$ $\frac{\sin 91°}{t} = \frac{\sin 40°}{20}$ $r \approx 23.5$ $\quad$ $t \approx 31.1$
special right triangle A 45° –45°–90° triangle or a 30°–60°–90° triangle.	**triángulo rectángulo especial** Triángulo de 45°–45°–90° o triángulo de 30°–60°–90°.	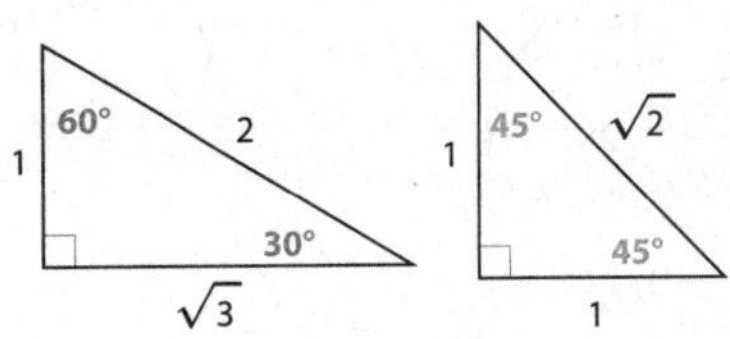
square matrix A matrix with the same number of rows as columns.	**matriz cuadrada** Matriz con el mismo número de filas y columnas.	$\begin{bmatrix} 1 & 2 \\ 0 & -3 \end{bmatrix}, \begin{bmatrix} 1 & -3 & 1 \\ 2 & 0 & -2 \\ 0 & 1 & 3 \end{bmatrix}$

ENGLISH	SPANISH	EXAMPLES

square-root function A function whose rule contains a variable under a square-root sign.

función de raíz cuadrada Función cuya regla contiene una variable bajo un signo de raíz cuadrada.

$f(x) = \sqrt{x}$

---

standard deviation A measure of dispersion of a data set. The standard deviation σ is the square root of the variance.

desviación estándar Medida de dispersión de un conjunto de datos. La desviación estándar σ es la raíz cuadrada de la varianza.

Data set: $\{6, 7, 7, 9, 11\}$
Mean: $\frac{6+7+7+9+11}{5} = 8$
Variance: $\frac{1}{5}(4+1+1+1+9) = 3.2$
Standard deviation: $\sigma = \sqrt{3.2} \approx 1.8$

---

standard form of a polynomial A polynomial in one variable is written in standard form when the terms are in order from greatest degree to least degree.

forma estándar de un polinomio Un polinomio de una variable se expresa en forma estándar cuando los términos se ordenan de mayor a menor grado.

$3x^3 - 5x^2 + 6x - 7$

---

standard form of a quadratic equation $ax^2 + bx + c = 0$, where a, b, and c are real numbers and $a \neq 0$.

forma estándar de una ecuación cuadrática $ax^2 + bx + c = 0$, donde a, b y c son números reales y $a \neq 0$.

$2x^2 + 3x - 1 = 0$

---

standard normal value A value that indicates how many standard deviations above or below the mean a particular value falls, given by the formula $z = \frac{x - \mu}{\sigma}$, where z is the standard normal value, x is the given value, μ is the mean, and σ is the standard deviation of a standard normal distribution.

valor normal estándar Valor que indica a cuántas desviaciones estándar por encima o por debajo de la media se encuentra un determinado valor, dado por la fórmula $z = \frac{x - \mu}{\sigma}$, donde z es el valor normal estándar, x es el valor dado, μ es la media y σ es la desviación estándar de una distribución normal estándar.

---

standard position An angle in standard position has its vertex at the origin and its initial side on the positive x–axis.

osición estándar Ángulo cuyo vértice se encuentra en el origen y cuyo lado inicial se encuentra sobre el eje x.

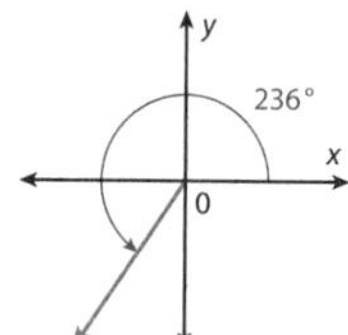

---

statistic A number that describes a sample.

estadística Número que describe una muestra.

---

step function A piecewise function that is constant over each interval in its domain.

función escalón Función a trozos que es constante en cada intervalo en su dominio.

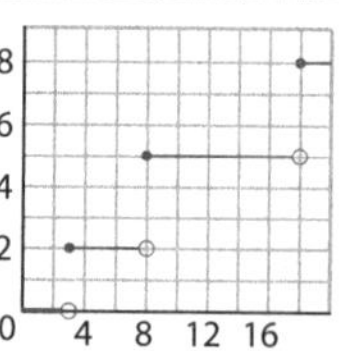

---

stratified sample A sample in which a population is divided into distinct groups and members are selected at random from each group.

muestra estratificada Muestra en la que población está dividida en grupos diferenciados y los miembros de cada grupo se seleccionan al azar.

Ms. Carter chose a stratified sample of her school's student population by randomly selecting 30 students from each grade level.

ENGLISH	SPANISH	EXAMPLES
stretch A transformation that pulls the points of a graph horizontally away from the *y*–axis or vertically away from the *x*–axis.	**estiramiento** Transformación que desplaza los puntos de una gráfica en forma horizontal alejándolos del eje y o en forma vertical alejándolos del eje *x*.	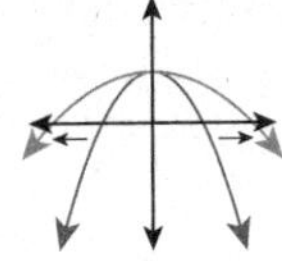
substitution A method used to solve systems of equations by solving an equation for one variable and substituting the resulting expression into the other equation(s).	**sustitución** Método utilizado para resolver sistemas de ecuaciones resolviendo una ecuación para una variable y sustituyendo la expresión resultante en las demás ecuaciones.	$\begin{cases} 2x + 3y = -1 \\ x - 3y = 4 \end{cases}$ Solve for x. $x = 4 + 3y$ Substitute into the first equation and solve. $2(4 + 3y) + 3y = -1$ $y = -1$ Then solve for x. $x = 4 + 3(-1) = 1$
summation notation A method of notating the sum of a series using the Greek letter $\sum$ (capital *sigma*).	**notación de sumatoria** Método de notación de la suma de una serie que utiliza la letra griega $\sum$ (SIGMA mayúscula).	$\sum_{n=1}^{5} 3k = 3 + 6 + 9 + 12 + 15 = 45$
synthetic division A shorthand method of dividing by a linear binomial of the form $(x - a)$ by writing only the coefficients of the polynomials.	**división sintética** Método abreviado de división que consiste en dividir por un binomio lineal del tipo $(x - a)$ escribiendo sólo los coeficientes de los polinomios.	$(x^3 - 7x + 6) \div (x - 2)$ $\begin{array}{r\|rrrr} 2 & 1 & 0 & -7 & 6 \\ & & 2 & 4 & 6 \\ \hline & 1 & 2 & -3 & 0 \end{array}$ $(x^3 - 7x + 6) \div (x - 2) = x^2 + 2x - 3$
synthetic substitution The process of using synthetic division to evaluate a polynomial $p(x)$ when $x = c$.	**sustitución sintética** Proceso que consiste en usar la división sintética para evaluar un polinomio $p(x)$ cuando $x = c$.	
system of equations A set of two or more equations that have two or more variables.	**sistema de ecuaciones** Conjunto de dos o más ecuaciones que contienen dos o más variables.	$\begin{cases} 2x + 3y = -1 \\ x^2 = 4 \end{cases}$
system of linear inequalities A system of inequalities in two or more variables in which all of the inequalities are linear.	**sistema de desigualdades lineales** Sistema de desigualdades en dos o más variables en el que todas las desigualdades son lineales.	$\begin{cases} 2x + 3y \geq -1 \\ x - 3y < 4 \end{cases}$
systematic sample A sample based on selecting one member of the population at random and then selecting other members by using a pattern.	**muestra sistemática** Muestra en la que se elige a un miembro de la población al azar y luego se elige a otros miembros mediante un patrón.	Mr. Martin chose a systematic sample of customers visiting a store by selecting one customer at random and then selecting every tenth customer after that.

T

tangent line A line that is in the same plane as a circle and intersects the circle at exactly one point.	**línea tangente** Línea que está en el mismo plano que un círculo y corta al círculo en exactamente un punto.	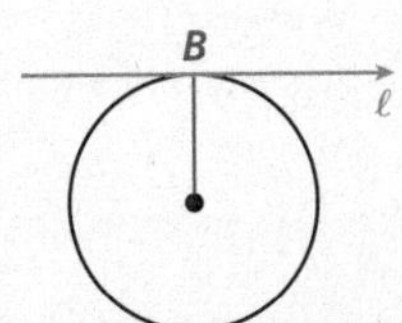

ENGLISH	SPANISH	EXAMPLES
term of a sequence An element or number in the sequence.	**término de una sucesión** Elemento o número de una sucesión.	5 is the third term in the sequence 1, 3, 5, 7, . . .
terminal side For an angle in standard position, the ray that is rotated relative to the positive *x*–axis.	**lado terminal** Dado un ángulo en una posición estándar, el rayo que rota en relación con el eje positivo *x*.	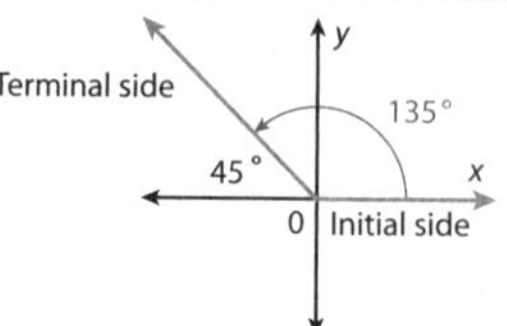
theoretical probability The ratio of the number of equally likely outcomes in an event to the total number of possible outcomes.	**probabilidad teórica** Razón entre el número de resultados igualmente probables de un suceso y el número total de resultados posibles.	The theoretical probability of rolling an odd number on a number cube is $\frac{3}{6} = \frac{1}{2}$.
third quartile The median of the upper half of a data set. Also called *upper quartile*.	**tercer cuartil** La mediana de la mitad superior de un conjunto de datos. También se llama *cuartil superior*.	Lower half: (18, 23, 28,) Upper half: (29, (36), 42) Third quartile
three-dimensional coordinate system A space that is divided into eight regions by an *x*–axis, a *y*–axis, and a *z*–axis. The locations, or coordinates, of points are given by ordered triples.	**sistema de coordenadas tridimensional** Espacio dividido en ocho regiones por un eje *x*, un eje *y* y un eje *z*. Las ubicaciones, o coordenadas, de los puntos son dadas por tripletas ordenadas.	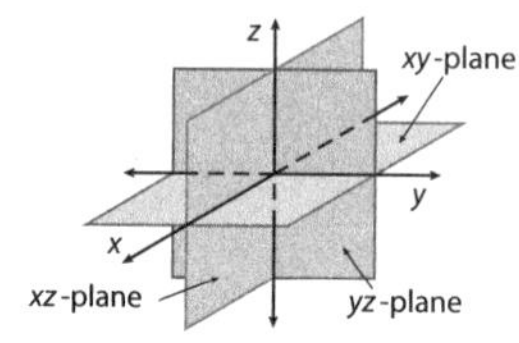
transformation A change in the position, size, or shape of a figure or graph.	**transformación** Cambio en la posición, tamaño o forma de una figura o gráfica.	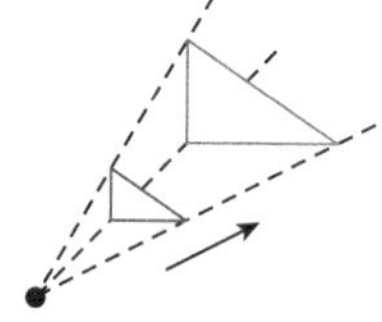
translation A transformation that shifts or slides every point of a figure or graph the same distance in the same direction.	**traslación** Transformación en la que todos los puntos de una figura se mueven la misma distancia en la misma dirección.	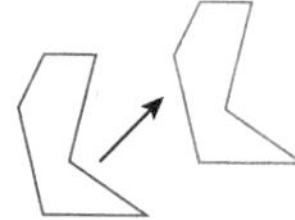
translation matrix A matrix used to translate points on the coordinate plane.	**matriz de traslación** Matriz utilizada para trasladar puntos en el plano cartesiano.	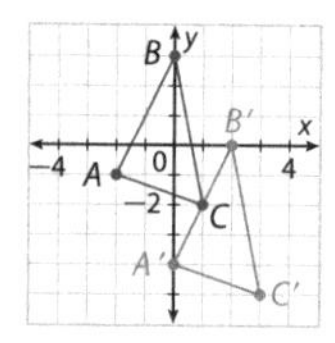 Matrix $\begin{bmatrix} -2 & -2 & -2 \\ 3 & 3 & 3 \end{bmatrix}$ is used to translate the figure 2 units left and 3 units up.
transpose A matrix that reverses the rows and columns of a matrix.	**transposición** Matriz que invierte las filas y columnas de una matriz.	$\begin{bmatrix} 1 & 2 \\ 3 & 4 \\ 5 & 6 \end{bmatrix}$ is the transpose of $\begin{bmatrix} 1 & 3 & 5 \\ 2 & 4 & 6 \end{bmatrix}$.

ENGLISH	SPANISH	EXAMPLES
transverse axis The axis of symmetry of a hyperbola that contains the vertices and foci.	**eje transversal** Eje de simetría de una hipérbola que contiene los vértices y focos.	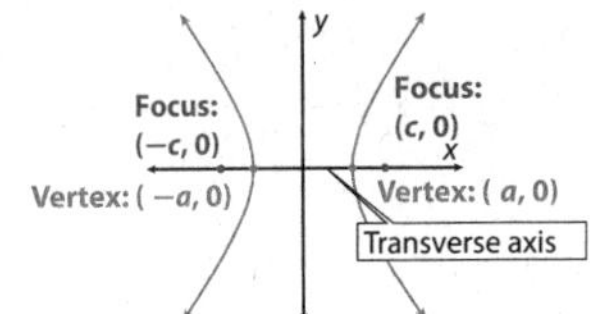
treatment group In a controlled experiment, the group that receives treatment.	**grupo experimental** En un experimento controlado, el grupo que está expuesto a la manipulación.	
trial In probability, a single repetition or observation of an experiment.	**prueba** En probabilidad, una sola repetición u observación de un experimento.	In the experiment of rolling a number cube, each roll is one trial.
trigonometric function A function whose rule is given by a trigonometric ratio.	**función trigonométrica** Función cuya regla es dada por una razón trigonométrica.	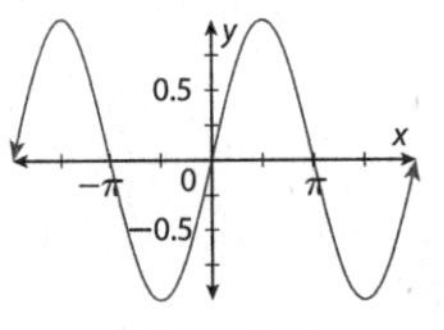 $f(x) = \sin x$
trigonometric ratio Ratio of the lengths of two sides of a right triangle.	**razón trigonométrica** Razón entre dos lados de un triángulo rectángulo.	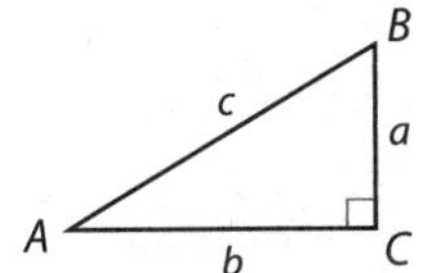 $\sin A = \frac{a}{c}, \cos A = \frac{b}{c}, \tan A = \frac{a}{b}$
trigonometry The study of the measurement of triangles and of trigonometric functions and their applications.	**trigonometría** Estudio de la medición de los triángulos y de las funciones trigonométricas y sus aplicaciones.	
trinomial A polynomial with three terms.	**trinomio** Polinomio con tres términos.	$4x^2 + 3xy - 5y^2$
turning point A point on the graph of a function that corresponds to a local maximum (or minimum) where the graph changes from increasing to decreasing (or vice versa).	**punto de inflexión** Punto de la gráfica de una función que corresponde a un máximo (o mínimo) local donde la gráfica pasa de ser creciente a decreciente (o viceversa).	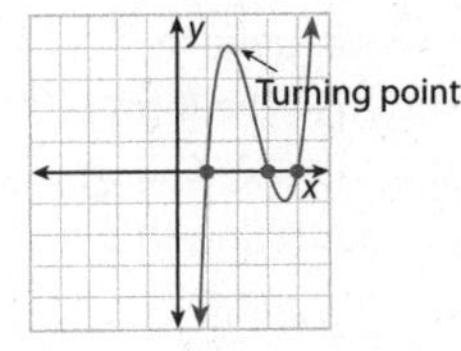

Glossary/Glosario

ENGLISH	SPANISH	EXAMPLES

U

unit circle A circle with a radius of 1, centered at the origin.

círculo unitario Círculo con un radio de 1, centrado en el origen.

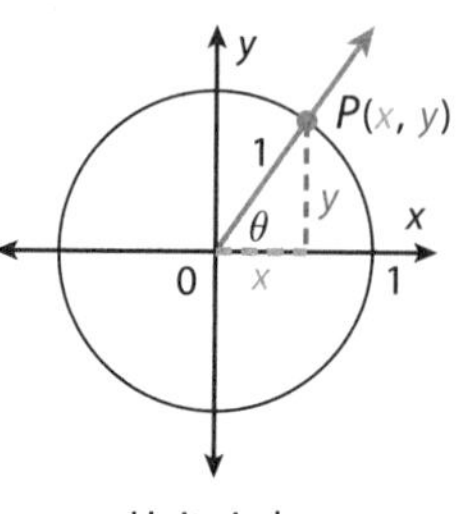

Unit circle

V

variable A symbol used to represent a quantity that can change.

variable Símbolo utilizado para representar una cantidad que puede cambiar.

$2x + 3$ ↑ variable

variable matrix The matrix of the variables in a linear system of equations.

matriz de variables Matriz de las variables de un sistema lineal de ecuaciones.

System of equations	Variable matrix
$\begin{cases} 2x + 3y = -1 \\ x - 3y = 4 \end{cases}$	$\begin{bmatrix} x \\ y \end{bmatrix}$

variance The average of squared differences from the mean. The square root of the variance is called the *standard deviation.*

varianza Promedio de las diferencias cuadráticas en relación con la media. La raíz cuadrada de la varianza se denomina *desviación estándar.*

Data set: is $\{6, 7, 7, 9, 11\}$

Mean: $\frac{6 + 7 + 7 + 9 + 11}{5} = 8$

Variance: $\frac{1}{5}(4 + 1 + 1 + 1 + 9) = 3.2$

Venn diagram A diagram used to show relationships between sets.

diagrama de Venn Diagrama utilizado para mostrar la relación entre conjuntos.

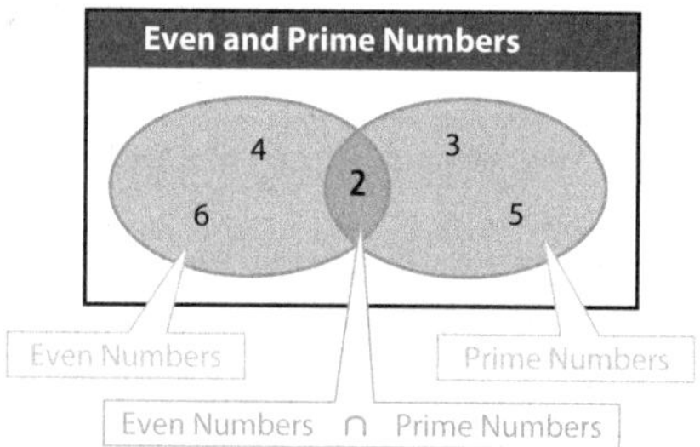

vertex form of a quadratic function A quadratic function written in the form $f(x) = a(x - h)^2 + k$, where a, h, and k are constants and (h, k) is the vertex.

forma en vértice de una función cuadrática Una función cuadrática expresada en la forma $f(x) = a(x - h)^2 + k$, donde a, h y k son constantes y (h, k) es el vértice.

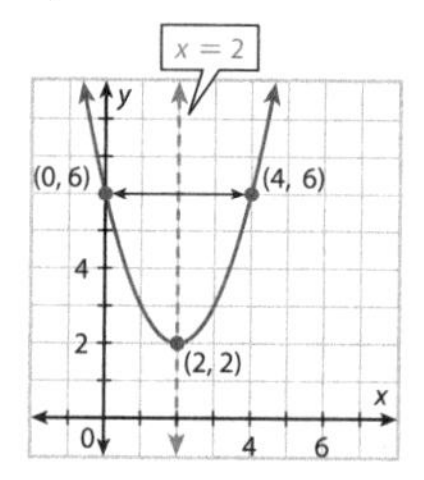

$f(x) = (x - 2)^2 + 2$

ENGLISH	SPANISH	EXAMPLES
vertex of a hyperbola (vertices) The endpoints of the transverse axis of the hyperbola.	**vértice de una hipérbola** Extremos del eje transversal de la hipérbola.	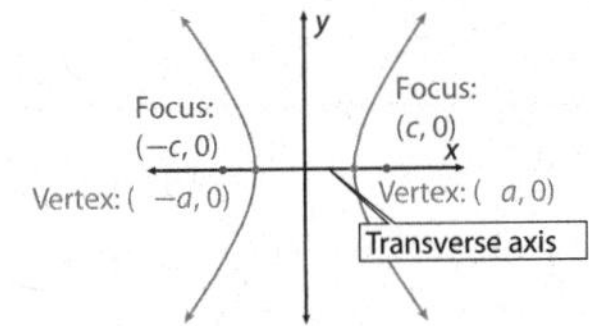
vertex of an absolute-value graph The point where the axis of symmetry intersects the graph.	**vértice de una gráfica de valor absoluto** Punto donde en el eje de simetría interseca la gráfica.	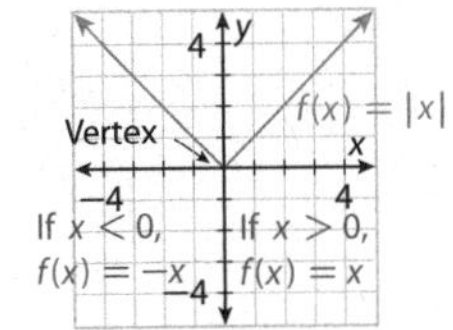
vertex of an ellipse (vertices) The endpoints of the major axis of the ellipse.	**vértice de una elipse** Extremos del eje mayor de la elipse.	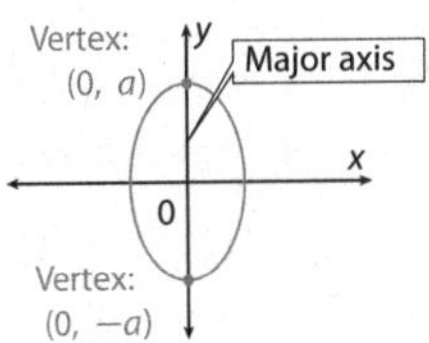
vertex of a parabola The highest or lowest point on the parabola.	**vértice de una parábola** Punto más alto o más bajo de una parábola.	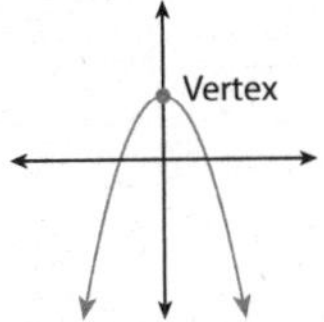
vertical line A line whose equation is $x = a$, where a is the x-intercept. The slope of a vertical line is undefined.	**línea vertical** Línea cuya ecuación es $x = a$, donde a es la intersección con el eje x. La pendiente de una línea vertical es indefinida.	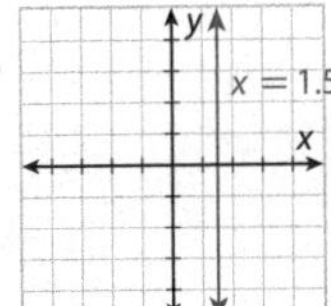
vertical-line test A test used to determine whether a relation is a function. If any vertical line crosses the graph of a relation more than once, the relation is not a function.	**prueba de la línea vertical** Prueba utilizada para determinar si una relación es una función. Si una línea vertical corta la gráfica de una relación más de una vez, la relación no es una función.	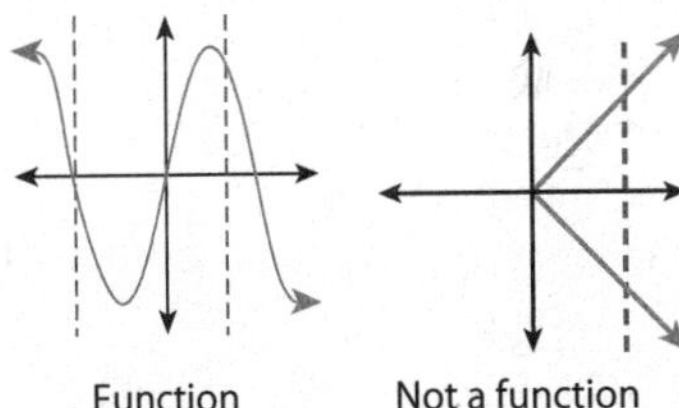

whole number The set of natural numbers and zero.	**número cabal** Conjunto de los números naturales y cero.	0, 1, 2, 3, 4, 5, …

X

x-intercept The x-coordinate(s) of the point(s) where a graph intersects the x-axis.	**intersección con el eje x** Coordenada(s) x de uno o más puntos donde una gráfica corta el eje x.	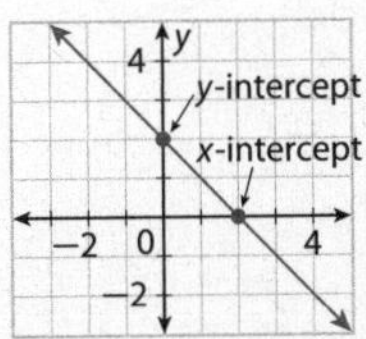

Glossary/Glosario

ENGLISH	SPANISH	EXAMPLES

Y

y-intercept The *y*-coordinate(s) of the point(s) where a graph intersects the *y*-axis.

intersección con el eje *y* Coordenada(s) de uno o más puntos donde una gráfica corta el eje *y*.

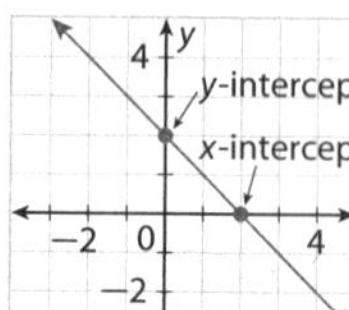

Z

***z*-axis** The third axis in a three-dimensional coordinate system.

eje *z* Tercer eje en un sistema de coordenadas tridimensional.

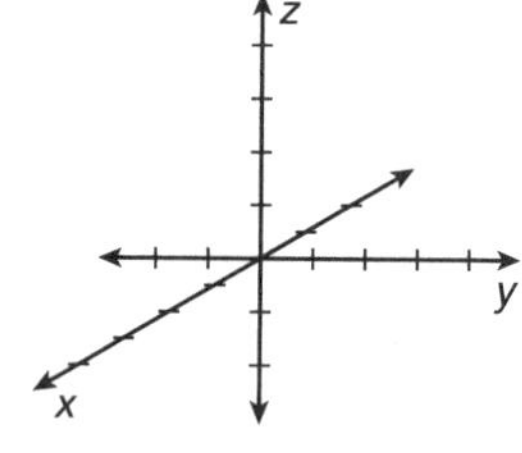

zero exponent For any nonzero real number x, $x^0 = 1$.

exponente cero Dado un número real distinto de cero x, $x^0 = 1$.

$5^0 = 1$

zero of a function For the function f, any number x such that $f(x) = 0$.

cero de una función Dada la función f, todo número x tal que $f(x) = 0$.

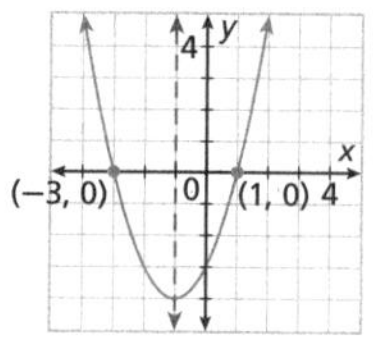

The zeros of $f(x) = x^2 + 2x - 3$ are −3 and 1.

Index

Index locator numbers are in Module. Lesson form. For example, 2.1 indicates Module 2, Lesson 1 as listed in the Table of Contents.

F

G

H

I

L

M

N

O

P

Q

R

Index

S

T

U

V

W

X

Z

Table of Measures

LENGTH

1 inch = 2.54 centimeters
1 meter = 39.37 inches
1 mile = 5,280 feet
1 mile = 1760 yards
1 mile = 1.609 kilometers
1 kilometer = 0.62 mile

MASS/WEIGHT

1 pound = 16 ounces
1 pound = 0.454 kilograms
1 kilogram = 2.2 pounds
1 ton = 2000 pounds

CAPACITY

1 cup = 8 fluid ounces
1 pint = 2 cups
1 quart = 2 pints
1 gallon = 4 quarts
1 gallon = 3.785 liters
1 liter = 0.264 gallons
1 liter = 1000 cubic centimeters

Symbols

Symbol	Meaning
$\neq$	is not equal to
$\approx$	is approximately equal to
10^2	ten squared; ten to the second power
$2.\overline{6}$	repeating decimal 2.66666...
$\lvert -4 \rvert$	the absolute value of negative 4
$\sqrt{\ }$	square root
π	pi: (about 3.14)
$\perp$	is perpendicular to
$\parallel$	is parallel to
$\overleftrightarrow{AB}$	line *AB*
$\overrightarrow{AB}$	ray *AB*
$\overline{AB}$	line segment *AB*
$m\angle A$	measure of $\angle A$

Formulas

FACTORING	
Perfect square trinomials	$a^2 + 2ab + b^2 = (a + b)^2$
	$a^2 - 2ab + b^2 = (a - b)^2$
Difference of squares	$a^2 - b^2 = (a - b)(a + b)$
Sum of cubes	$a^3 + b^3 = (a + b)(a^2 - ab + b^2)$
Difference of cubes	$a^3 - b^3 = (a - b)(a^2 + ab + b^2)$

PROPERTIES OF EXPONENTS	
Product of powers	$a^m a^n = a^{(m+n)}$
Quotient of powers	$\frac{a^m}{a^n} = a^{(m-n)}$
Power of a power	$(a^m)^n = a^{mn}$
Rational exponent	$a^{\frac{m}{n}} = \sqrt[n]{a^m}$
Negative exponent	$a^{-n} = \frac{1}{a^n}$

QUADRATIC EQUATIONS	
Standard form	$f(x) = ax^2 + bx + c$
Vertex form	$f(x) = a(x - h)^2 + k$
Parabola	$(x - h)^2 = 4p(y - k)$ $(y - k)^2 = 4p(x - h)$
Quadratic formula	$x = \frac{-b \pm \sqrt{b^2 - 4ac}}{2a}$
Axis of symmetry	$x = \frac{-b}{2a}$